FAMILIES, CRIME AND CRIMINAL JUSTICE

CONTEMPORARY PERSPECTIVES IN FAMILY RESEARCH

Series Editor: Felix M. Berardo

Volume 1: Through the Eyes of the Child: Revisioning Children as Active Agents of Family Life

Volume 2: Families, Crime and Criminal Justice

CONTEMPORARY PERSPECTIVES IN FAMILY RESEARCH
VOLUME 2

FAMILIES, CRIME AND CRIMINAL JUSTICE

EDITED BY

GREER LITTON FOX
Department of Child & Family Studies, University of Tennessee, USA

MICHAEL L. BENSON
Department of Sociology, University of Tennessee, USA

Emerald

United Kingdom – North America – Japan
India – Malaysia – China

Emerald Group Publishing Limited
Howard House, Wagon Lane, Bingley BD16 1WA, UK

British Library Cataloguing in Publication Data
A catalogue record for this book is available from the British Library

ISBN: 978-0-76-230737-1
ISSN: 1530-3535 (Series)

Printed and bound by CPI Group (UK) Ltd, Croydon, CR0 4YY

CONTENTS

PART II. THE FAMILY'S ROLE IN THE PRODUCTION, MEDIATION, AND PREVENTION OF CRIME

PART III. FAMILIES IN THE CRIMINAL JUSTICE SYSTEM

PART IV. OPPORTUNITIES FOR SYSTEM REFORM

LIST OF CONTRIBUTORS

Robert Agnew	Department of Sociology, Emory University, Atlanta, GA 30322, USA
Ronet Bachman	Division of Criminal Justice, University of Cincinnati, Cincinnati, OH 45221, USA
Michael L. Benson	Department of Sociology, University of Tennessee, Knoxville, TN 37996, USA
Felix Berardo	Department of Sociology, University of Florida, Gainesville, FL 32611, USA
Gary L. Bowen	School of Social Work, The University of North Carolina at Chapel Hill, Chapel Hill, NC 27599, USA
Natasha K. Bowen	School of Social Work, The University of North Carolina at Chapel Hill, Chapel Hill, NC 27599, USA
Timothy Brezina	Department of Sociology, Tulane University, New Orleans, LA 70118, USA
Rand D. Conger	Institute for Social and Behavioral Research, Department of Sociology, Iowa State University Ames, Iowa 50010, USA

Patricia G. Cook	School of Social Work, The University of North Carolina at Chapel Hill, Chapel Hill, NC 27599, USA
Francis T. Cullen	Division of Criminal Justice, University of Cincinnati, Cincinnati, OH 45221, USA
Alfred DeMaris	Department of Sociology, Bowling Green State University, Bowling Green, OH 43403, USA
Greer Litton Fox	Department of Child and Family Studies, University of Tennessee, Knoxville, TN 37996, USA
Stephen M. Gavazzi	Department of Human Development and Family Science, The Ohio State University, Columbus, OH 43210, USA
Dr. Jaana Haapasalo	Department of Psychology, University of Jyväskylä, FIN-40351, Jyväskylä, Finland.
Marvin D. Krohn	Department of Sociology, University at Albany, State University of New York, Albany, NY 12222, USA
Suzanne Kurth	Department of Sociology, University of Tennessee, Knoxville, TN 37996, USA
Alan J. Lizotte	University at Albany Clinic, School of Criminal Justice, State University of New York, Albany, NY 12222, USA
Jeff Maahs	Division of Criminal Justice, University of Cincinnati, Cincinnati, OH 45221, USA
Karen A. Mason	Criminal Justice Program, Washington State University, Pullman, Washington, USA

Paul Mazerolle	Division of Criminal Justice, University of Cincinnati, Cincinnati, OH 45221–0389, USA
Cynthia Perez McCluskey	School of Criminal Justice, Michigan State University, East Lansing, MI 48824, USA
Sara McLanahan	Centre for Research on Child Wellbeing, Princeton University, Princeton, NJ 08544, USA
Anne Nurse	Department of Sociology and Anthropology, College of Wooster, Wooster, Ohio 44691, USA
Charles Partridge	Department of Human Development and Family Science, The Ohio State University, Columbus, OH 43210, USA
Cesar Rebellon	Department of Sociology, Emory University, Atlanta, GA 30322, USA
Jacquelyn L. Sandifer	Divison of Social Science, Campbellsville University, Campbellsville, KY 42718, USA
Ronald L. Simons	Institute for Social and Behavioral Research, Department of Sociology, Iowa State University, Ames, Iowa 50010, USA
Carolyn A. Smith	School of Criminal Justice, University at Albany, State University of New York, Albany, NY 12222, USA
Ryan E. Spohn	Department of Sociology, The University of Iowa, Iowa City, IA 52242, USA
Eric A. Stewart	Department of Sociology, Georgia State University, Atlanta, GA 30302, USA

Magda Stouthamer-Loeber Western Psychiatric Institute & Clinic, University of Pittsburgh, Pittsburgh, PA 15213, USA

Sherod Thaxton Department of Sociology, Emory University, Atlanta, GA 30322, USA

Margaret Vandiver Department of Criminology and Criminal Justice, University of Memphis, Memphis, TN 38152, USA

Judy Van Wyk Department of Sociology and Anthropology, University of Rhode Island, Kingston, RI 02881, USA

Deborah Wasserman Department of Human Development and Family Science, The Ohio State University, Columbus, OH 43210, USA

Anne Weiher Boulder, CO 80304, USA

Bruce Western Department of Sociology, Princeton University, Princeton, NJ 08544, USA

John D. Wooldredge Division of Criminal Justice, University of Cincinnati, Cincinnati, OH 45221, USA

John Paul Wright Division of Criminal Justice, University of Cincinnati, Cincinnati, OH 45221, USA

Courtney Yarcheck Department of Human Development and Family Science, The Ohio State University, Columbus, OH 43210, USA

PREFACE

Felix M. Berardo, Series Editor

This second volume of our series on Contemporary Perspectives in Family Research deals with a wide spectrum of important and sometimes controversial issues involving the interconnections between, family, crime, and the criminal justice system. As our guest editors note, for too long researchers from these various sub-specialties operated rather independently from one another, despite many overlapping empirical interests and theoretical approaches, and exhibited little in the way of collaborative endeavors. A similar observation was made some years ago by Lincoln & Straus (1985). In commenting on the high crime rate and changes that occurring in the family, they noted that, with some exceptions, especially with regards to juvenile delinquency, these . . . two issues are studied independently by social scientists. One group of researchers studies crime and another group studies families . . . there does not seem to be a recognition of the general principle that crime and the family are interrelated, nor is there a group of researchers who take 'crime and the family' as the focus of their work . . . family researchers have tended to ignore the possibility that crime may be as much a part of the family as it is a part of other spheres of life, and criminologists have tended to ignore the possibility that families may be involved in crime as much as any other group or type of person. They went on to suggest that one reason for this lack of focus may be that we resist recognizing the connection between family and crime because it would force us to accept a less than perfect image of this revered institution. It would run contrary to what Straus & Steinmetz (1971, p. 7) had earlier characterized as the myths or ideologies of 'familial love and gentleness' and 'family consensus and harmony.' The 'perceptual blindness' of the 'dark side' of the family, for example, in relation to violence among its members, prevents correcting such problems and impedes development of a more realistic picture of marriage and family life. A second reason suggested for the absence of crime and the family as distinct field of study may be the absence of a conceptual framework which could be used to think about the issue as well as a guide to research.

The Lincoln & Straus book, which in many ways was a precursor to the present volume, was aimed at creating such a framework, one that would be helpful in grasping the central issues in crime within the family, crime by the family, and crime against the family. Its systematic perspective involved an integration of ideas and theories from traditional criminology, including the emerging field of victimology, and family science. "Family specialists might be able to achieve a better understanding of families if they can use the theories that have developed to explain criminal behavior, and criminologists might achieve a better understanding of crime if they can use some of the theories that have been developed to explain family behavior."

The present volume represents the latest extension of this interdisciplinary approach. The addition of the criminal justice system in the equation prompts new and more complex questions concerning its connections with crime and the family and, hopefully, to the formulation of testable propositions about those interrelationships. This complexity reinforces the caution by the current editors, and others, that to comprehend criminal behaviors one must additionally examine factors beyond family structure and process alone. Indeed, the resolution, prevention, and treatment of such behaviors mandates attention be given to a broader, extra-familial framework. Fox & Benson note the growing recognition of the importance of contextual effects. "Families are no longer considered in isolation from the neighborhoods and communities in which they are located. While what goes on behind the closed doors of a family's domicile obviously is important, it does not occur in a vacuum. Families invariably and unavoidably are touched by what goes on around them on the street, in the school, and in the houses of other families." The view here characterizes families as semi-permeable or open social systems which interface and interact with other social systems or institutions in a myriad of ways, especially with respect to violence and crime.

The advantages gained through periodically assessing and drawing upon commonalities between the separately developed literatures of these different fields cannot be over-emphasized. As Fox & Benson note in their opening chapter, this strategy increases the potential for each area to identify issues that may have gone unnoticed until the other's disciplinary lens is applied. Moreover, as the contributors to this volume demonstrate, such cross-fertilization enhances the research potential for each of the disciplines involved, including a number of methodological gains, for example, through sharing different research designs and alternative approaches to measurement. Such methodological sophistication is essential in order to deal with the growing vitality of alternative family forms which challenge the prevailing family paradigm. None the less important, however, are the practical advantages of such collaboration, not only at the empirical and theoretical levels, but in the subsequent application of findings as well. Such findings can lead to reforms in the justice system, new public policies, and programs designed to reduce the incidence of criminal behaviors and their impact on families, communities, and society.

REFERENCES

Lincoln, A. J., & Straus, M. A. (1985). *Crime and the Family*. Illinois: Charles C. Thomas.
Steinmetz, S. K., & Straus, M. A. (1974). *Violence in the Family*. New York: Dodd, Mead.

FAMILIES, CRIME AND CRIMINAL JUSTICE: CHARTING THE LINKAGES

Greer Litton Fox and Michael L. Benson

INTRODUCTION

To some observers a focus on linkages between families, crime, and the justice system may seem not to require any justification whatsoever. After all, the idea that the family plays an important role in producing criminal behaviors would come as no surprise to many. As Wilson & Herrnstein (1985, 213) have noted, "if asked why some persons are more likely than others to commit crimes, the average parent would very likely attribute these individual differences to family experiences." A long tradition of research suggests that parents are not mistaken in this belief. Deficits in family structure and functioning appear to have a lot to do with why some fathers and sons or mothers and daughters, but not others, are crime-prone. From a different perspective, though, there is something troubling about linking crime with families. The "family has always been regarded through a murk of sentimentality" (Collins, 1975, 225). Idealized cultural images of the family emphasize consensus, harmony, and stability and have no place for the ugly, painful, and conflict-ridden realities of crime and criminal justice. Well-functioning families do not harbor criminals, and they are not involved with the criminal justice system. Thus, an attempt to examine the linkages of families – the building blocks of a democracy – with crime and deviance begs an explanation.

Families, Crime and Criminal Justice, Volume 2, pages 1–21.

ISBN: 0-7623-0737-4

In this volume we bring together the research of more than forty criminologists and family scientists who address different aspects of the interconnections between families, crime, and criminal justice. This collection of chapters is timely and appropriate for several reasons. First, despite considerable overlap in their concerns and subject matter, and in their theoretical frameworks, the fields of criminology and family studies as sub-specialties of sociology have developed in relative isolation from each other. There are advantages in taking stock from time to time of commonalties between these separately developed literatures.

Second, bringing together specialists from two areas heightens the potential for each area to identify problematic issues that may have escaped notice until the other's disciplinary lens is applied. An example of this possibility is the realization that the criminal justice system often functions as a potent source of secondary victimization of families of crime victims, an effect not evident until victims were viewed within the context of their familial memberships (Elias, 1986). Extending this line of research beyond the families of crime victims, several chapters in this volume consider the equally consequential effects of criminal justice processing on the families of perpetrators.

Third, and perhaps most important, is the potential for improvement in the conduct of research in both disciplines. Theoretically, there is potential for improvement in model specification, that is, in the identification of relevant predictors and mediating processes that foster criminal outcomes or that moderate the impact of potentially criminogenic processes in early developmental settings. For example, it is advantageous to draw upon research guided by theoretical frameworks in family studies, such as family stress theory, to better understand what questions to ask, what variables to include, and how to operationalize such critical concepts as stressor, crisis definition, and family coping processes. These and other conceptually similar variables are often used in an unsophisticated manner by criminologists investigating crime and the family. Several chapters in this volume show the power of drawing from domain-linked theoretical perspectives to examine families and crime. Other potential gains to methodology include exposure to different research designs, such as, for example, the longitudinal panel designs that are common in developmental studies of families but that until recently were less common in criminological research. Identifying alternative approaches to measuring variables shared across the two disciplines is also a potential advantage of aligning the two fields more closely. For example, more precise ways of measuring family configurations enhances the validity of research on the impact of family structures and processes on delinquency outcomes, as exemplified by several chapters in this volume. Further benefits from considering the interconnections between the study of families and crime are evident as we sketch various points of convergence in theoretical and empirical approaches to families and crime.

POINTS OF CONVERGENCE IN THEORETICAL APPROACHES TO FAMILIES AND CRIME

The Inherited Parsonian Theoretical Tradition

In classical sociological tradition, most especially that of Parsonian structure-functionalism, socialization and social control are seen as complementary processes in well functioning social systems. The family system is the primary social institution entrusted with the task of socialization (Parsons & Bales, 1955). In this model, families encourage individuals through their social ties as members of a family unit to "come to want to do what they must do." Because of their intimacy, mutual and reciprocated feelings of love and concern, and continual access to one another, families were thought to be the most salient and effective social group for the continuous task of socialization. Families were also expected to engage in mild forms of social control – discipline, punishment, social regulation – when these were needed as corrective actions to keep family members in line with societal expectations.

One of the most important tasks of any social system, and one of the socialization functions entrusted to the family, was the establishment of a sense of trust in the social order. Indeed, the Parsonian sociologists found it hard to imagine that a society could function if its members lacked a deeply held, internalized sense of the moral order. Establishing this sense of justice is an early function of family socialization, one that is undergirded by larger institutional systems as well.

The socialization children initially received within the family was supposed to be supplemented by other social institutions such as churches, schools, and media, all functioning as redundant socializers to reinforce societal expectations and mores governing appropriate and expected behaviors and to ensure the maintenance of socially guided actions. The criminal justice system in particular has as its major societal purpose the reinforcement of the social order learned originally within the context of family interaction.

Even Parsonian sociologists recognized, however, that socialization failures were possible. Such failures were seen as more likely when the form of the family deviated from what was felt to be most efficacious, that is, a two-parent family in which one parent specialized in paid labor in the market economy and the other specialized in the unpaid labor involved in family and home maintenance. Socialization failures were also more likely when the adult members were disaffected or disengaged from major social institutions, such as the economy, the political system, and religious institutions.

To accommodate families that failed to socialize and regulate their members, the justice system was posited as the fallback control mechanism. That is, a system of civil and criminal justice, complete with the regulatory and police powers of the state, would provide an extra-familial source of authority for monitoring and reinforcing social norms of appropriate behavior. Together the family and justice systems were engaged in what Parsons labeled 'pattern maintenance', or the reinforcement and continuation of socially appropriate patterns of individual thought, affect, and behavior. Aggregated across all members of a society, these patterns ensured the continual functioning of the group or social system.

This theoretical framework has been instrumental in depicting the relationship between the family and justice systems. It has also inspired the bulk of the extant research literature on the role of families in the production of crime and delinquency as well as the juvenile justice practice literature that is based upon it. Among the most important impacts that this theoretical framework has had on the fields of criminology and family science are the assumptions that criminal behavior represents a failure of socialization and that those interested in the emergence of delinquency should look first at the family for evidence of socialization failure.

It has also directed researchers and practitioners to a focus on 'family breakdown', or departures from what were considered normative family structures and characteristics. The emphasis that crime and delinquency researchers have placed on such factors as single parent families, the employment of mothers outside the home, parental lapses in the supervision and regulation of their children, the unemployment of fathers, weak parental commitments to conventional values, and absence of respect and affective ties across generations – all flow from this theoretical framework.

Of equal consequence is that this framework made it easy for researchers to ignore dysfunctional processes in structurally normative families, such as covert marital conflict, harsh parental discipline, and coercion and violence among family members – all aspects of family functioning that more recent research has found to be significant antecedents of antisocial behaviors. Moreover, it also made it more difficult to see that under certain circumstances criminal behaviors among youth could represent familial socialization successes rather than failures. That is, the focus of this theoretical framework on accounting for the production of normative conventionality failed to acknowledge that the very same structures and processes that characterize well functioning families also may be present in families who socialize their children into criminal careers. It was left to other theorists to point out that to understand the emergence of criminal behaviors required a focus on factors beyond family structure and process alone.

Control Theory

A second theoretical scheme that focuses on the family to account for the emergence of non-conventional behaviors, including delinquency and crime, is control theory, articulated most effectively by Hirschi (1969). Control theories start with less sanguine assumptions about the naturalness of the social order than does structural-functionalism. Control theories assume that individuals are likely to engage in deviance, delinquency, and crime unless they are somehow prevented from doing so. Often referred to as social bond theory, Hirschi's theory of juvenile delinquency stressed the development of the social bond as the primary way in which delinquency was prevented. The elements of the social bond include attachment to others, commitment to conformity, involvement in conventional activities, and belief in mainstream norms and values. Youths' development of attachment to conventional role models, such as parents and teachers, was a matter of prime importance. The extent to which a family can encourage and reinforce a youth's commitment to conventionality is a function of family affective and regulatory processes, which are themselves affected by family composition and structure. Social bond theory has been used extensively as a theoretical model driving research on the development and maintenance of several categories of youth 'problem behaviors', such as smoking, sexual behaviors including pregnancy, rape, and prostitution, drug use, conduct disorders, and criminal behaviors of varying seriousness. It also underlies a considerable body of research on normative adolescent development investigating the relative influences of parents versus peers in accounting for youth outcomes in values, attitudes, and behaviors.

The most recent and influential extension of control theory is self-control theory, promulgated by Gottfredson & Hirschi (1990). According to self-control theory, crime and delinquency result from low self-control. Low self-control refers to a behavioral tendency to pursue one's own self-interest with little concern for others or for long-term consequences. Levels of self-control vary over individuals, but remain more or less constant within individuals over time. Although families play an important role in social bond theory as objects of attachment, they play an even more crucial role in self-control theory. The theory assumes that self-control will not develop unless it is instilled very early on by parents or other primary caregivers. Thus, to prevent delinquency, families must at a minimum: "(1) monitor the child's behavior; (2) recognize deviant behavior when it occurs; and (3) punish such behavior" (Gottfredson

& Hirschi, 1990, 1997). Although obviously related to social bond theory, self-control theory places much less emphasis on social relationships than does its theoretical predecessor. In self-control theory, the family does not function as a place of warmth and emotional nourishment in which children come to avoid deviance because they love their parents and want to please them. Rather, parents primarily are conceived to operate as watchful disciplinarians, ready to pounce on any and all of their children's deviations.

While the importance of discipline cannot be ignored, self-control theory may go too far in its emphasis on parental control. Recently, important modifications to self-control theory have been articulated by criminologists who argue that to produce well-adjusted and law-abiding children parents must do more than simply recognize and punish deviance. They must also provide emotional and social support for their offspring (Cullen, 1994). As several of the articles in this volume illustrate, it is just as important for parents to share time with and support their children as it is for them to supervise and punish them.

Strain Theory of Crime and the Family Stress Model

A third theoretical linkage is found in modern strain theory (Agnew, 1992), which posits that some criminal behaviors result from life stresses and frustrations, which may often originate in unsatisfactory performance in normatively mandated family roles, such as provider role failure. For example, from the perspective of strain theory, male-to-female intimate violence is more common among men with frequent bouts of unemployment and job loss because these events are sources of acute embarrassment and loss of status for the male partner. Some men attempt to overcome embarrassment and regain status through violent domination of their intimate partners. Similarly, fraud, embezzlement, and other white collar crimes that seem otherwise inexplicable may be understood as motivated by a perpetrator's desire to provide for his or her family more adequately or in response to a family crisis (Benson, 1985; Daly, 1989).

Strain theory in criminology has a counterpart in family stress theory, a dominant theoretical framework in family science. First articulated by Hill (1949), family stress theory postulated a model of family crises in terms of three factors: a stressor event, the family's resources for coping with that stressor, and the family's definition of the stressor situation as manageable. Interestingly, one of the examples Hill (1949) originally used to illustrate family stress theory was the criminal activity and subsequent imprisonment of a family breadwinner as a stressor event.

Later refinements of family stress theory recognized that familial reactions to stressors occur over time, adding to the model such concepts as stressor pileup and the build up or depletion of familial coping resources (Boss, 1988; McCubbin et al., 1980). Empirical research has found that two dimensions of family functioning, adaptability and cohesion, are useful indicators of family resources for dealing with potential stressor events. Family adaptability is the ability to change or revise organizational procedures in the face of new situations. Cohesion refers to the degree of attachment and bonding among family members. How families characteristically define situations and develop perceptions of events has been referred to as family paradigms, or family value and belief systems. An especially salient familial value orientation in coping with stressors is that of mastery/activity versus passivity/fatalism. Mastery families are more likely to move toward problem resolution, to shape circumstances and take control over their situation. Fatalistic families, by contrast, are less likely to take action to solve their problems and accept rather than shape their situations (Boss, 1988).

An extensive body of empirical work on family stress and coping suggests that families who score high on measures of cohesion, adaptability and mastery orientation manage stress more effectively, are less likely to devolve from stress into crisis, and more likely to recover from stressful events, situations or states of crisis (Boss, 1988; Figley, 1983; Ochberg, 1988). Somewhat surprisingly, however, neither criminal perpetration by nor victimization of a family member has been investigated as a critical event in this literature. Nonetheless, we can speculate that mastery families will pose special challenges to personnel in the criminal justice system, just as they do in medical settings (Heimer & Staffen, 1995). Such families will be likely to question police practices, become restive at the dilatory pace of criminal proceedings, and refuse to be mollified by standard assurances that their case is being accorded due process. A mastery orientation to crisis is often at odds with the business-as-usual orientation of personnel in such large-scale, people-processing bureaucracies as the criminal justice system.

Life Course Theory and Developmental Processes

Life course theories of criminal behavior offer a final example of theoretical linkages between families and crime. The life course approach arose originally in developmental psychology and developmental psychopathology to account for the indeterminancy of child outcomes based on predictive models that included only variables that described aspects of a child's origins, including family structure and process indicators. Risk and resilience models were postulated to explain why some children are able to overcome origins that seemed to stack the deck against positive developmental outcomes, while other children in the same circumstances fared so predictably poorly.

The potential complexity and sophistication of risk and resilience models proffer distinct conceptual advantages in the study of negative child outcomes, such as juvenile delinquency, with its multiple potential antecedents, pathways of influence, and potential outcomes. Factors that increase or decrease a child's vulnerability to harm can be identified within a child's social ecology, including family, neighborhood, school, and other social settings within which the child's daily life is lived. Such vulnerability and protective factors can be enduring and stable or transient and short-term. In risk and resilience models, a child's living in a violence-marred family, for example, can be identified as a threat to child wellbeing as a family origin factor that may predispose the child to seek affective ties outside the family, such as through close association with a sports team coach or through gang membership. The relationship between risk factors and outcomes, however, is not a deterministic one; the correlation is not perfect. Certain aspects of a child's living environment and prior experience may increase or decrease the child's risk of poor outcome resulting from his/her exposure to domestic violence. For example, the family's residence in a neighborhood with a high degree of social disorganization and normlessness, by itself a risk factor for negative child outcomes, might increase the degree of harm ensuing from parental marital violence alone, just as residence in a neighborhood with high social efficacy can counteract or buffer the negative potential of parental marital violence by reducing the potential for gang formation and persistence.

There is increasing reliance upon risk and resilience conceptual models in research on children and adolescence. Such models are able to take into account multiple sources of influence on child outcomes, influences that interact with other influences in highly intricate ways. This flexibility helps produce a more complete picture of the complexity of life course trajectories that appear to lead some children inexorably into criminal pathways and allow others to chart more conventional life courses.

Since the mid 1980s, life course and developmental approaches have become increasingly popular in criminology (for an overview, see Thornberry, 1997). The most important and influential work done in the life course perspective is Robert Sampson's & John Laub's (1993) reanalysis of the famous data on 1,000 lower class boys that Sheldon & Elenore Glueck gathered over half a century ago (1950). The work by Sampson & Laub (1993) indicates that families influence crime in different ways at different points in the life course. In early childhood, families may promote later delinquency by their disciplinary practices. Children who experience harsh and erratic discipline from their parents are at greater risk of developing into serious delinquents than those whose parents use other means of controlling their children. However, later in the life course, families can serve as an important route out of crime. Delinquents who marry and develop strong affective ties with their spouses are much more likely to desist from crime than their counterparts who do not marry or who enter into troubled relationships with women (Laub, Nagin & Sampson 1998). Thus, good marriages can serve as an important transition in life course trajectories in crime.

Major government sponsored research efforts are now underway that seek to chart the lives of high-risk youth from early childhood on to adulthood. In the Program on Research on the Causes and Correlates of Delinquency, over 4,000 children, parents, and others at research sites located in Rochester, New York, Denver, Colorado, and Pittsburgh, Pennsylvania have been interviewed at regular intervals for over a decade. This large-scale multi-disciplinary project has produced more publications and bulletins than can be listed here. This project is important not only because of the wealth of material it has produced on how early childhood experiences in the family influence the risk of juvenile delinquency in adolescence, but also because it addresses how teenage delinquency affects the formation and maintenance of families by young people (Thornberry et al., 2000). The more teens are involved in delinquency and drug use the more likely they are to become teenage fathers and mothers.

Researchers in the United States do not have a monopoly on the life course approach to crime. Major longitudinal projects have been underway for some time in England (West & Farrington, 1973; 1977) and New Zealand (Henry et al., 1993). These international studies have produced findings that echo those of American-based studies on the development of delinquency. Early childhood experiences in the family matter a great deal.

EMPIRICAL LINKAGES BETWEEN FAMILIES AND CRIME

We turn now to the empirical linkages between families and crime, that is, the obvious and not so obvious ways in which real families are touched by crime and the justice system. The chapters in this book illustrate in different ways the empirical connection between families and crime. As shown in the Table of Contents, we have organized the chapters in this book in four parts, focusing on families as sites of criminal perpetration and victimization, families as sources of crime protection or vulnerability to children, the impact on families of involvement with the criminal justice system, and the potential for system reform.

The Family as a Site of Offending and Victimization

Ordinary family life is often enacted in the shadow of violations of the law. The practice of marital contraception, for example, was illegal until the U.S.

Supreme Court affirmed a conjugal right to privacy in 1965. The strategies couples devise for management of their post-divorce, co-parental relationships often violate the details of the judicial decrees governing child custody, visitation, and child support. Informal adoptions sometimes barely skirt laws proscribing the sale of children. With some regularity parents of teens often are knowingly or unwittingly guilty of facilitating violations of curfews, laws regulating operations of vehicles, and laws governing the underage consumption of alcohol. And yet, despite their ubiquity, few persons beyond a small circle of family members and close friends know – or care much – about the commission or prosecution of such crimes. Such crimes as these are only rarely treated with concern, and only rarely brought to the attention of authorities.

Such was the case until recently for crimes of violence between family members. Although Pleck (1989) has documented considerable variability over time in the level of public concern about domestic violence in our national history, scholars suggest that the most recent phase of heightened concern can trace its beginnings to the 1960s when pediatricians began to take serious note of the phenomenon of child abuse. Prior to the 1960s, family violence was not widely recognized as an important social or legal problem. Indeed, until the latter half of the nineteenth century, violence between family members enjoyed a tradition of legal legitimation. The law largely ignored or condoned violence in the family, particularly when family members held in higher social regard victimized their less powerful kin (Gelles 1982; Goode 1971). Thus, husbands could assault their wives and parents could abuse their children with a degree of severity that now would be regarded as unacceptable and trigger legal intervention.

The social and legal contexts of domestic violence have changed dramatically since the 1960s. Violence in families now is recognized as an important public problem, one to which significant legal resources are devoted. All fifty states have established mandatory reporting laws for child abuse, and police in many communities are now required to make arrests in cases of domestic violence between spouses and cohabiting partners. Underpinning and driving the changes in law and law enforcement were broad cultural and social forces that began to gather force in the 1960s. According to Straus (1974), family violence began to shift from a private issue to a high priority public issue in the 1960s for three reasons: (1) a growing public and scientific awareness of and sensitivity to violence, (2) the rise of the women's movement, which uncovered and brought into public view the plight of battered women, and (3) a paradigmatic change in social science thinking in which the heretofore dominant consensus model of society was supplanted by conflict and social action models.

These forces continue to operate today. The amount of public and research attention devoted to the problem of violence is now enormous. Even small increases or decreases in national violent crime rates garner headlines in newspapers and featured story status on nightly newscasts. Interest in violence has not been confined to the traditional street crimes of rape, robbery, aggravated assault and murder. Public and scientific concern also extends to forms of violence that previously were hidden from view – child abuse, violence between intimates, and abuse of the elderly. Recognizing the serious problem posed by domestic violence and in response to pressure from women's groups, Congress passed the Violence Against Women Act in 1994. It directed the National Research Council to develop a research agenda on violence against women. The Panel on Research on Violence Against Women (1996) subsequently produced an extensive review of the research literature and detailed recommendations for researchers. Many of the articles in this section reflect this agenda.

Although domestic violence between adults and child abuse are now widely acknowledged as important social problems, other forms of family violence have received less notice. Timothy Brezina investigates one of these forms, that of violence by teenagers against their parents. He shows that the violence that parents bestow upon their children in the form of corporal punishment sometimes is reciprocated by the children. Male adolescents who are physically punished by their parents are more likely to reciprocate than are teens whose parents use non-physical approaches to discipline.

The article by Paul Mazzerole, Jeffrey Maahs, and Ronet Bachman and the one by Jaana Haapasalo address an issue that has become a perennial topic of concern among family researchers and criminologists: the cycle of violence. A consistent conclusion of domestic violence research is that children who grow up in violent and abusive homes are at greater risk to become child and spouse abusers as adults than are children who grow up in homes with little or no violence (Widom, 1989). Haapasalo's research on men in prison in Finland indicates that the cycle of violence is a general pattern in family relationships and not a phenomenon that is unique to the United States.

One of the important issues that the Panel on Violence Against Women noted needed further investigation was the social location of domestic violence. The Panel recommended that researchers should focus on "the influence of socio-economic, cultural, and ethnic factors" on domestic violence. Michael L. Benson, Greer L. Fox, Alfred DeMaris and Judy Van Wyk attempt to address this issue in their article. Using a data set that was constructed by merging wave 2 of the National Survey of Families (NSFH) and Households with tract level data from the 1990 U.S. Census, they investigate the complex intersection of race, poverty, and community context.

The Family's Role in the Production, Mediation, and Prevention of Crime

As we noted above, the issue of violence in families became an important area of public concern in part because of an increased sensitivity to violence and because of the growing political power of women. There is also another reason why violence in families has increasingly attracted public and scientific attention. We now understand that domestic violence and traditional street crime violence are strongly linked together. Conflict and violence between spouses and between parents and children may spill over into violence outside the home. After a long hiatus during the 1950s, 60s and 70s, the idea that families are an important cause of crime rebounded in popularity in the 1980s. With the rise of developmental and life course approaches to criminality in the 1980s, research on early family experiences as precursors to juvenile delinquency and adult crime exploded. Influential theoretical statements by Wilson & Herrnstein (1985) and Gottfredson & Hirschi (1990) were accompanied by comprehensive research reviews by Loeber & Stouthammer-Loeber (1986) and Farrington (1978). These and other works provided theoretical cover for delinquency researchers who wanted to turn their attention away from peers and toward families.

Contemporary research on families and crime is distinguished from earlier research in several ways. One important and much needed development is the growing attention being paid to delinquent girls. Although teenage boys remain heavily over-represented in official crime statistics and continue to dominate the research literature, girls are no longer completely ignored. John Hagan's power control theory represents perhaps the most systematic attempt to account for gender differences in delinquency (Hagan, Gillis & Simpson, 1985). In this volume, Ryan E. Spohn investigates gender differences in the effects of child maltreatment on criminal activiy.

Another sea change in family research has been the rise of developmental and life course approaches. The life course approach has sensitized researchers to the importance of continuity and discontinuity in behavioral trajectories. With the advent of the life course approach and the growing use of longitudinal research designs, it has become clear that teenage delinquency is often preceded by problem behavior in early childhood and followed by serious crime in adulthood. Yet, many children who engage in delinquency during their teenage years do not have histories of problem behavior as young children, and many, indeed most, delinquents do not go on to become adult criminals. These individuals display discontinuity in their behavior. Identifying the factors which contribute to continuity or which facilitate discontinuity has become perhaps the central question in delinquency research. Being raised in a troubled family is one of the factors that consistently is associated with individual continuity in anti-social behavior.

In this volume, Robert Agnew, Cesar Rebellon and Sherod Thaxton carefully delineate the different ways in which troubles in families, particularly those involving parent-child relations, may lead to strain in children. They explore how children may respond to strain by engaging in crime and delinquency. Approaching the connection between parents and children from a different perspective, John Paul Wright, Francis T. Cullen, and John D. Wooldredge explore the role of parental support in mitigating and reducing the extent to which juveniles engage in delinquency.

The life course approach has helped us understand that it is useful to conceive of crime and delinquency as independent as well as dependent variables. Researchers are now increasingly aware that involvement in crime has collateral consequences for individuals in other domains of life, most notably employment and family life. For example, persons who compile serious criminal records as juvenile delinquents appear to have reduced employment opportunities when they reach their twenties compared to their less criminal counterparts. Early involvement in crime and delinquency also predicts difficult marital relations in the future. Thus, crime is no longer viewed as simply a dependent variable that we try to explain by reference to factors that precede it in temporal order. Rather, crime is now considered as one event in a chain of events that have reciprocal effects on one another over time. The article by Carolyn A. Smith, Marvin D. Krohn, Alan J. Lizotte, Cynthia Perez McCluskey, Magda Stouthamer-Loeber, and Anne Weiher addresses this issue. They explore how involvement in delinquency and substance abuse may lead to what they call precocious transitions to adulthood, such as leaving home and causing pregnancies.

A final important development in family research and criminology is the growing recognition of the importance of contextual effects. Families are no longer considered in isolation from the neighborhoods and communities in which they are located. While what goes on behind the closed doors of a family's domicile obviously is important, it does not occur in a vacuum. Families invariably and unavoidably are touched by what goes on around them on the street, in the school, and in the houses of other families. After a long period in limbo, social disorganization theory has been reinvigorated and given new life (Bursik, 1988; Sampson, 1997, 1999). Researchers are beginning to understand how neighborhoods vary in collective efficacy, which refers to a neighborhood's sense of active engagement in solving problems and stimulating the informal social control and mutual support of children (Sampson et al., 1999).

Two articles in this volume illustrate this theme. Gary L. Bowen, Natasha K. Bowen, and Patty Cook investigate how neighborhood characteristics influence the parenting practices of single mothers. Greer L. Fox explores how families help or fail to help children deal with exposure to extreme violence, an aspect of the families and crime connection that has received little attention from researchers. This issue is important because of the localization of violence. The chances that a child who grows up in an economically and socially disadvantaged area will witness violence in person during their childhood is much greater than it is for children who grow up in middle class suburbs. In a provocative analysis, Fox explores class differences in responses to violence.

Families in the Justice System

According to the National Crime Victimization Survey, in 1994 there were over 42,000,000 criminal victimizations in the United States, including over 10,000,000 crimes of violence. That same year, there were over 911,000 felony convictions in state and federal courts (U.S. Department of Justice 1997a). In 1996, 1,630,940 persons were incarcerated in federal, state, and local institutions (U.S. Department of Justice 1997b). When these numbers are reconfigured to include the related family members of victims and offenders, the number of persons touched by crime and the criminal justice system is very large indeed.

The scholarly literatures in criminology and family sciences, including clinical family social work, are surprisingly thin in investigations of patterns by which families manage their experiences of direct and indirect victimization by the criminal involvement of family members as offenders or victims. Most victimization research concerns the reactions and responses of individuals, not families, to traumatic experiences and their aftermath. Yet, as research on other kinds of stressors shows, the involvement of families can be critical in moderating or buffering the potentially long-term negative effects of stress and crisis, such as the experience of criminal victimization, subsequent involvement with the criminal justice system, or incarceration.

One of the most important questions left unanswered by recent victimization research is how families come to view justice in the abstract after experiencing the reality of crime and the criminal justice system. This society has a vested interest in the development and maintenance of a 'sense of justice' among all citizens. Viewed sociologically, crime victimization or perpetration represents an assault on the internalized normative order, the sense that the world is a 'just world'. Importantly, the sense of trust in the social order is not just a construction of the individual psyche. Rather, in the truest Durkheimian sense, it is a social phenomenon; it resides in the domain of the collective conscience. In sociological terms, it is the social order which is violated by acts of crime. Destruction of trust in and commitment to the social order is one of the most important outcomes of crime; reconstruction of the sense of the world as a just and orderly place is one of the most important outcomes of criminal justice (Garofalo, 1981; Greenberg & Ruback, 1984; Janoff-Bulman & Frieze, 1983). The sense of justice is sorely strained for families touched by crime, whether as families of the victim or of the offender or both, and the criminal justice system is a major contingency in the process of family reaction to criminal involvement, whether families are those of the offenders or victims. Of special concern is the impact of their experiences with the criminal justice system on the family's subsequent orientation to the normative order. The criminal justice system can reinforce or undermine a family's sense of trust and commitment to the system. By viewing the criminal justice system through the eyes of inadvertent participants, that is, family members of perpetrators and victims, several chapters in this section suggest where it erodes or sustains their sense of fairness, equity and justice.

The chapter by Eric Stewart, Ron Simons, and Rand Conger examines the aftermath of involvement with the criminal justice system on parent-adolescent relationships, by looking at parental behaviors *after* a youth's delinquent behavior has resulted in police and court contact. Thus, these researchers reverse the usual research focus on the impact of parenting styles on children's delinquency outcomes. By so doing, they add to our understanding of the multiple pathways through which youth are embedded in a cycle of increasing disadvantage, in this case through the unintended and deleterious effects of legal sanction on the perpetuation and exacerbation of harsh, and counterproductive, parental disciplinary styles.

The next two chapters focus on young fathers who are or have been incarcerated during their children's early years of development. Anne Nurse's ethnography explores the transitions of young men upon release from prison and their often futile attempts at reintegration into the lives of families left behind. Unrealistic expectations, dashed hopes, broken promises, and unrelenting difficulties with custody and child support dog many of these young fathers, rendering them especially vulnerable to a return to the criminal activities that separated them from their families in the first place. The chapter by Bruce Western and Sara McLanahan complements Nurse's analysis by providing an exacting assessment of the impact of incarceration on family formation among low-income parents, using data from the Fragile Families study, a multi-site investigation of young families. The authors argue that incarceration of fathers discourages the formation and stability of young families both directly by making it difficult for fathers to live with their children during their prison terms and indirectly through decreasing their employment and earnings prospects upon release. They point to the growth in the prison population, especially among young men and women of color, as a large contributor to the rise of family instability and single parent families among minority groups.

The final two chapters in this section focus on families of offenders. Karen Mason's chapter describes the familial reactions and adjustment processes among middle-class families as reported by a small sample of previously imprisoned white-collar offenders. Her analysis highlights the impact of community ostracism and stigma in creating strain for these families. She also elaborates how the indirect experience of criminal justice processes through the treatment of their family member invites contempt of the justice system and its officers, and gives rise to negative sentiments about the legitimacy of the law. The final chapter in this section, that of Margaret Vandiver and Felix Berardo, focuses on the families of the more than 3,500 inmates currently condemned and living on death row. Using a comparative framework, Vandiver and Berardo contrast the experiences of these families with those of inmates incarcerated for less serious offenses, with families of homicide victims, and with other families who have endured loss. The authors also review the available research on executions, including such difficult issues as the effect of executions on witnesses, on families of victims, and on families of the condemned prisoner.

Opportunities for System Reform

Paradoxically, the involvement of increasing numbers of people in the criminal justice system, both victims and offenders and their respective families, has opened doors to changes in the system. The victimization movement, for example, has spawned interest in the ways certain crimes have traditionally been conceptualized by law enforcement agencies. Homicide, for example, until recently was described in such publications as the Uniform Crime Reports as a crime that was unpredictable and unpreventable, and therefore, not responsive to law enforcement procedures. However, once it was noted that certain intra-familial homicides followed a discernible pattern, that is, they followed closely a battered wife's determination to leave a batterer, it became possible to determine homicide risk and to alter police practices to prevent homicides in such cases. Similarly, the very large numbers of young men involved in gun-related crimes as both victims and perpetrators spurred the involvement of public health workers and emergency room physicians in seeking system changes. Their involvement led to the application of a public health perspective in a re-examination of the nature of youth violence, and the role of guns in such violence (Prothrow-Stith, 1992). This in turn has brought about increased regulatory attention to gun manufacture, marketing, and distribution as supplemental policies to the previous exclusive reliance upon punitive measures against young people to control gun violence.

Vehicular homicide provides an example of a crime in which the involvement of victims' families has sparked system reform. Although in many jurisdictions vehicular homicide continues to be treated in the criminal codes and criminal justice system as a mere accident with unfortunate and tragic consequences, in other states there has been enormous reform. Unlike most crimes of violence, such as street crimes, which are largely concentrated among persons of lower economic and social standing, vehicular homicide is a more democratic crime. Its victims are drawn across a broader socioeconomic spectrum. Vehicular homicide and other 'democratic crimes' that are not selective by socioeconomic status bring more articulate families into the criminal justice system. Such families are both more capable of and more likely to translate their anguish and outrage into demands for system reform and to carry out campaigns for such reform both inside and outside the system. The success of MADD, Mothers Against Drunk Driving, in reshaping attitudes about the culpability of alcohol-impaired drivers for their crimes is an example of a family-based victims group. Similar victims advocacy groups, such as Parents of Murdered Children and the National Organization for Victims Assistance, have been instrumental in altering the standing of victims' families in the criminal prosecution of homicide perpetrators.

Currently the rate of incarceration in the United States is among the highest rates of incarceration in the world. Among black men 18–34 years old, the incarceration rate reaches 1 in 10, and an even larger proportion – 1 in 3 – of young black men are admitted to prisons or jails in the course of a year (Donziger, 1996, 35). Whether the incarceration of such large proportions of young men and women, often for crimes that were not considered serious enough for incarceration a mere decade ago, will lead to changes in law, prosecutorial practice, and criminal justice remains to be seen. Several chapters in this volume have pointed out the unintended consequences of such practices not just for the families directly involved, but for the larger society as well. The long term consequences of our current practices are projected to exacerbate the amount of criminal activity in the future in no small part because they undermine the ability of families to form, survive, and care adequately for their members.

One of the two chapters in this section offers a perspective on the need for system reform because of the consequences for children of the explosive growth in the population of incarcerated mothers. The chapter by Jacquelyn Sandifer and Suzanne Kurth provides an overview of the large numbers of women who have been caught in the increasingly wide net of arrest, prosecution, conviction, and mandatory sentencing with incarceration for drug-related offenses. They argue that the children of incarcerated women are not only unrecognized and underserved, but are at high risk for disruptions in care and custody. Two strengths of their analysis are their review of policies and practices of both the criminal justice and social welfare bureaucracies that facilitate and hinder the maintenance of mother-child ties during arrest and incarceration and their suggestions for supplement and change in those systems.

The chapter by Stephen Gavazzi, Courtney Yarcheck, Deborah Wasserman, and Charles Partridge reviews the limitations of retribution or punishment as a goal of juvenile justice. As an alternative, the authors review work based upon the Balanced and Restorative Justice (BARJ) model, which emphasizes the offender's responsibility for restitution. The BARJ model focuses on the harm that the crime has caused, not solely to the victim, but also to the community at large and to the offender, himself or herself, and the offender's role in making amends and restoring losses. The authors describe an innovative series of family-based interventions developed to implement the BARJ model, which offer a potentially fruitful alternative to punitive responses to juvenile crime.

CONCLUSION

We close with the observation that the past 50 years in American society has witnessed the gradual processes of deinstitutionalization of the family system and delegitimation of the justice system. As it pertains to the family, there is plentiful demographic evidence that Americans are making different choices than did persons of their parents' generation in terms of marriage and family formation and dissolution. And it is equally evident that roles and relationships within intimate settings have changed in both dramatic and minor ways for women and men. Moreover, fewer men and women feel compelled to justify their life style choices, ignoring the heretofore strong normative pressures to live in conformity with the prevailing family paradigm of the two-parent, one breadwinner, heterosexual, child-present, lifelong family.

As more people have turned to other institutions and settings to meet needs for affection, guidance, sustenance, and care that once were met primarily or exclusively in family settings, we continue to learn about the essential importance of participation and embeddedness in caring familial relationships to optimal human development across the life course. There are measurable human costs of a social system that fails to provide the pervasive structural supports necessary for a strong family system.

Similarly, the justice system itself faces increasing public resistance and disillusionment as it struggles to meet demands for swift and sure justice. Part of the struggles of justice can be attributed to an increase in numbers of cases that need to be processed at all levels of the system; the system is inefficient and thereby unfair. But part of its troubles stem from the difficulty in the articulation and implementation of justice ideals amid a diverse public among whom there is decreasing consensus on the meaning of justice. However inefficiently, whatever justice is finally delivered will be perceived by a portion of that public as incomplete, unsure, and thereby illegitimate.

The complementary relationship between the institutions of the family and justice envisioned by the parsonian sociologists no longer holds. The justice system is increasingly called upon not simply to act as an agent of social control and social regulation, but as a family institution as well. As a larger proportion of the population spends ever increasing amounts of its life course incarcerated within the justice system, that system inadvertently becomes a substitute for adult family roles and settings for its inmates and a destroyer of the very family relationships that, in the past, have offered the surest pathways away from crime. Given the continual unfolding of these social processes, we urge that connections between families and crime, and the institutional interface between family and justice systems, be given the serious scholarly attention they deserve. This book is one step in that direction.

REFERENCES

Benson, M. L. (1985). Denying the Guilty Mind: Accounting for Involvement in White-Collar Crime. *Criminology, 23*, 583–607.

Boss, P. (1988). Family Stress Management. Newbury Park, CA: Sage.

Bursik, R. (1988). Social Disorganization Theories of Crime and Delinquency: Problems and Prospects. *Criminology, 26*, 519–551.

Cullen, F. T. (1994). Social Support as an Organizing Concept for Criminology: Presidential Address to the Academy of Criminal Justice Sciences. *Justice Quarterly, 11*, 527–559.

Daly, K. (1989). Gender and Varieties of White-Collar Crime. *Criminology, 27*, 769–794.

Donziger, S. R. (1996). *The Real War on Crime: The Report of the National Criminal Justice Commission.* New York: HarperCollins.

Elias, R. (1986). *The Politics of Victimization: Victims, Victimology, and Human Rights.* New York: Oxford University Press.

Farrington, D. P. (1978). The Family Background of Aggressive Youth. In: L. A. Hersov, M. Berger, & D. Shaffer (Eds), *Aggression and Antisocial Behavior in Childhood and Adolescence.* Oxford: Pergamon Press.

Figley, C. R. (1983). Catastrophes: An Overview of Family Reactions. In: C. R. Figley, & H. I. McCubbin (Eds), *Stress and the Family: Vol.2. Coping with Catastrophe.* New York: Bruner/Mazel.

Garofalo, J. (1981). The Fear of Crime: Causes and Consequences. *The Journal of Criminal Law & Criminology, 72*, 839–857.

Gelles, R. J. (1982). Domestic Criminal Violence. In: M. E. Wolfgang, & N. A. Weiner (Eds), *Criminal Violence*, (201–235). Beverley Hills, CA: Sage Publications.

Glueck, S., & Glueck, E. (1950). *Unraveling Juvenile Delinquency*. Cambridge, MA: Harvard University Press.

Goode, W. J. (1971). Force and Violence in the Family. *Journal of Marriage and the Family, 33*, 624–636.

Gottfredson, M. R., & Hirschi, T. (1990). *A General Theory of Crime*. Stanford, CA: Stanford University Press.

Greenberg, M. S., & Ruback, R. B. (1984). Criminal Victimization: Introduction and Overview. *Journal of Social Issues, 40*, 1–8.

Hagan, J., A. R. G., & Simpson, J. (1985). The Class Structure of Gender and Delinquency: Toward a Power-Control Theory of Common Delinquent Behavior. *American Journal of Sociology, 90*, 1151–1178.

Heimer, C. A., & Staffen, L.R. (1995). Interdependence and reintegrative social control: Labeling and reforming 'inappropriate' parents in neonatal intensive care units. *American Sociological Review, 60*, 635–654.

Henry, B., Moffitt, T., Robins, L., Earls, F., & Silva, P. (1993). Early Family Predictors of Child and Adolescent Behavior: Who Are the Mothers of Delinquents? *Criminal Behavior and Mental Health, 3*, 97–118.

Hill, R. (1949). *Families Under Stress: Adjustment to the Crises of War Separation and Reunion*. New York: Harper.

Janoff-Bulman, R., & Frieze, I. H. (1983). A Theoretical Perspective for Understanding Reactions to Victimization. *Journal of Social Issues, 39*, 1–17.

Laub, J., Nagin, D. S., & Sampson, R. J. 1998. Trajectories of Change in Criminal Offending: Good Marriages and the Desistance Process. *American Sociological Review, 63*, 225–238.

Loeber, R., & Stouthamer-Loeber, M. (1986). Family Factors as Correlates and Predictors of Juvenile Conduct Problems and Delinquency. In: M. Tonry, & N. Morris (Eds), *Crime and Justice: An Annual Review of Research, 7*, 29–149. Chicago: University of Chicago Press.

McCubbin, H. I., Constance, B. J., Cauble, A. E., Comeau, J. K., Patterson, J. M., & Needle, R. H. (1980). Family Stress and Coping: A Decade Review. *Journal of Marriage and the Family, 42*, 855–871.

Ochberg, F. M. (1988). Post-traumatic therapy and victims of violence. In: F. M. Ochberg (Ed.), *Post-traumatic Therapy and Victims of Violence*. New York: Bruner/Mazel.

Panel on Research on Violence Against Women. (1996). *Understanding Violence Against Women*. Washington, D. C: National Academy Press.

Parsons, T. C., & Bales, R. F. (1955). *Family, Socialization and Interaction Process*. New York: Free Press.

Pleck, E. (1989). Criminal Approaches to Family Violence, 1640–1980. In: L. Ohlin, & M. Tonry (Eds), *Family Violence*, 1640–1980. Chicago, IL: University of Chicago Press.

Prothrow-Stith, D. (1991). *Deadly Consequences: How Violence Is Destroying Our Teenage Population and A Plan to Begin Solving the Problem* New York: Harper Collins.

Sampson R. J., & Laub, J. H. (1993). *Crime in the Making: Pathways and Turning Points Through Life*. Cambridge, MA: Harvard University Press.

Sampson, R. J., Raudenbush, S. W., & Earls, F. (1997). Neighborhoods and Violent Crime: A Multilevel Study of Collective Efficacy. *Science, 277*, 918–924.

Sampson, R. J., Morenhoff, J. D., & Earls, F. (1999). Beyond Social Capital: Spatial Dynamics of Collective Efficacy for Children. *American Sociological Review, 64*, 633–660.

Straus, M. A. (1974). Leveling, Civility, and Violence in the Family. *Journal of Marriage and the Family, 36*, 13–29.

Thornberry, T. P. (Ed.). (1997). *Developmental Theories of Crime and Delinquency*. New Brunswick, NJ: Transaction Publishers.

Thornberry, T. P., Wei, E. H., Stouthamer-Loeber, M., & Van Dyke, J. (2000). Teenage Fatherhood and Delinquent Behavior. *Juvenile Justice Bulletin*, January 2000. Washington DC: U.S. Department of Justice.

U.S. Department of Justice, Bureau of Justice Statistics. (1997a). *Felony Sentences in the United States 1994*. Bulletin NCJ-165149. Washington, DC: Department of Justice.

U.S. Department of Justice, Bureau of Justice Statistics. (1997b). *Prison and Jail Inmates at Midyear 1996*. Bulletin NCJ-162843. Washington, DC: Department of Justice.

West, D. J., & Farrington, D. P. (1973). *Who Becomes Delinquent?* London: Heinemann.

West, D. J., & Farrington, D. P. (1977). *The Delinquent Way of Life*. London: Heinemann.

Widom, C. S. (1991). The Cycle of Violence. *Science, 244,* 160–166.

Wilson, J. Q., & Herrnstein, R. J. (1985). *Crime and Human Nature*. New York: Simon and Shuster.

PART I

THE FAMILY AS A SITE OF OFFENDING AND VICTIMIZATION

CORPORAL PUNISHMENT AS A CAUSE OF TEENAGE VIOLENCE TOWARD PARENTS: DATA FROM A NATIONAL SAMPLE OF MALE ADOLESCENTS

Timothy Brezina

Available estimates indicate that adolescent-to-parent violence represents a significant social problem in the United States. In particular, national estimates reveal that roughly 10% of all adolescents hit their parents each year (Cornell & Gelles, 1982; Peek, Fischer & Kidwell, 1985; Straus, Gelles & Steinmetz, 1980). Although this percentage may appear small, when projected to the total population of adolescents it translates into roughly 3 million cases of adolescent-to-parent assault annually. Moreover, child-initiated assaults are believed to contribute to much psychological distress for family members and at least some of these attacks result in physical injury (e.g., cuts, welts, bruises, major lacerations, and broken bones) (Agnew & Huguley, 1989; Charles, 1986; Harbin & Madden, 1979).

Yet the issue of adolescent-to-parent violence has received only limited attention from researchers. Studies on domestic violence focus predominantly on

---

Author's Note: An earlier version of this chapter was presented at the 1999 meetings of the American Sociological Association, Chicago, IL. I wish to thank Murray A. Straus, Michael L. Benson, and Greer Litton Fox for their helpful comments and suggestions.

Families, Crime and Criminal Justice, Volume 2, pages 25–43.

ISBN: 0-7623-0737-4

violence between spouses or violence by parents against children. Only a handful of studies give systematic treatment to parent battering by adolescent children, and several of these investigations are limited to the examination of small clinical samples (for exceptions, see Agnew & Huguley, 1989; Brezina, 1999; Browne & Hamilton, 1998; Cornell & Gelles, 1982; Peek et al.1985; Straus et al., 1980).

The relative dearth of research on parent assault is surprising and not fully understood, although several explanations have been proffered. First, unlike cases of child abuse or spousal assault which "typically involve attacks by *more* powerful individuals on *less* powerful individuals," parent assault appears, from the perspective of an outside observer, to involve attacks by less powerful individuals on more powerful others (Agnew & Huguley, 1989, 699, emphasis in original). Consequently, few researchers may suspect that parent assault is an extensive problem. Second, research on parent assault may have been discouraged by both the taboo nature of the topic (see Straus et al., 1980) and its lack of visibility in the public arena. For example, parents of abusive children often try to conceal evidence of their victimization and relatively little effort has been made to bring their plight to the attention of the general public[1] (Agnew & Huguley, 1989; Charles, 1986; Gelles 1997; Harbin & Madden 1979).

Whatever the causes of neglect, existing studies indicate that the problem of adolescent-to-parent assault is extensive, it is therefore deserving of scientific scrutiny, and many researchers agree that more research on the topic is needed (e.g., Agnew & Huguley, 1989; Brezina, 1999; Cornell & Gelles, 1982; Peek et al., 1985). The need for additional research is especially apparent in light of the unique correlates of parent assault. According to the results of prior research, many of the variables affecting adult-initiated family violence (such as socio-economic status, stress, and social isolation) do not appear to have the same (or any) impact on child-initiated family violence (Agnew & Huguley, 1989; Cornell & Gelles, 1982; Peek et al., 1985). As Peek et al. (1985, 1057) observe, "teenage violence toward parents may not be cut from the same cloth as adult-initiated family violence."

Although limited in number, existing studies document the importance of variables highlighted in the juvenile delinquency as opposed to family violence literature, such as parental attachment and parental punitiveness (Agnew & Huguley, 1989; Cornell & Gelles, 1982; Peek et al., 1985). As several researchers emphasize (Peek et al., 1985; Straus et al., 1980), the control strategies exercised by parents appear to be of special significance. Reliance on physical punishment (e.g., slapping and hitting) is one such strategy and is believed to have a particularly important connection to parent assault.

The fact that corporal punishment may contribute to adolescent-to-parent violence is of special interest because it is likely to have important policy implications. If a causal relationship exists between corporal punishment and teenage violence toward parents, then it may be possible to prevent this form of family violence by reducing the use of corporal punishment. The causal status of the relationship between corporal punishment and parent assault, however, remains unclear. Existing studies on the link between corporal punishment and parent assault are plagued by a number of methodological limitations. Although the preponderance of such studies observe an association between corporal punishment and parent assault, most fail to control for other important aspects of parenting (e.g., the overall quality of the parent-child relationship and consistency of discipline) and almost all rely on cross-sectional data. The possibility exists, then, that the observed relationship is spurious.

The purpose of this study is to clarify the relationship between corporal punishment and teenage violence toward parents. Longitudinal data from a national survey of male adolescents are used to examine the effect of corporal punishment on the occurrence of adolescent-to-parent assault. To avoid the methodological problems identified above, the analysis includes controls for prior level of assault and a range of other parenting characteristics such as parent-child attachment, parental consistency, and emotional maltreatment. The following section provides a more detailed review of theory and research on corporal punishment and its possible connection to parent assault.

THE LINK BETWEEN CORPORAL PUNISHMENT AND PARENT ASSAULT

Definition of Corporal Punishment

To be consistent with several recent studies, corporal punishment is defined in this chapter as the "use of physical force with the intention of causing a child to experience pain but not injury, for purposes of correction or control of the child's behavior" (Straus & Donnelly, 1993, 420; Straus, Sugarman & Giles-Sims, 1997). Examples of corporal punishment defined in these terms include spanking, slapping, grabbing, and hitting the child with the aid of a belt or paddle. Although it is believed that corporal punishment is used mainly to discipline young children (Straus & Donnelly, 1993), evidence described below indicates that parents frequently rely on corporal punishment to control adolescent children as well. In addition, there are strong indications that corporal punishment is a contributing factor to adolescent-to-parent violence.

Prevalence and Chronicity

Data from several large and nationally representative surveys indicate that the corporal punishment of adolescents is highly prevalent in the United States. Drawing on data from a national survey of male adolescents, Bachman (1975) reports that 60% had been slapped by their parents at least once during their teens. Drawing on more recent data, Straus & Donnelly (1993) report 'lower bound' corporal-punishment estimates of 44% for female adolescents and 58% for male adolescents. These data suggest, then, that many female adolescents and the *majority* of male adolescents in the United States have been subject to physical punishment administered by their parents.

The data presented by Straus & Donnelly (1993) also provide information on the chronicity (or incidence) of corporal punishment during adolescence. Their findings show that, of the parents who hit their adolescent children, "70% to 90% did it more than once during the 12 months" preceding the survey interview (Straus & Donnelly, 1993, 431). On average, such parents slapped or hit their adolescent children 6 to 8 times during the 12-month period, although a significant percentage of these parents (at least 10%) slapped or hit them 21 times or more. In short, when parents of adolescent children resort to corporal punishment, it is usually not limited to a single or isolated incident.

Theoretical Grounds for Concern

The above findings are noteworthy because of concerns about the harmful effects of corporal punishment on adolescent development, especially its possible contribution to violent behavior. As suggested above, the consequences of corporal punishment may involve an increased likelihood of parent assault. Moreover, there are strong theoretical grounds for anticipating such a relationship.

Social-psychological strain theorists, for example, assume that negative treatment by parents is a leading cause of juvenile delinquency and aggression. Negative parental treatment in the form of frequent physical or verbal attacks (e.g., slapping, hitting, yelling, screaming) is said to generate negative affect and thus create pressures for 'corrective action' (Agnew, 1985, 1992; for empirical validation see Brezina, 1996). Delinquency is one possible response and may represent a means to cope with negative treatment or 'strain,' especially when legitimate coping strategies are taxed or unavailable. As Agnew (1992, 60) contends, delinquency or aggression "may be a method for alleviating strain . . . for protecting or retrieving valued stimuli, or for terminating or escaping negative stimuli."

From the standpoint of social-psychological strain theory, frequent exposure to corporal punishment would be expected to increase the likelihood of delinquency and aggression, including child-to-parent assault. Parent assault is one means of attacking the source of strain, in the hopes of terminating noxious stimuli. This type of response may be especially likely among youths who are not otherwise able to avoid or escape physical punishment, and who therefore strike back at their parents in anger or frustration (Agnew & Huguley, 1989).

Social learning theorists also predict a relationship between corporal punishment and parent assault, but for different reasons. Social learning theorists stress that the use of corporal punishment may teach children to value violence as a mode of conflict resolution. A favorable orientation toward violence is, in turn, likely to affect how children respond to conflict with their parents and is expected to promote child-to-parent interactions of an aggressive nature:

> children tend to model their behavior after that of their parents. When a parent hits a child, it is a powerful learning experience, teaching the moral rightness of the use of force between family members (Straus et al., 1980, 121).

The work of Gerald Patterson and his associates appears to be particularly relevant to the issue of child-to-parent violence. Patterson also adopts a social learning framework, but emphasizes the processes of reinforcement that are believed to shape and sustain aggressive behavior patterns in the context of dysfunctional family dynamics (Patterson, Littman & Bricker, 1967; Patterson, 1982). According to Patterson's (1982) 'coercion model', provocations or attacks by parents or other family members (e.g., teasing and slapping) tend to elicit aggressive counterattacks from the child. When sustained with sufficient force or persistence, these counterattacks are often successful in terminating the initial attack and, when this occurs, the child comes to learn the value of aggression as a protective response. As Patterson (1982, 151) describes:

> The key assumption is that coercive [i.e., aggressive] child behavior may serve the function of terminating aversive intrusions by other family members . . . The major reinforcer for attack-instigated aggression is termination of the attack.

In short, parent assault is a possible if not likely response to corporal punishment because the display of violent counter-attack by the child can serve to intimidate parents and force them to 'back off' (for relevant evidence on this point, see Brezina, 1999). Although Patterson's model is based on observations of young children, one might suspect that adolescents would be particularly prone to this type of display. Adolescent children are, after all, more likely to possess the capacity for physical retaliation. As Wolman (1998) observes, awareness of increasing physical size and strength tempts teens to "rebel against adults who *were* stronger and dominated them, especially parents and teachers" (p. 88, emphasis in original). Likewise, as Abrams (1981) observes, children develop resources of their own as they grow up. In response to physical punishment, they can "hit back and choose devious means of behavior as they grow larger" (p. 254). The observations of Straus et al. (1980, 121) suggest that some parents intuitively recognize this fact:

> Parents of older children often say they don't hit them any more because "they're too big now." This is often said in the sense of retaliation being dangerous, rather than because they think it is wrong to hit children of that age.

Of course, not all teens are prone to violence toward their parents. Not all teens are subjected to corporal punishment and, among those that are subjected, violent counter-attack cannot be described as an inevitable response. As stressed by Wolman (1998), adolescents are also aware of their dependence on parents and this fact usually serves to inhibit the most extreme forms of rebellion. Nevertheless, the theoretical statements reviewed above suggest that, for various reasons, corporal punishment is likely to increase the *odds* of adolescent-to-parent assault, especially when administered in a frequent or persistent manner and, perhaps, among children who are not particularly dependent on their parents, emotionally or otherwise.

Studies Linking Corporal Punishment to Parent Assault

Consistent with theoretical expectations, indications of a possible connection between corporal punishment and child-to-parent violence are provided by both clinical studies and studies based on survey data. Some of these studies focus on children of all ages, and some include measures that may conflate corporal punishment with other forms of negative treatment (e.g., yelling and screaming). Because such studies are nonetheless informative, they are described below. However, studies that focus exclusively on adolescents and/or corporal punishment are emphasized.

Drawing on a small and clinical sample of assaultive children and their parents, Harbin & Madden (1979) observe that at least some cases of parent battering appear to be precipitated by the actions of the parent, including the delivery of hostile statements and/or corporal punishment. In one such case, "An 11-year-old boy whose mother had spanked him for disobeying orders pushed her into a door, breaking her coccyx, and then kicked her in the face while she was on the floor" (Harbin & Madden, 1979, 1289).

Consistent with the above observations are findings from four studies based on nationally representative survey data. Based on their analysis of interview data collected from 1,146 families with children between the ages of 3 and 17,

Straus et al. (1980, 119) conclude that, "the more often parents have hit the child, the higher the probability that the child has hit the parent." After restricting the analysis of these same data to families with adolescent children (10 to 17 years of age), Cornell & Gelles (1982) observe the same relationship: the physical assault of adolescents by parents increases the likelihood of parent assault. A third study, conducted by Peek et al. (1985) and based on data from the *Youth in Transition Survey* (Bachman, 1975), finds the use of a 'violent parenting style' to be positively correlated with teenage violence toward parents. Their measure of 'violent parenting style' included three items: (1) parental slapping, (2) threatening to slap, and (3) yelling, shouting, or screaming at the child. A fourth study, conducted by the present author (Brezina, 1999) also relies on the *Youth in Transition* data, but draws on several waves of these data to examine longitudinal effects. Although the main purpose of this study was to examine the *consequences* rather than causes of parent battering, the findings nonetheless show that parental slapping has a positive effect on subsequent parent assault, controlling for prior level of assault.

Methodological Limitations

The above studies are highly suggestive of a causal link between corporal punishment and teenage violence toward parents. However, as most of the authors recognize, all of these studies suffer from important methodological limitations. First, all but one of these studies (Brezina, 1999) are based on cross-sectional analyses, precluding "a definitive conclusion that striking a child will cause that child to . . . strike his or her parent" (Cornell & Gelles, 1982, 13). Second, at least one of the above studies conflates physical punishment with other forms of aversive treatment such as verbal threats and shouting.

Third, *all* of the above studies fail to control for an adequate range of parenting variables. Children exposed to corporal punishment may exhibit a greater propensity to assault their parents because corporal punishment may serve as a rough proxy for other problems in the parent-child relationship – not because of corporal punishment per se (see also Simons, Johnson & Conger, 1994). It is important, therefore, to control for other dimensions of parenting such as parental attachment, verbal/emotional maltreatment, and other variables indexing the quality of the parent-child relationship. In fact, Agnew & Huguley (1989) failed to observe a significant relationship between corporal punishment and parent assault after controlling for parental attachment – although it should be noted that their study is also limited by a reliance on cross-sectional data.

In addition to the quality of the parent-child relationship, it is believed that *consistency* of parental discipline is another important factor in the development of adolescent antisocial behavior. Data show that the inconsistent or erratic application of punishment is a significant correlate of delinquency in general, and some researchers argue that the consistency of punishment is a more important consideration than form of punishment (Agnew, 1983; Peterson & Becker, 1965). Furthermore, parents who rely on physical punishment to control their children may also have a tendency to apply discipline in an inconsistent manner and engage in other inept parenting practices (Simons et al., 1994). The observed relationship between corporal punishment and parent assault, then, may have more to do with parental inconsistency than with the use the physical punishment per se.

The above considerations point to the need for a more rigorous and thorough examination of the link between corporal punishment and parent assault. To provide a more conclusive assessment, it will be necessary to estimate the effect of corporal punishment on subsequent parent assault while controlling for prior assault and a wide range of parenting variables. The purpose of this study is to conduct such an analysis. In the following sections, the details and results of this analysis are described.

DATA AND METHODS

Data

To examine the relationship between corporal punishment and adolescent-to-parent assault, a weighted-least-squares regression analysis was conducted using data from the first and second waves of the *Youth in Transition* (YIT) survey (Bachman, 1975). The YIT survey employed a panel design and is based on a nationwide sample of male public high school students. The survey was conducted by the Institute for Social Research, University of Michigan, and is available to member universities through the Inter-University Consortium of Political and Social Research (ICPSR).

During the first wave of data collection, personal interview and questionnaire data were obtained from 2,213 male youths as they were entering the 10th grade (modal age = 15), clustered in 87 schools. The schools and boys were selected through the use of multistage probability sampling and, according to the authors of the study, the resulting sample constitutes "an essentially bias-free representation of tenth-grade boys in public high schools throughout the United States" (Bachman, 1975: 1).

A second wave of data was collected from 1,886 of these youths (85% of the time 1 sample) approximately 1.5 years later, when they were completing the 11th grade. Data presented by Bachman, O'Malley & Johnston (1978) indicate that the survey results are not seriously biased by either panel attrition or repeated measurement effects.

The YIT survey contains several measures relevant to this study, including measures of physical punishment (slapping) and adolescent-to-parent assault (hitting parent). This is also the same data set used in several prior studies of parent assault and/or corporal punishment (Agnew, 1983; Brezina, 1999; Peek et al., 1985). Moreover, secondary analyses of the YIT data have resulted in numerous other contributions to the research literature on juvenile delinquency (e.g., Agnew, 1985, 1989, 1990; Brezina, 1996, 1998; Felson et al., 1994; Rosenberg, Schooler & Schoenbach, 1989; Wiatrowski, Griswold & Roberts, 1981).

It is important to note that the YIT data are not without certain limitations. In particular, the data are several decades old, they are restricted to male adolescents of roughly the same age, the measures are based on self-reports, and certain measures are based on a single indicator. However, as discussed below, these limitations are not viewed as serious problems for the particular purposes of this study.

Measures

Physical Punishment. During the first wave of data collection, youths were asked, "How often do your parents actually slap you?" Responses to this item (recoded) range from 1 ('never') to 5 ('always'). Approximately 60% of the respondents report some level of physical punishment, with 9% (n = 193) stating that they are 'often' or 'always' slapped by their parents (see Appendix for descriptive statistics and a correlation matrix of the study variables). Physical punishment may, of course, assume other forms than slapping. Data presented by Gelles (1979, 87), however, indicate that slapping is the most common form of corporal punishment during adolescence. Also, while one may question adolescent reports of parental behavior, a prior study by McKenry, Price-Bonham & O'Bryant (1981, 333) observes no significant difference between parental and adolescent reports of physical punishment, and little difference for most other parenting practices.

Adolescent-to-Parent Assault. During both waves of data collection, the YIT survey also asked respondents to report the number of times they have: (1) hit their mother, and (2) hit their father in the recent past (the recall period was 3 years at time 1, and 1.5 years at time 2). Responses to each item (recoded) range from 1 ('never') to 5 ('five or more times'). Following Peek et al. (1985), responses to these items were summed to form additive scales that index the overall level of adolescent-to-parent assault (Chronbach's alpha time 1 = 0.48, time 2 = 0.67). While most respondents report that they have never hit either of their parents, approximately 11% at time 1 (n = 239) and 8% of the respondents at time 2 (n = 146) report that they have hit at least one of their parents one or more times during the specified recall period.[2]

Control Variables. A number of variables are entered as controls in the data analysis, including a wide range of parenting factors such as parental attachment, parental consistency, parental control, parental reasoning with the respondent, and a measure of verbal/emotional maltreatment. Respondents who score high on a 4-item *Parental Attachment* scale report that they (1) feel close to their father, (2) feel close to their mother, (3) want to be like their father, and (4) want to be like the kind of person their mother is. Due to different response formats, the individual items were standardized before scale construction (alpha = 0.73).

Following Agnew (1983), two separate measures are used to index various dimensions of parental consistency, including a measure of intermittent discipline and a measure of 'inconsistent demands'. The first measure of consistency, called *Intermittent Discipline*, is indexed by asking respondents, "How often do your parents disagree about punishing you?" As Agnew (1983, 228) observes, "If parents disagree often, it seems likely that punishment will be intermittent. While one parent may be dominant and always win out in the disagreements, there are likely to be many disciplinary situations where only one parent is present." The second measure, called *Inconsistent Demands*, taps the overall level of parental consistency concerning matters of child-rearing and is indexed by asking respondents, "How often do your parents disagree with one another when it comes to raising you?" This measure is distinct from Intermittent Discipline because, while parents may administer discipline/punishment in a consistent or continuous manner, they may not clearly or consistently indicate behavioral expectations and, consequently, the types of behaviors that are necessary to avoid punishment. Responses to each of the consistency items (recoded) range from 1 ('never') to 5 ('always').

High scorers on a 5-item scale called *Parental Control* say their parents often decide how late they can stay out, how much spending money they can have, what friends they can 'go around with', what movies and parties they can attend, and what other extra-curricular activities they can have (alpha = 0.72). High scorers on a 4-item scale called *Lack of Parental Reasoning* state that they have 'little or no influence' in family decisions that directly affect them and further report that their parents seldom 'listen to their side of the argument', consult them on important decisions, or act fair and reasonable in what they ask of the respondent (alpha = 0.70). High scorers on a 3-item *Verbal/Emotional Maltreatment* scale state that their parents often: (1) threaten to hit them, (2) yell, shout, or scream at the them, and (3) give out undeserved blame and criticism (alpha = 0.69).

In addition, the analysis also controls for the academic grades of the respondent, deviant beliefs, attitudes toward aggression, family structure, race, and socio-economic status. *Academic Grades* is based on a single item that asks respondents, "What is the average grade you got in your classes last year?" High scorers on a 3-item scale of *Deviant Beliefs* say it is good to: (1) get hold of a final exam copy, (2) borrow money and not expect to pay it back, and (3) charge bills without knowing how to pay them (alpha = 0.74). High scorers on a 3–item scale called Approval of Aggression devalue non-aggressive responses to personal attacks and wrongdoing, saying that is *not* good to: (1) turn the other cheek and forgive others when they harm you, (2) reply to anger with gentleness, and (3) be kind to people "even when they do things against one's own beliefs" (alpha = 0.61). This same measure has been used in previous research to index adherence to a 'subculture of violence' (Felson et al., 1994).

Family Stability is a dichotomous variable that is scored '1' if the respondent lives with both his mother and father. Race is also a dichotomous variable and is scored '1' for non-white respondents. *Socioeconomic Status* is measured by a 6-item index constructed by the original investigators (Bachman, 1975), which combines information on father's occupational status, father's education, mother's education, number of rooms per person in the home, number of books in the home, and a checklist of other possessions in the home (e.g., a map or globe, a set of encyclopedias, a camera). The mean of the six items constitutes the scale score.

Analysis

The regression estimates are obtained using the weighted-least-squares (WLS) method available in the LISREL 8 program by Joreskog & Sorbom (1993). The WLS method produces standardized estimates based on the analyses of a polychoric correlation matrix and a weight matrix computed with the aid of PRELIS 2 (Joreskog & Sorbom, 1993). It is a distribution-free method of estimation and was selected as the most appropriate strategy given the highly skewed nature of the parent assault distributions and the ordinal nature of other measures in this study.

RESULTS

The WLS estimates are presented in Table 1. As seen in Table 1, physical punishment has a significant and positive standardized effect of 0.15 ($p < 0.05$)

Table 1. The Effect of Physical Punishment on Subsequent Adolescent-to-Parent Assault, Controlling for Prior Assault and Other Variables (WLS Analysis)

Time 1 Independent Variables	Dependent Variable: Time 2 Parent Assault Beta	t-value
Physical Punishment	**0.15**	**2.75***
Control Variables:		
Parental Attachment	**–0.09**	**–2.40***
Intermittent Discipline	**0.06**	**2.68***
Inconsistent Demands	–0.01	–0.19
Parental Control	0.02	0.76
Lack of Parental Reasoning	0.04	0.82
Verbal/Emotional Maltreatment	0.01	0.03
Approval of Aggression	0.05	0.43
Deviant Beliefs	0.03	1.40
Academic Grades	–0.03	–0.16
Family Stability (1 = intact)	0.03	0.99
Race (1 = nonwhite)	**0.21**	**10.65***
SES	–0.00	–0.05
Time 1 Parent Assault	**0.44**	**10.64***
R Squared	0.36	

N (listwise deletion) = 1,528
Note: Standardized effects shown (significant effects in bold)
* $p < 0.05$ (two-tailed test)

on subsequent parent assault, controlling for prior assault, parental attachment, measures of parental consistency, and other variables. In other words, a one standard-deviation increase in corporal punishment is associated with a 0.15 standard-deviation increase in parent battering, controlling for other factors. These results, then, are consistent with the hypothesis that corporal punishment contributes to the development of teenage violence toward parents.

The only other variables that predict parent battering include Parental Attachment (-0.09), Intermittent Discipline (0.06), Race (0.21) and prior assault (0.44, $p < 0.05$). Not unexpectedly, respondents who report high levels of parental attachment are less likely to assault their parents (see also Agnew & Huguley, 1989). The odds of parent assault increase when parents administer discipline/punishment in an intermittent or inconsistent manner (see also Agnew, 1983). The effect of race indicates that non-white respondents are more likely to engage in subsequent parent assault – a finding that appears to be consistent with Elliott's (1994) observation that non-whites have higher rates of continuity in violent behavior. In addition, respondents who have hit their parents in the past are prone to hit them again in the future, with prior assault having the single largest effect in the equation. This effect is consistent with the well-established finding in criminology that prior delinquency is the best known predictor of subsequent delinquency (Nagin & Paternoster, 1991).

Additional analyses (not shown) were conducted to test for possible interaction effects between physical punishment and the other study variables (e.g., Physical Punishment × Parental Attachment; Physical Punishment × Intermittent Punishment; Physical Punishment × Inconsistent Demands; etc.). An interaction term involving physical punishment and parental attachment was the only such term to achieve statistical significance ($p < 0.05$). The nature of this interaction suggests that physical punishment has a particularly strong effect on parent assault when levels of parental attachment are low (one or more standard deviations below the mean) and a relatively weak effect when parental attachment is high (one or more standard deviations above the mean). Further inspection of the data, however, suggests a need for caution. In particular, very few cases ($n = 25$) exist in which respondents report high levels of both physical punishment and parental attachment (little more than 1% of the effective sample size). Due to the few number of cases falling into this category, and hence the potential unreliability of the estimate, no conclusion is drawn from the observed interaction effect.

To sum up, the results of the WLS analysis suggest that the relationship between corporal punishment and parent assault, observed in prior studies, is not spurious and that corporal punishment may indeed be a causal factor in the development of parent assault (i.e., it increases the likelihood of parent assault). These results, however, should be interpreted with the following considerations in mind. First, it is important to note that the above results do not *prove* causation because, as with all such studies, it is always possible that the analysis failed to control for all relevant antecedent variables. Yet, to the author's knowledge, this study provides the most compelling evidence to date of a causal link between corporal punishment and teenage violence toward parents.

Second, it should be stressed that the sample is restricted to adolescent males of roughly the same age. The above results may not generalize to female adolescents or to younger children. Previous research reveals at least one gender difference related to parent assault: while data indicate that male and female adolescents hit their parents at roughly the same rate, assaults by males are more likely to be non-trivial in nature, involving punching, pushing, and the use of weapons such as sticks or chairs (Agnew & Huguley, 1989). Future research will be required to examine the consequences of corporal punishment among these other populations.

Third, many of the measures in this study are based on a single indicator and, consequently, it was not possible to control for measurement error or to correct for possible serial correlation. In the case of measurement error, however, the most likely result is the attenuation of parameter estimates. In the case of serial correlation, the most likely result is the overestimation of stability effects, and the *underestimation* of other effects in the model, such as the effect of physical punishment on parent assault (Kessler & Greenberg, 1981).

A fourth consideration involves the age of the data. As stated above, the data are several decades old and it is possible that various social processes may have changed as a function of time, including the meaning of both corporal punishment and adolescence. The significance of age in this case is an empirical question, although there is room for reasoned speculation. To the extent that corporal punishment has lost cultural legitimacy (see Straus & Mathur, 1996), it is possible that corporal punishment may have a *stronger* effect on the behavior of adolescents today than it did several decades ago, when corporal punishment was more likely to have been viewed as a legitimate exercise of parental authority. In other words, there is reason to believe that the YIT data may provide a conservative estimate of the effect of corporal punishment on teenage violence toward parents.

DISCUSSION

The analysis described in the present study corrects for some of the more serious methodological limitations plaguing prior studies on the relationship between corporal punishment and teenage violence toward parents. Consequently, the results of the present study allow for a more definitive assessment of this relationship. In particular, the findings indicate that corporal punishment increases the likelihood of subsequent parent assault, controlling for prior assault, parental attachment, parental consistency, and other important variables. The results, then, provide relatively compelling (albeit tentative) empirical support for the assertion that corporal punishment is a causal factor in the development of teenage violence toward parents.

These results are important because they have implications for the control of teenage violence toward parents. It may be possible to reduce the occurrence of parent battering by reducing corporal punishment. In addition, the findings also provide tentative justification for interventions not specifically designed to address corporal punishment. Given the observed negative effect of parental attachment on parent assault (see above), it may also be possible to control or reduce parent battering by improving the quality of parent-child relationships. In this vein, Paulson, Coombs & Landsverk (1990) recommend parent training programs designed to facilitate affectionate and mutually rewarding interactions (see also Harbin & Madden, 1979).

The results of this study are also important because they appear to have implications for corporal punishment research in general. Certain previous studies have failed to observe an association between corporal punishment and the anti-social behavior of adolescents (e.g., Simons et al., 1994). The failure to observe an effect, however, may be related to the use of broad and inclusive outcome measures. Reliance on such measures (which have included items indexing a range of behaviors such as shouting, throwing things, getting into fights, skipping school, drinking alcohol, and theft) may have obscured the effect of corporal punishment. Perhaps it is in a relatively specific domain of behavior – in the physical interaction between parent and adolescent child – where the direct behavioral effects of corporal punishment are most likely to be observed. This may be especially true for the effect of corporal punishment on adolescent violence. Research indicates that many if not most violent attacks arise from disputes between family members, intimates, and other associates and, in the eyes of the offender, are intended as a *response* to provocations delivered by the victim or target (Black, 1983; Luckenbill, 1977). The failure to examine adolescent-to-parent violence as a specific outcome variable, then, may represent a significant oversight of much corporal punishment research.

To explore this possibility, an additional analysis (not shown) was conducted in which a measure of general delinquency served as the dependent variable (the measure included such items as theft, arson, robbery, general assault, and shoplifting). In this case, the effect of corporal punishment on subsequent behavior was insignificant (0.01, $p > 0.05$), controlling for prior delinquency and the other study variables. These results appear to affirm the importance of discriminating between general and specific outcome measures in corporal punishment research (complete set of results available from the author on request).

In terms of future research on the link between corporal punishment and parent assault, it will be worthwhile to attempt replication of the study results with more recent data, data drawn from a more diverse sample of respondents (e.g., samples that include both male and female adolescents), and data containing multiple indicators of theoretical constructs to allow for the correction of measurement error and autocorrelation. In addition, future investigations might also benefit from the examination of more detailed data on corporal punishment and parent assault (e.g., data sets that contain information on the use of weapons or the occurrence of injury). Because of the potential implications of corporal punishment research for the control of parent battering, research advances on the etiological contribution of corporal punishment are needed and should be of special interest to practitioners. The present study can be viewed as a significant step in this direction.

It is hoped that this study might also stimulate more general interest in the problem of teenage violence toward parents. As stated earlier, this problem remains a relatively neglected dimension of family violence. Although we now possess limited knowledge of the family correlates and possible causes of parent assault, much less is known about the long-term consequences of this behavior once it appears. The occurrence of parent assault represents the establishment of an 'inverse dominance hierarchy' and, once established in the family of origin, it is possible that the behaviors associated with this kind of hierarchy carry over to other environments and influence interactions with non-family members (Loeber & Stouthamer-Loeber, 1998; Loeber, Weissman & Reid, 1983). For example, children who assault their parents may also display a tendency to intimidate peers, threaten teachers, and batter their future partners. Parent assault may, in fact, be a 'missing link' in the cycle of violence (Agnew & Huguley, 1989; Straus et al., 1980). If consequences of this nature can be verified, it may be possible to disrupt the cycle of violence by developing interventions that target parent abusers.

Even less is known about the extent to which parent battering represents a growing problem or, perhaps, a uniquely American problem. On the one hand, data indicate that the rate of corporal punishment in the United States decreased over the past several decades – although it remains high (Straus & Mathur, 1996). For this reason, one would suspect that the rate of parent assault has also decreased. On the other hand, factors that are said to inhibit violent forms of rebellion are believed to have diminished or weakened over the same time period, such as the legitimacy of parental authority and the dependence of adolescent children on their parents (Felson, 1998). For other reasons, then, one might suspect an *increase* in the rate of parent battering. Research designed to estimate actual trends in parent assault would be of substantial interest, because such trends could serve as a kind of 'barometer' for gauging the health of America's families.

The above considerations raise yet another issue. As one reviewer of this chapter noted, parental assault by retaliatory teens would be 'unthinkable' in other parts of the world where deference to elders is firmly established, including cultures known for their authoritarian approach to child rearing. It is quite possible that the extant research on parent assault, based almost exclusively on American samples (this study included), is 'culture bound' (but see Browne & Hamilton, 1998). In other words, the patterns observed in existing studies could be specific to American culture (or, at least, post-industrial societies), where the dependence of teens on their parents is fairly limited and cultural scripts emphasize the value of personal autonomy. To the author's knowledge, cross-cultural comparisons of child-to-parent violence have yet to be conducted.

Long-term consequences, trends, and cultural variation – these are just a few issues that remain to be explored in relation to teenage violence toward parents. From a sociological or research standpoint, these issues point toward a fruitful and intriguing line of inquiry. From a policy standpoint, these issues involve a problem that, in our own society, is too large to ignore (Browne & Hamilton, 1998).

NOTES

1. To be sure, the issue of parricide has received a fair amount of attention from researchers (e.g., Heide, 1992). Simple assault, however, is a far more typical manifestation of child-to-parent violence.

2. Consistent with the results of prior research (Agnew & Huguley, 1989), the adolescent males in this sample – in their mid- to late teens – are somewhat more likely to hit their fathers than their mothers. (Other studies indicate that the pattern is reversed for males under the age of 15, who are more likely to assault their mothers). For several reasons, the parent-specific assault items were combined in this study. The combined scale, which measures the *overall* level of parent assault, maximizes variation in the dependent variable. The combined scale is also more appropriate, given that the available measure of physical punishment is not itself parent specific. In any event, exploratory analyses based on the parent-specific assault items produced a nearly identical pattern of results – although the effects of certain variables, such as physical punishment, were slightly more robust in the case of father versus mother assault (results available from the author on request).

REFERENCES

Abrams, J. R. (1981). Adolescent Perceptions of Parental Discipline and Juvenile Delinquency. In: R. J. Hunner, & Y. E. Walker (Eds), *Exploring the Relationship Between Child Abuse and Delinquency*, (252–265). Montclair, NJ: Allanheld, Osmun, and Co.

Agnew, R. (1983). Physical Punishment and Delinquency: A Research Note. *Youth & Society, 15*, 225–236.

Agnew, R. (1985). A Revised Strain Theory of Delinquency. *Social Forces, 64*, 151–167.

Agnew, R. (1989). A Longitudinal Test of the Revised Strain Theory. *Journal of Quantitative Criminology, 5*, 373–387.

Agnew, R. (1990). Adolescent Resources and Delinquency. *Criminology, 28*, 535–565.

Agnew, R. (1992). Foundation for a General Strain Theory of Crime and Delinquency. *Criminology, 30*, 47–87.

Agnew, R., & Huguley, S. (1989). Adolescent Violence Toward Parents. *Journal of Marriage and the Family, 51*, 699–711.

Bachman, J. G. (1975). *Youth in Transition: Documentation Manual.* Ann Arbor, MI: Institute for Social Research.

Bachman, J. G., OMalley, P. M., & Johnston, J. (1978). *Youth in Transition, Vol. 6: Adolescence to Adulthood.* Ann Arbor, MI: Institute for Social Research.

Black, D. (1983). Crime as Social Control. *American Sociological Review, 48*, 34–45.

Brezina, T. (1996). Adapting to Strain: An Examination of Delinquent Coping Responses. *Criminology, 34*, 39–60.

Brezina, T. (1998). Adolescent Maltreatment and Delinquency: The Question of Intervening Processes. *Journal of Research in Crime and Delinquency, 35*, 71–99.

Brezina, T. (1999). Teenage Violence Toward Parents as an Adaptation to Family Strain *Youth & Society, 30*, 416–444.

Browne, K. D., & Hamilton, C. E. (1998). Physical Violence Between Young Adults and Their Parents: Associations with a History of Child Maltreatment. *Journal of Family Violence, 13*, 59–79.

Charles, A. V. (1986). Physically Abused Parents. *Journal of Family Violence 4*, 343–355.

Cornell, C. P., & Gelles, R. J. (1982). Adolescent to Parent Violence. *Urban and Social Change Review, 15*, 8–14.

Elliott, D. S. (1994). Serious Violent Offenders: Onset, Developmental Course, and Termination. *Criminology, 32*, 1–21.

Felson, M. (1998). *Crime and Everyday Life* (2nd ed.). Thousand Oaks, CA: Pine Forge.

Felson, R. B., Liska, A. E., South, S. J., & McNulty, T. L. (1994). The Subculture of Violence and Delinquency: Individual vs. School Context Effects. *Social Forces, 73*, 155–73.

Gelles, R. J. (1979). *Family Violence.* Beverly Hills, CA: Sage

Gelles, R. J. (1997). *Intimate Violence in Families.* Thousands Oaks, CA: Sage.

Harbin, H. T., & Madden, D. J. (1979). Battered Parents: A New Syndrome. *American Journal of Psychiatry, 136*, 1288–1291.

Heide, K. M. (1992). *Why Kids Kill Parents.* Columbus: Ohio State University Press.

Joreskog, K. G., & Sorbom, D. (1993). *LISREL 8.* Hillsdale, NJ: Lawrence Erlbaum.

Kessler, R. C., & Greenberg, D. F. (1981). *Linear Panel Analysis: Models of Quantitative Change.* New York: Academic Press.

Loeber, R., & Stouthamer-Loeber, M. (1998). Juvenile Aggression at Home and at School. In: D. S. Elliott, B. A. Hamburg, & K. R. Williams (Eds), *Violence in American Schools*, (94–126). New York: Cambridge University Press.

Loeber, R., Weissman, W., & Reid, J. B. (1983). Family Interactions of Assaultive Adolescents, Stealers, and Nondelinquents. *Journal of Abnormal Child Psychology, 11*, 1–14.

Luckenbill, D. F. (1977). Criminal Homicide as a Situated Transaction. *Social Problems, 25*, 176–186.

McKenry, P. C., Price-Bonham, S., & OBryan, S. L.. (1981). Adolescent Discipline: Different Family Members Perception. *Journal of Youth and Adolescence, 10*, 327–337.

Nagin, D. S., & Paternoster, R. (1991). On the Relationship of Past to Future Participation in Delinquency. *Criminology, 30*, 501–523.

Patterson, G. R. (1982). *A Social Learning Approach, Volume 3: Coercive Family Process.* Eugene, OR: Castalia.

Patterson, G. R., Littman, R. A., & Bricker, W. (1967). Assertive Behavior in Children: A Step Toward a Theory of Aggression. *Monographs of the Society for Research in Child Development, 32*(5), Serial No 113

Paulson, M. J., Coombs, R. H., & Landsverk, J. (1990). Youth who Physically Assault Their Parents. *Journal of Family Violence, 5*, 121–133.

Peek, C. W., Fischer, J. L., & Kidwell, J. S. (1985). Teenage Violence Toward Parents: A Neglected Dimension of Family Violence. *Journal of Marriage and the Family, 47*, 1051–1058

Peterson, D. R., & Becker, W. C. (1965). Family Interaction and Delinquency. In: H. C. Quay (Ed.), *Juvenile Delinquency: Research and Theory* (36–99). Princeton, N.J.: Van Nostrand.

Rosenberg, M., Schooler, C., & Schoenbach, C. (1989). Self-esteem and Adolescent Problems: Modeling Reciprocal Effects. *American Sociological Review, 54,* 1004–1018.

Simmons, R. L., Johnson, C., & Conger, R D. (1994). Harsh Corporal Punishment Versus Quality of Parental Involvement as an Explanation of Adolescent Maladjustment. *Journal of Marriage and the Family, 56,* 591–607.

Straus, M. A., & Donnelly, D. A. (1993). Corporal Punishment of Adolescents by American Parents. *Youth and Society, 24,* 419–442.

Straus, M. A., Gelles, R. J., & Steinmetz, S. K. (1980). *Behind Closed Doors: Violence in the American Family.* Garden City, NY: Anchor Books.

Straus, M. A., & Mathur, A. K. (1996). Social Change and the Trends in Approval of Corporal Punishment by Parents from 1968 to 1994. In: D. Frehsee, W. Horn, & K.-D. Bussmann *Family Violence Against Children: A Challenge for Society,* (91–105). New York: Walter de Gruyter.

Straus, M. A., Sugarman, D. B., & Giles-Sims, J. (1997). Spanking by Parents and Subsequent Antisocial Behavior of Children. *Archives of Pediatrics and Adolescent Medicine, 151,* 761–767.

Wiatrowski, M. D., Griswold, D. B., & Roberts, M. K. (1981). Social Control Theory and Delinquency. *American Sociological Review, 46,* 525–41.

Wolman, B. B. (1998). *Adolescence: Biological and Psychosocial Perspectives.* Westport, CT: Greenwood Press.

EXPOSURE TO VIOLENCE IN THE FAMILY: UNPACKAGING THE LINKAGES TO INTIMATE PARTNER VIOLENCE*

Paul Mazerolle, Jeff Maahs and Ronet Bachman

INTRODUCTION

Recent events raise concern over the level of exposure to violence by children and adolescents in America. Indeed, acts of violence appear widespread in our society. Violence is regularly depicted in movies and news media (Berry, 1993) as evidenced by the popularity of certain television programs and sporting events (e.g. 'pro wrestling'). Lethal acts of violence are too often reported in our schools as evidenced by unfortunate events in Jonesboro, Arkansas and Columbine, Colorado among others; and violence within families continues to have negative health consequences for women and children. Despite the fact that family processes and interactions consistently shape human behavior in positive ways, the family is a major source of violence and conflict. Evidence reveals that over 3 million children observe inter-parental violence (Jaffe et al., 1992) and approximately 2.9 million have been reported as being maltreated

*This project was supported by Grant No. 98-WT-VX-0007 awarded by the National Institute of Justice, Office of Justice Programs, U.S. Department of Justice. Points of view in this document are those of the authors and do not necessarily represent the official position or policies of the U.S. Department of Justice.

Families, Crime and Criminal Justice, Volume 2, pages 45–71.

ISBN: 0-7623-0737-4

annually in the United States (U.S. Department of Health and Human Services, National Center of Child Abuse and Neglect, 1996). Unfortunately, many children are exposed to multiple forms of conflict and violence within the family. In sum, given the extent of this social problem as well as the deleterious consequences of violence, research that examines the links between early exposure to violence and various negative outcomes is important on theoretical, empirical and public policy grounds.

Evidence consistently reveals that early exposure to violence in the family has several negative consequences for children (Loeber & Stouthamer-Loeber, 1986; Widom, 1989a; Smith & Thornberry, 1995; Singer et al., 1999). For example, early exposure to violence has been linked to depression (Cerezo-Jimenez & Frias, 1994), post traumatic stress disorders (Famularo et al., 1994), substance abuse (Duncan et al., 1996), delinquency (Smith & Thornberry, 1995; Widom, 1989b), and a host of other internalizing and externalizing problems. In addition to these negative outcomes, early exposure to violence has been linked with an increase in arrests for violence in adulthood (Rivera & Widom, 1990) as well as an increased likelihood of violence in adult relationships (Cappell & Heiner, 1990; Carter et al., 1988).

Despite the accumulating knowledge on the consequences of early exposure to violence in the family, many of the mechanisms are not fully understood. Moreover, previous research has not always examined whether experiences with different kinds of violence in the family give rise to unique outcomes. For example, despite the fact that experiencing physical child abuse and witnessing inter-parental violence have been linked with negative life outcomes (Rivera & Widom, 1990; Smith & Thornberry, 1995; Cerezo-Jimenez & Frias, 1994), such experiences might be differentially related to intimate partner violence later in adulthood. Observing inter-parental violence in childhood, despite its other negative consequences, might be more consistently related to intimate partner violence in adulthood through social learning (Bandura, 1977) or social cognitive processes (Bandura, 1986). Physical child abuse is perhaps not as strongly related to intimate partner violence despite the number of associated negative outcomes (Jonson-Reid, 1998). In sum, despite the important research conducted to date on these issues, there is still a need to disentangle unique experiences with violence in the family to assess whether they are related differentially to adult outcomes such as, intimate partner violence.

In the current study, we examine two main questions related to the linkages between early exposure to violence and later life outcomes. First, in exploring the consequences of early exposure to violence, we examine relationships between cumulative exposure to violence in childhood with a series of negative life outcomes including early delinquent participation, violence, alcohol abuse, and illicit drug use. Second, recognizing that being exposed to physical child abuse as opposed to observing inter-parental violence may have different consequences, we examine the independent effects of these influences on intimate partner violence through a series of multivariate analyses.

CONSEQUENCES OF EXPOSURE TO VIOLENCE IN THE FAMILY

A wealth of past research has uncovered a number of deleterious consequences that are related to childhood exposure to violence (Thornberry, 1994; Smith & Thornberry, 1995; Dodge et al., 1990; Widom, 1989a; Singer et al., 1999; Duncan et al., 1996; Malinosky-Rummell & Hansen, 1993). Exposure to physical child maltreatment has been associated with a host of negative outcomes including depression (Duncan et al., 1996), running away (Kurtz et al., 1991; Kaufman & Spatz-Widom, 1999), substance abuse (Duncan et al., 1996), delinquency (Widom 1989b, Smith & Thornberry, 1995, Zingraff et al., 1993), and violent crime (Singer et al., 1999; Thornberry, 1994; Rivera & Widom, 1990). Additionally, observing parental violence during childhood is problematic and has been linked with increases in violence (Spaccerelli et al., 1995; Thornberry, 1994) as well as an increased likelihood of intimate partner violence (Choice et al., 1995; Cappell & Heiner, 1990; Carter et al., 1988; Hotaling & Sugarman, 1986).

Some research has examined the consequences of the combined impact of witnessing partner violence and experiencing maltreatment. For example, research by Hughes and colleagues (1989) finds that children who witness partner violence and are abused have many more problems than youth in non-violent or less violent comparison groups. Additionally, Hotaling and colleagues (1989) find that children who experience multiple exposure to violence in the family (witness parental violence and experience abuse) are six times more likely than children from non-violent homes to assault other children.

Given the range of negative consequences associated with childhood exposure to different types of violence, there is a need to examine whether exposure to specific types of violent experiences give rise to specific problems later in adulthood. Moreover, it remains unclear the extent to which different levels of exposure to violence in childhood impact later life outcomes, such as intimate partner violence. In sum, it is important to consider whether the cumulative exposure to violence in childhood gives rise to a range of problems later in adulthood, such as intimate partner violence, and whether there are specific effects when the context of these experiences are disaggregated.

DOES EXPOSURE TO VIOLENCE IN CHILDHOOD DIFFERENTIALLY AFFECT INTIMATE PARTNER VIOLENCE?

There are reasons to believe that exposure to physical child abuse and observing inter-parental violence in childhood might be differentially related to intimate partner violence later in adulthood. To be sure, there are many pathways and precursors to intimate partner violence, but one of the consistent risk factors is having witnessed inter-parental violence while growing up (Hotaling & Sugarman, 1986). Social learning (Bandura, 1977) and social cognitive (Bandura, 1986) mechanisms have been identified as the main processes accounting for the strong relationship between witnessing inter-parental violence and perpetration in adulthood. Individuals exposed to inter-parental violence in childhood are expected to recall many of the contexts and patterns that were involved in their parents' violent incidents. Such experiences set up expectations for behavioral scripts between partners. Within this context, males are likely to embrace physical means of social interaction and dispute resolution within their marital relationships. Consistent with this belief, research reveals that batterers are more likely to have poor communication and social skills (Holtzworth-Munroe & Stuart, 1994) and educational attainment (Hotaling & Sugarman, 1986). Moreover, further research examining the direct and indirect linkages between witnessing inter-parental violence and partner violence finds that ineffective conflict resolution skills and marital distress mediate the relationship between exposure to inter-parental violence and subsequent partner violence perpetration. Using data from the Second National Family Violence Survey (Gelles & Straus, 1989), Choice and colleagues (1995) find support for the mediating effects of ineffective conflict resolution skills and partner violence. However, their results also reveal direct relationships between exposure to parental violence and subsequent partner violence by males. In sum, there is good evidence that exposure to inter-parental violence in childhood is one of the strongest and most consistent predictors of male perpetrated intimate partner violence. Additionally, there are reasons to believe that some of these effects are direct and some are more indirect (see also Downs et al., 1996).

Research suggests that exposure to physical child abuse increases the probability of violent behavior (Thornberry, 1994; Rivera & Widom, 1990; Hotaling et al., 1989; Wolfe et al., 1998). However, there are reasons to suspect that the effects of childhood exposure to physical abuse are less strongly related to intimate partner violence than observing inter-parental violence. First, past exposure to physical child abuse has not been identified as a consistent risk factor for intimate partner violence (Hotaling & Sugarman, 1986; Kalmuss, 1984). Second, past research linking physical child abuse with partner violence has not consistently isolated the independent effects of exposure to physical child abuse and observing inter-parental violence (Dutton, 1988; Kalmuss, 1984). As a result, many of these relationships may be over-stated or confounded because the criminogenic effects of observing inter-parental violence were not isolated and controlled. Finally, despite research revealing that exposure to physical child maltreatment leads to physical conflict in dating relationships (Wolfe et al., 1998), such relationships are arguably not as strong when directly compared to the independent effects of observing inter-parental violence. In sum, in light of past research, we expect exposure to inter-parental violence during childhood to be more strongly related to male perpetrated intimate partner violence in adulthood than is exposure to physical child abuse.

THE CURRENT STUDY

In the current study, we examine relationships between early exposure to violence and a range of negative outcomes using data collected from an 'at risk' sample of male parolees in New York state (Blane et al., 1994). Our first hypothesis is that exposure to violence in the family is associated with a range of problematic outcomes including substance abuse, delinquency, and violence. Moreover, we examine whether there is a consistent 'dose response' between exposure to violence in childhood and problems in adulthood. In short, we expect greater exposure to violence to be associated with more problems later in adulthood.

A second hypothesis to be tested in the current analysis concerns whether exposure to certain types of violence in childhood lead more strongly to specific outcomes, such as intimate partner violence. Past research has not consistently examined whether exposure to multiple types of violence in childhood is differentially related to partner violence. Therefore, our analysis examines the independent effects of childhood exposure to physical child abuse and observing inter-parental violence on intimate partner violence through a series of multivariate models.

METHOD

Sample

The data used in this study are drawn from a previous study on the inter-relationships between substance abuse, and criminal and domestic violence across three generations (Blane et al., 1994). Specifically, subjects were drawn from a stratified random sample of all parolees residing in the greater Buffalo, New York, area between January 1987 and June 1987.[1] Of the 259 subjects sampled, 194 (75%) agreed to participate and were interviewed. The parolees who chose not to participate in the study did not differ from parolees participating on a range of measures including age, race, marital status, criminal history, drug/alcohol problems, or instant offense variables. However, one difference was observed in that parolees who refused to be the interviewed were more likely to have emotional problems.

Information collected in the interview covered several domains including self-reported delinquency and crime, early exposure to different types of violence, and drug/alcohol use. Additionally, a revised version of the Conflict Tactics Scale (Straus, 1979, hereinafter CTS) was given to parolees who were in a relationship in the year prior to their most recent incarceration. Items comprising the CTS were used to gauge varying forms of intimate partner violence as well as exposure to physical child abuse and observing inter-parental violence in childhood.

The sub-sample of the larger study used for the present analysis consists of 111 males who reported that they were in a relationship in the year prior to their incarceration. Female parolees (N = 14) were excluded from the analysis due to their small group size. To investigate whether male parolees who were in a relationship prior to prison (N = 111) differed from those who were not (N = 69), we compared the groups on demographic and criminal history information as well as their drug/alcohol use patterns and level of exposure to violence in childhood. Most of the comparisons revealed no differences between the groups although two differences were observed. Male parolees in a relationship were significantly older and had more prior incarcerations.[2]

Measures

Early Exposure to Violence. Childhood exposure to violence was assessed using revised versions of the Conflict Tactics Scale that gauged inter-parental violence and physical child abuse. More specifically, for each respondent the measures assessed: (a) each of their parental figures' physical violence toward them as a child; and (b) each of their parental figures' violence towards each other.[3] For each of these relationships, subjects were asked to report the frequency of twelve specific acts of violence, ranging from pushing or shoving to beating and choking, that they had observed or experienced as a child (see specific items listed in the Appendix). Because these reports were retrospective and the response categories were somewhat ambiguous, the specific items were dichotomized and summed to form an index.[4] This practice, which yields a 'variety index,' represents a more accurate gauge for measuring self-reports of past experiences with violence.

Child Abuse. The physical child abuse measure was created by summing respondents' reports of whether their mother, father, or step-parent committed any of twelve specific violent acts toward them. The specific acts ranged in severity from slapped, pushed, grabbed, or shoved to very serious events such as beat up, choked, or used knife or gun. All of the specific acts of violence are reported in the Appendix (Cronbach's alpha = 0.80).

Exposure to Parental Violence. This scale represents a summated index of violence perpetrated by each of the subjects' parents (or step-parents) toward each other (e.g. mother toward father, and father toward mother). The specific items comprising the scale are consistent with items used in the previously discussed child abuse scale. Items comprising the scale, as previously noted, range in severity from slapping, pushing, grabbing to more serious events including choking, beat up, and use of a knife or gun (Cronbach's alpha = 0.94).

Global Exposure to Violence. To assess cumulative exposure to violence during childhood, we combined the scales measuring exposure to inter-parental violence and physical child abuse into an overall index assessing global exposure to violence (Cronbach's alpha = 0.89).

Negative Life Outcomes. Childhood exposure to violence may have consequences across many life-course domains. We examine several outcome measures, including early delinquent onset, alcohol-related problems, drug use, self-reported acts of violence, and intimate partner violence.

Delinquent Onset. Early onset to delinquency is a strong and consistent correlate of adult offending (Loeber & Dishion, 1983; Wolfgang et al., 1972). Moreover, according to recent theory and research, early delinquent participation is related to several criminal career dimensions including violence, enhanced offending frequency, versatility, persistence in crime, and offending chronicity (Moffitt, 1993; Patterson & Yoerger, 1993; Blumstein et al., 1986; Piquero et al., 1999; Tolan, 1987). Respondents are classified as having early onset status in the present analysis if they report engaging in one of four criminal acts (theft, burglary, assault, or robbery) prior to the age of 13.[5]

Alcohol-Related Problems. This measure reflects the number of alcohol-related problems experienced by the respondent (out of twenty-one problems) over their lifetime. Examples of problems include job or school difficulties, drinking while ill, neglecting responsibilities because of drinking binges. A complete list of alcohol-related problems included in the scale is reported in the Appendix (Cronbach's alpha = 0.83).

Drug Use. A measure for illicit drug use was included which indexes the number of different illicit drugs (heroin, barbiturates, cocaine, marijuana, LSD, and PCP) that the respondent used on a regular basis. Regular illicit drug use was defined as weekly use of a substance for at least a month (Cronbach's alpha = 0.67).

Self-Reported Violence. Respondents reported whether or not they had ever committed a number of violent criminal acts. The measure reflects the respondent's participation in a number of different types of violent crimes including assault, kidnapping, robbery, armed robbery, and rape.

Intimate Partner Violence. Similar to measures for childhood exposure to violence, the measure for intimate partner violence (derived from a revised version of the CTS) reflects the number of violent acts perpetrated by the parolee toward his intimate partner in the year prior to his current incarceration. Unlike the CTS measures for early childhood violence, the male parolees were asked to report on nine (rather than twelve) specific violent acts including items ranging in severity from slapped, hit with something to even more serious acts such as used knife or gun, choked, and forced sex.[6] The specific acts comprising the scale are reported in the Appendix.

Demographic Information. Information on demographic characteristics is included for age (measured in years) and race (codes as 1 = white, 0 = non-white). Descriptive statistics for all measures used in the analysis are reported in Table 1.

ANALYSIS PLANS

There are two main components of our analysis. First, we examine the relationship between childhood exposure to violence and various negative outcomes, such as substance abuse and violence. Our measure of childhood exposure to violence gauges cumulative exposure to physical child abuse and observing inter-parental violence. We divided the respondents into three categories of childhood exposure to violence in the family (high, medium and low levels of exposure to violence). High, medium and low levels of exposure to violence in childhood were assessed by selecting groups at the 25th and 75th percentile.[7] We conduct a series of one-way analysis of variance (ANOVA) models to assess mean level differences in lifetime alcohol problems, regular drug use, early onset to delinquency, self-reported violence, and intimate partner violence across the three groups of respondents differentiated by their level of exposure to violence in childhood. As mentioned previously, our hypothesis is that greater levels of exposure to violence during childhood is expected to be associated with more negative outcomes in adulthood.

The second component of our analysis involves estimating a series of Tobit models to examine the independent effects of different types of childhood

Table 1. Descriptive statistics (N = 111)

Variable	Mean	Sd	Range
Demographics			
Age	33.67	7 79	20–72
Race (white = 1)	0.36	0.48	0–1
Childhood Exposure to Violence			
Child Abuse	6.75	4.50	0–24
Observe Parents Violence	5.78	6.24	0–23
Global Violence	12.32	8.63	0–37
Negative Outcomes			
Lifetime Alcohol Problems	5.93	6.13	0–21
Number of Drugs Used Regularly	2.00	1.68	0–6
Early Delinquent Onset	0.52	0.50	0–1
Self Report Violence	1.40	1.12	0–5
Intimate Partner Violence	0.55	1.03	0–5

exposure to violence on male perpetrated intimate partner violence in adulthood. Tobit regression models are estimated because of the censored nature of our dependant variable. Many of the responses were left-hand censored at zero and, not surprisingly, the range of positive outcomes is restricted. In this context Ordinary Least Squares estimation is not appropriate.

Our multivariate analysis proceeds through a series of steps. First, we estimate a baseline model that includes control variables, such as age and race as well as various risk factors for partner violence including alcohol problems, substance use and early delinquent participation. The second model introduces cumulative exposure to violence in childhood to assess whether it is a significant predictor of partner violence when these other influences are statistically controlled. In subsequent models, we assess whether different types of exposure to violence in childhood, such as exposure to physical child abuse and observing inter-parental violence are independently related to intimate partner violence. We assess these relationships independently as well as directly control for these influences in our final model.

RESULTS

Assessing the Effects of Exposure to Violence on Negative Life Outcomes

The relationship between different levels of exposure to cumulative levels of violence in childhood and various negative outcomes are displayed in Figures

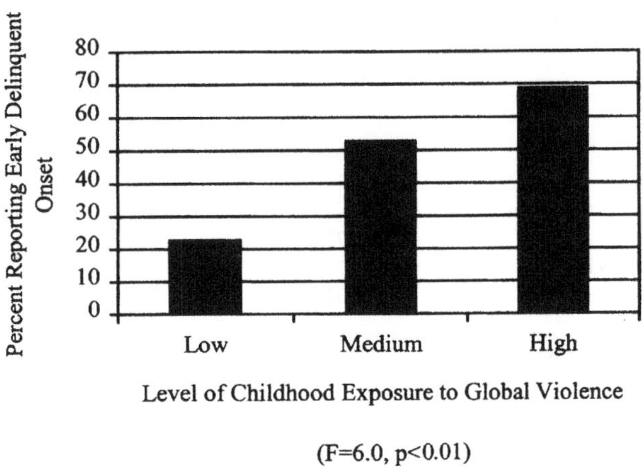

Figure 1. The Relationship Between Childhood Exposure to Violence and Early Onset of Delinquency.

1 through 5. Our first comparison examines the relationship between childhood exposure to violence and early delinquent participation. Early onset to delinquency is often viewed as a proxy for high criminal propensity (Moffitt, 1993; Paternoster et al., 1997). Moreover, available research reveals early delinquent participation to be positively correlated with offending frequency and versatility (Tolan & Lorion, 1988; Tolan, 1987; Piquero et al., 1999) as well as chronic offending (Loeber & LeBlanc, 1990; Wolfgang et al., 1972). Given the negative consequences associated with childhood exposure to violence in the family, we expect higher exposure to violence to lead to more problems, such as early delinquent conduct. Results from Figure 1 are consistent with this expectation. According to Figure 1, early onset to delinquency is much more prevalent (69%) among subjects with high levels of exposure to violence relative to those with medium (52%) or low (22%) levels of exposure. The results reveal substantive and significant differences in mean levels of early delinquent onset across groups ($F = 6.0$, $p < 0.01$). In sum, these results are consistent with our hypothesis that greater exposure to violence during childhood is associated with more problems, such as early delinquent participation.

Exposure to violence in childhood is also expected to lead to problems with alcohol and substance abuse later in life (Duncan et al., 1996). Results displayed in Figures 2 and 3 allow us to examine this issue further. The findings are largely consistent with our expectations. For example, high levels of exposure to childhood violence is associated with a greater number of problems with

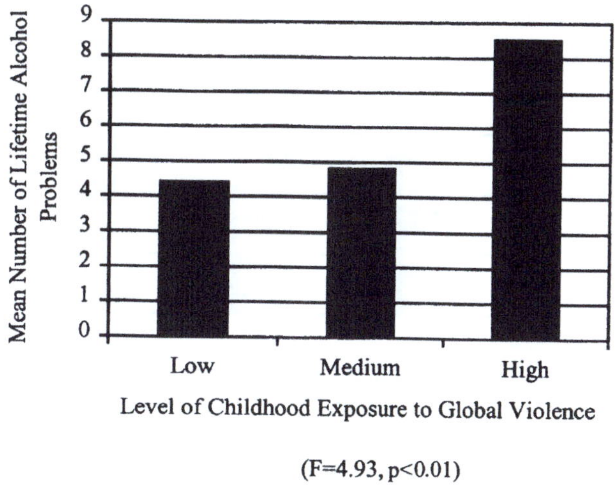

Figure 2. The Relationship Bteween Childhood Exposure to Violence and Lifetime Alcohol Problems.

alcohol use (mean = 8.5) than for respondents with medium (mean = 4.8) and low levels (mean = 4.4) of exposure. Additionally, the results reveal support for rejecting the null hypothesis of there being no mean level differences in alcohol use problems across groups ($F = 4.9$, $p < 0.01$).[8]

The results comparing mean level differences in drug abuse across respondents differentiated by varying levels of exposure to violence during childhood are displayed in Figure 3. The results are substantively consistent with theoretical expectations. Exposure to high levels of violence during childhood is associated with higher levels of regular drug use in adulthood. For example, the mean number of illicit drugs used regularly was 2.4 for parolees with high exposure to violence in childhood compared to a mean of 1.4 drugs used regularly for parolees with low levels of exposure to violence during childhood. While these differences are not dramatic (and not statistically different in this comparison, $F = 2.5$, $p > 0.05$), the general pattern of results is consistent with expectations that greater exposure to violence in the family during childhood is associated with more illicit drug use problems in adulthood.[9]

Exposure to violence during childhood is also expected to be strongly associated with violent behavior in adulthood (Thornberry, 1994; Rivera & Widom, 1990). The results in Figures 4 and 5 highlight the nature of this relationship for self-reported violent crime and intimate partner violence. The findings are largely consistent with our expectations. For example, the results in Figure 4

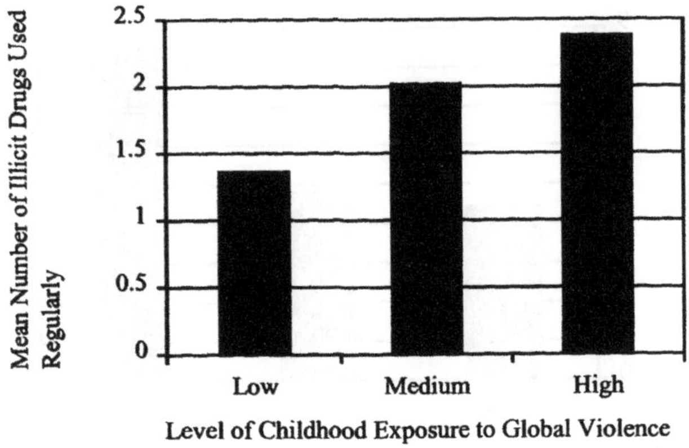

(F=2.50, p>0.05)

Figure 3. The Relationship Between Childhood Exposure to Violence and Number of Drugs Used Regularly.

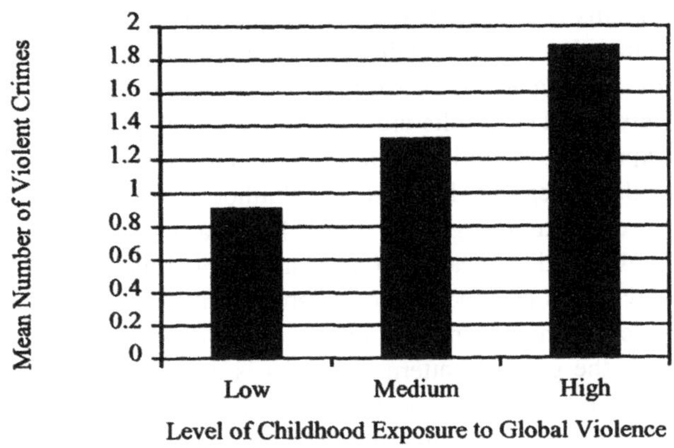

(F=6.42,p<0.01)

Figure 4. The Relationship Between Childhood Exposure to Violence and Self-Reported Violence.

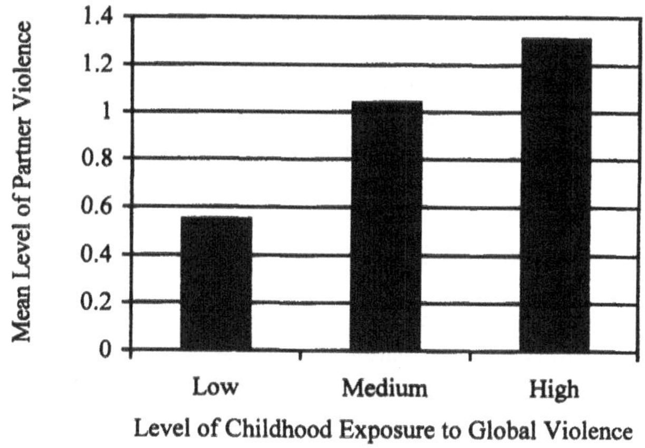

(F=2.27,p>0.05)

Figure 5. The Relationship Between Childhood Exposure to Violence and Intimate Partner Violence.

reveal significant differences in mean levels of self reported violent crime for parolees with varying levels of exposure to violence during childhood ($F = 6.4$, $p < 0.01$). Parolees with high levels of exposure to violence during childhood reported over twice as many violent crimes as parolees with low levels of exposure.

Finally, the relationship between childhood exposure to violence in the family and intimate partner violence perpetration in adulthood (in the year prior to current incarceration) is reported in Figure 5. The results reveal a general pattern that is consistent with our expectations that higher childhood exposure to violence is associated with higher mean levels of partner violence. However, despite the fact parolees with high levels of childhood exposure to violence report over twice the level of partner violence as parolees with low levels of childhood exposure to violence, the relationship is not statistically significant ($F = 2.3$, $p > 0.05$).

Thus far our results reveal a consistent pattern. Greater exposure to violence in the family during childhood is associated with more problems in adulthood. While some mean level differences between groups were not large enough to reject the null hypothesis, all of the relationships were substantive and in the theoretically expected direction. It is important to acknowledge that these comparisons did not control for extraneous influences. Additionally, these comparisons did not examine whether childhood exposure to different types of violence differentially affects certain outcomes, such as partner violence. We now examine whether childhood exposure to physical child abuse and observing parental violence have independent influences on intimate partner violence through a series of multivariate models.

Predictors of Intimate Partner Violence: Multivariate Analyses

In the next stage of the analyses, we examine whether exposure to physical child abuse and observing inter-parental violence are independently related to intimate partner violence through a series of Tobit regression models. As previously mentioned, Tobit models are used because of the limited range of positive responses for the dependant variable, intimate partner violence. The results for these models are presented in Table 2.

The results of a base model predicting partner violence are presented in the first (left hand) column of Table 2. The model includes the following variables: age, race, early delinquent participation, alcohol problems, and regular drug use. The results reveal that age, alcohol problems, regular drug use, and early delinquent onset are significantly related to partner violence in the predicted direction. Specifically, parolees with more alcohol-related problems, parolees who use illicit drugs regularly, parolees who are younger, and those with early onset to delinquency are more likely to engage in partner violence. These results are consistent with past research (Duncan et al., 1996).

In the next column (listed under model 1), cumulative or global exposure to violence in the family during childhood is introduced to assess whether it is

Table 2. Tobit models predicting intimate partner violence (n = 111)

Variable	BASE	Model 1	Model 2	Model 3	Model 4
Age	–0 023*	–0.028**	–0.021	–0.027**	–0.022*
Race	–0.362	–0.316	–0 032	–0.190	–0.170
Early Onset	0.771*	0.666	0.815*	0.573	0.651
Alcohol problems	0.065*	0.058*	0.067**	0.049	0.053
Regular Drug Use	0.237*	0.227*	0.240*	0.204*	0.208*
Global Violence	–	0.023	–	–	–
Child Abuse	–	–	–0.019	–	–0.045
Parents' Violence	–	–	–	0.069**	0.077**
-2 Log Likelihood	323.0	322.8	322.8	318.4	317.6

*p < 0.05; **p < 0.01, one-tailed test

related to partner violence perpetration when other risk factors (alcohol problems, etc.) are statistically controlled.

Recall that exposure to violence during childhood is expected to give rise to several negative outcomes (e.g. substance abuse). Thus, it is important to examine whether exposure to violence during childhood is related to partner violence perpetration over and above its effects on other outcomes, such as drug abuse. Moreover, it is important to assess whether certain factors mediate the effects of prior exposure to violence during childhood on intimate partner violence in adulthood. The results displayed in the second column of Table 2 reveal that global exposure to violence during childhood is not a significant predictor of male perpetrated intimate partner violence in adulthood when other influences are controlled. As in the base model, partner violence is significantly related to age, alcohol problems, and regular illicit drug use. Younger parolees and those with more alcohol-related problems and who use illicit drugs regularly have a greater likelihood of partner violence. While the results are largely unchanged from the base model, some of the earlier effects are somewhat attenuated. Additionally, the finding that early delinquent onset increased the likelihood of intimate partner violence was not observed.

Based on our earlier results reported in Figures 1 through 5, there are at least two explanations to account for these null findings. First, exposure to violence during childhood has several negative consequences (e.g., substance abuse) and a large part of its effects on partner violence perpetration operate indirectly through other influences (i.e., a mediating process).

The precise mechanisms are displayed in Figure 6. Results from our earlier bivariate analysis as well as the regression results (see base model and model 1) suggest that cumulative exposure to violence during childhood is not directly related to intimate partner violence in adulthood (note the absence of a significant, direct linkage). The results do suggest however, that the criminogenic impact of childhood exposure to violence in the family operates through and is mediated by alcohol and illicit drug. The fact that early onset to delinquency is no longer significantly related to intimate partner violence in adulthood (as indicated by the broken line) illustrates that early exposure to violence during childhood has a number of negative outcomes.

A second explanation for the results is that distinct experiences with exposure to violence during childhood are differentially related to partner violence perpetration in adulthood. Thus, it is important to disentangle the independent effects of unique experiences with violence. We turn to this issue in additional models reported in Table 2.

In model 2, reported in the third column of Table 2, exposure to physical child abuse is entered to examine its independent effects on intimate partner

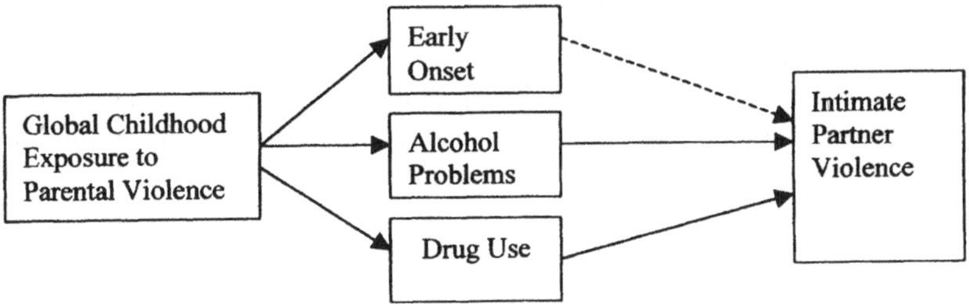

Figure 6. Theoretical Model Linking Childhood Exposure to Violence, Negative Life Outcomes, and Intimate Partner Violence.

violence perpetration in adulthood. The results from this analysis are virtually unchanged from previous models. Experiencing physical abuse in childhood does not increase the likelihood of partner violence perpetration in adulthood among male parolees in this sample. The results reveal that partner violence is significantly related to early delinquent participation, alcohol problems, and regular drug use.

The results from a model examining the independent effects of observing inter-parental violence on partner violence perpetration in adulthood is reported in the fourth column of Table 2. The findings offer a stark contrast to our previous results. Observing inter-parental violence during childhood surfaces as a strong predictor of intimate partner violence in adulthood even when other risk factors are controlled. Additionally, the effects of observing partner violence during childhood appear to reduce the criminogenic effects of other predictors in the model. While age and illicit drug use still remain as significant predictors of partner violence, the effects are diminished. Moreover, early delinquent participation and alcohol abuse are no longer significantly related to partner violence. In sum, the results from this model highlight the consequences of observing parental violence in childhood, both as a predictor of partner violence perpetration in adulthood and as an influence on other criminogenic factors (early delinquent onset, alcohol abuse, etc).

Results from models 2 and 3 are illustrated further in Figure 7. The model linking exposure to physical child abuse and intimate partner violence in adulthood reveals the absence of a direct relationship, whereas the model linking observing partner violence illustrates both direct and indirect effects on intimate partner violence in adulthood. Thus, the results highlight the importance of clearly specifying the context of early experiences with violence in the family.

In our final model, we consider whether child abuse and observing partner violence have significant effects on partner violence when both are included in

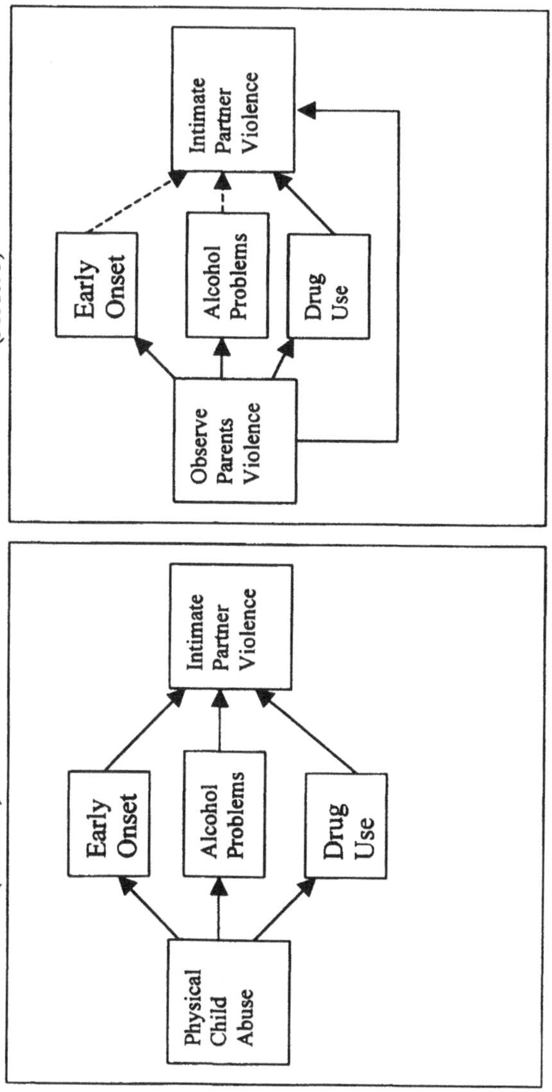

Figure 7. Theoretical Models Linking Physical Child Abuse and Exposure to Inter-parental Violence with Negative Life Outcomes and Intimate Partner Violence.

a model.[10] The results, reported in the last column of Table 2, are unchanged from the previous model. Male parolees who are younger, who regularly use illicit drugs, and who observe inter-parental violence during childhood are more likely to engage in intimate partner violence in adulthood. Exposure to physical child abuse during childhood is not significantly related to partner violence perpetration in adulthood when the effects of observing inter-parental violence are also considered. These results are supportive of our theoretical expectations.[11]

While the results from our multivariate analyses suggest that exposure to physical child abuse is not a salient predictor of partner violence, this does not necessarily mean that physical child abuse is not consequential. Rather, it might simply highlight that certain relationships between early exposure to violence and subsequent partner violence perpetration are stronger or more magnified. In short, we do not view these relationships in discrete either/or terms but rather we view them probabilistically in which differences in the effects of exposure to physical child abuse and observing inter-parental violence reflect differences in degree. To illustrate this issue further we examine specific bivariate relationships between these two types of exposure to violence and intimate partner violence in adulthood for respondents with varying levels of exposure (high, medium, and low) during childhood. The results of one way ANOVAs are displayed in Figure 8. The results clearly reveal that greater exposure to different types of violence in childhood is associated with higher mean levels of partner violence in adulthood. While the general form of the relationships is similar across types of exposure, the results are most magnified for observing inter-

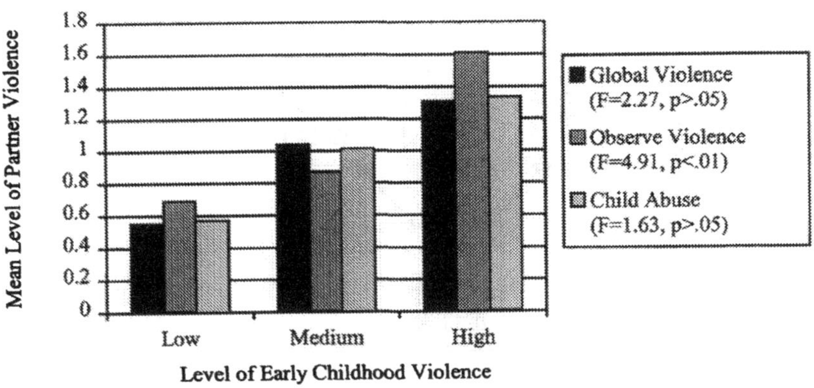

Figure 8. The Relationship Between Different Types of Childhood Exposure to Violence and Intimate Partner Violence.

parental violence which is consistent with our multivariate findings. Not surprisingly therefore, observing inter-parental violence in childhood is the only comparison that reveals a statistically significant difference in mean levels of partner violence in adulthood across respondents with varying levels of exposure (high medium, and low) to partner violence in childhood ($F = 4.91$, $p < 0.01$). Thus, these results suggest that observing inter-parental violence during childhood has strong and statistically significant effects on intimate partner violence in adulthood even though other relationships appear substantive. Combining these influences into a global index of exposure to violence in childhood could lead one to fail to reject the null hypotheses of there being no differences in mean levels of partner violence perpetration (in adulthood) across groups. Clearly, such a conclusion would be in incorrect because the criminogenic influences of observing inter-parental violence would be masked. In sum, these results highlight the importance of isolating the independent effects of specific types of exposure to violence during childhood.

DISCUSSION AND CONCLUSION

Using data collected from a sample of male parolees in New York State, our analysis examined two main issues. First, we examined whether exposure to violence in the family during childhood was related systematically to negative outcomes, such as alcohol abuse later in life. Second, we examined whether exposure to physical child abuse and observing inter-parental violence during childhood were differentially related to male perpetrated intimate partner violence in adulthood. Our review of prior research lead us to expect that greater exposure to violence in the family would be associated with more problems, such as violent crime, and illicit drug use later in life. Additionally, based on our reading of the literature, we expected that observing inter-parental violence during childhood would be more strongly related to intimate partner violence in adulthood than being exposed to physical child abuse. The results of our analysis were consistent with our research expectations. A series of mean level comparisons across negative life outcomes (e.g. substance abuse, violence, etc.) revealed that male parolees with greater exposure to violence in the family during childhood consistently reported more negative outcomes. For example, respondents with greater exposure to violence during childhood were more likely to self report early delinquent onset, substance abuse, and violent behavior than respondents with lower levels of exposure. In short, our results suggest that there are negative and diverse consequences associated with exposure to violence in the family during childhood.

Results from our multivariate models reveal that observing inter-parental violence during childhood is a strong and consistent predictor of intimate partner violence in adulthood even when other criminogenic influences are controlled. Moreover, the findings reveal that observing inter-parental violence during childhood is consistently a significant predictor of intimate partner violence in adulthood even when the independent effects of exposure to physical child abuse are considered. These results highlight the fact that observing inter-parental violence during childhood represents a salient influence on intimate partner violence in adulthood and are consistent with reviews of previous research on family violence (Hotaling & Sugarman, 1986). Moreover, our findings highlight the need to disentangle the effects of specific violent experiences in childhood on intimate partner violence in adulthood so that certain influences are not masked or misconstrued.

There are some limitations to our analysis that might impact how the results are interpreted. First, given the high-risk nature of the sample of parolees, our results are not generalizable to more representative populations. Second, the relatively small sample size for the analysis may have presented problems in uncovering significant relationships. Third, the cross-sectional and retrospective aspects of the data collection could be viewed as problematic given the potential problems associated with accurately recalling past incidents of violence. Because of the cross-sectional design of the study, temporal ordering between key variables of interest may have been inconsistent in some cases. For example, prior exposure to violence in the family may not have preceded early onset to delinquency for all parolees, although our interpretation of the results are consistent with extant theory and prior research. Fourth, some of the measures available in the data set have limitations. For example, measures of family violence were based on a revised version of the CTS. While the CTS has been widely used in research on family violence, critics have identified its methodological shortcomings (Dobash et al., 1992). To the extent that measures of past exposure to violence, such as the CTS, fail to fully uncover previous violent incidents (e.g., some types of incidents might be overlooked), baseline estimates will be reduced and relationships between key variables of interest could be attenuated. Given the potential for these measurement deficiencies, our results should be viewed as providing a conservative test of some relationships between exposure to violence in the family during childhood and various negative outcomes in adulthood including intimate partner violence.

It is important not to overstate the methodological problems with our data. On a more positive note, the data set includes a rich interview schedule that allows for a detailed examination of the intergenerational transmission of violence. Moreover, despite the limitations of certain measures, it is important to note that our findings are largely consistent with past research.

As with many studies, the results of our analysis highlight a number of important directions for future research. First, future research needs to examine the precise linkages – both direct and indirect – between observing inter-parental violence during childhood and engaging in intimate partner violence later in adulthood. Such information would prove valuable in making comparisons to alternative pathways, such as for exposure to physical child abuse, so that areas of similarity and dissimilarity could be identified. Second, there is an important need to assess whether the processes linking early exposure to violence in the family are invariant across different subgroups of the population. Most research on partner violence does not consider how certain contexts, related to, for example, social class, race, and age can shape various processes linking early exposure to violence in the family with intimate partner violence in adulthood.

Finally, future research needs to examine whether and to what extent exposure to alternative types of violence and conflict in the family give rise to differences in the processes leading to intimate partner violence in adulthood. Our analysis compared the independent effects of exposure to physical child abuse with observing inter-parental violence on intimate partner violence in adulthood. Clearly, children are exposed to many forms of conflict and violence during their formative years. From a research and public policy perspective, it is important to uncover whether exposure to verbal abuse, physical abuse, neglect, sibling violence, and other forms of violence in the community and in schools follow unique processes leading to partner violence. At that time, a more comprehensive understanding of the complex linkages between early exposure to violence in the family and intimate partner violence will be realized.

NOTES

1. In selecting the sample, the parolees were stratified into violent and non-violent groups based on their last conviction offense. Parolees whose last conviction offense could not be clearly classified as either violent or non-violent (e. g. DWI, reckless endangerment) were excluded from the sample. While this sample selection criteria suggests that offenders may specialize in particular types of crimes, such as violence, a wealth of information reveals that offenders, especially serious offenders, exhibit versatile offending patterns (Wolfgang et al., 1972; Tracy & Kempf-Leonard, 1996, Hindelang, 1971). While our analysis does not examine this issue, we acknowledge that the sample selection criteria may mistakenly impose violent and non-violent classifications upon parolees in the sample.

2. There were no differences between groups when a prevalence measure of prior incarceration was used. Thus, the groups differed only in the frequency of their prior incarcerations.

3. Information was not limited to biological parents as respondents also reported the activities of their step-parents.

4. The original response categories were: never, happened/frequency unknown, once or twice, several times, about once a month, about once a week or more.

5. While our selection of early onset prior to age 13 appears arbitrary, past research has utilized similar criteria (Tolan, 1987). Using this age cut-off is desirable because it clearly precedes and is not confounded with adolescent-limited delinquency (see Moffitt, 1993). Additional analyses using alternative measures of early delinquency onset status revealed similar results.

6. We acknowledge that couple reports (e. g. combining both partner's reports) may represent a more valid indicator of partner violence (Szinovacz, 1983). However, limitations with the data preclude using them from the current analysis. While the data do include CTS scores reported by the parolees' intimate partners, for the year prior to prison, partner reports were only available for 29 of the 111 male parolees. Therefore, we rely on male parolee reports of partner violence during the year prior to their current incarceration which past research demonstrates are acceptable under certain research conditions (see Moffitt et al., 1997).

7. This procedure was repeated by selecting one standard deviation above and below the mean, and produced substantively similar results. These results are available upon request.

8. Post hoc pairwise comparisons indicate that individuals exposed to high levels of childhood violence experience significantly more alcohol related problems as adults than those with low levels of exposure to violence during childhood.

9. Failing to find a statistical difference in this comparison is perhaps not surprising given the relatively small size of the sample as well as its 'high risk' composition.

10. As expected, exposure to physical child abuse and observing inter-parental violence are significantly correlated ($r = 0.32$, $p < 0.01$). However, the strength of this relationship suggests that multicollinearity was not problematic in the model.

11. It is also important to consider whether exposure to physical child abuse and observing inter-parental violence interact to affect partner violence in adulthood. We estimated one final model that included standardized measures for exposure to physical child abuse and observing inter-parental violence as well as their product term (i.e. interaction term) as predictors. The results revealed that exposure to physical child abuse and observing interparental violence did not interact to increase the likelihood of partner violence in adulthood. These results are available upon request.

REFERENCES

Bandura, A. (1977). *Social Learning Theory*. Englewood Cliffs, NJ: Prentice Hall.

Bandura, A. (1986). *Social Foundations of Thought and Action: A Social Cognitive Theory*. Englewood Cliffs, NJ: Prentice Hall.

Berry, D. (1993). Screen Violence and Americas Children. *Spectrum: The Journal of State Government, 66*, 37–42.

Blane, H. T., Miller, B., & Leonard, K. (1994). *Intra and Intergenerational Aspects of Serious Domestic Violence and Alcohol and Drug Abuse in Buffalo, 1987.* (Computer file). Los Altos, CA: Sociometrics Corporation, 1992 (producer). Ann Arbor, MI: Inter-university Consortium for Political and Social Research (distributor).

Blumstein, A., Cohen, J., Roth, J., & Visher, C. (1986). *Criminal Careers and Career Criminals.* Washington D.C.: National Academy Press.

Cappell, R., & Heiner, C. (1990). The Intergenerational Transmission of Family Aggression. *Journal of Family Violence, 5*, 135–152.

Carter, J., Stacey, W., & Shupe, A. (1988). Male Violence Against Women: Assessment of the Generation Transfer Hypothesis. *Deviant Behavior, 9*, 259–263.

Cerezo-Jimenez, M. A., & Frias, D. (1994). Emotional and Cognitive Adjustment in Abused Children. *Child Abuse and Neglect, 18*, 923–932.

Choice, P., Lamke, L. K., & Pittman, J. (1995). Conflict Resolution Strategies and Marital Distress as Mediating Factors in the Link Between Witnessing Interparental Violence and Wife Battering. *Violence and Victims, 10*, 107–119.

Dobash, R. P., Emerson, D. R., Wilson, M., & Daly, M. (1992). The Myth of Sexual Symmetry in Marital Violence. *Social Problems, 39*, 71–91.

Dodge, K. A., Bates, J., & Petit, G. (1990). Mechanisms in the Cycle of Violence. *Science, 250*, 1678–1683.

Downs, W. R., Smyth, N., & Miller, B. (1996). The Relationship Between Childhood Violence and Alcohol Problems Among Men who Batter: An Empirical Review and Synthesis. *Aggression and Violent Behavior, 1*, 327–344.

Duncan, R., Saunders, B., Kilpatrick, D., Hanson, R., & Resnick, H (1996). Childhood Physical Assault as a Risk Factor for PTSD, Depression, and Substance Abuse: Findings from a National Survey. *American Journal of Orthopsychiatry, 66*, 437–448.

Dutton, D. G. (1988). Profiling of Wife Assaulters: Preliminary Evidence for a Trimodel Analysis. *Violence and Victims, 3*, 5–29.

Famularo, R., Fenton, T., Kinscherff, R., Ayoub, C., & Barnum, R. (1994). Maternal and Child Posttraumatic Stress Disorder in Cases of Child Maltreatment. *Child Abuse and Neglect, 18*, 27–36.

Gelles, R. J., & Straus, M. (1989). *Physical Violence in American Families, 1985 (computer file).* Durham, NH: University of New Hampshire Family Research Laboratory, 1988 (producer). Ann Arbor, MI: Inter-University Consortium for Political and Social Research #9211 (distributor).

Hindelang, M. (1971). Age, Sex, and the Versatility of Delinquent Involvement. *Social Problems, 18*, 522–535.

Holtzworth-Munroe, A., & Stuart, G. L. (1994). Typologies of Male Batterers: Three Subtypes and the Differences Among Them. *Psychological Bulletin, 116*, 476–497.

Hotaling, G., & Sugarman, D. (1986). An Analysis of Risk Markers in Husband to Wife Violence: The Current State of Knowledge. *Violence and Victims, 1*, 101–124.

Hotaling, G., Straus, M., & Lincoln, A. J. (1989). Intra Family Violence and Crime and Violence Outside the Family. In: L. Ohlin, & M. Tonry (Eds) *Family Violence. Crime and Justice: A Review of Research, 11*, (315–375). Chicago University of Chicago Press.

Hughes, H. M., Parkinson, D. & Vargo, M. (1989). Witnessing Spouse Abuse and Experiencing Physical Abuse: A Double Whammy? *Journal of Family Violence, 4*, 197–209.

Jaffe, P., Sudermann, M. & Reitzel, D. (1992). Child Witnesses of Marital Violence. In: R Ammerman, & M. Hersen (Eds), *Assessment of Family Violence: A Clinical and Legal Sourcebook.* New York: John Wiley & Sons, Inc.

Jonson-Reid, M. (1998). Youth Violence and Exposure to Violence in Childhood: An Ecological Review. *Aggression and Violent Behavior, 3*, 159–179.

Kalmuss, D. (1984). The Intergenerational Transmission of Marital Aggression. *Journal of Marriage and the Family, 46*, 11–19.

Kaufman, G., & Spatz-Widom, C. (1999). Childhood Victimization, Running away, and Delinquency. *Journal of Research in Crime and Delinquency* (forthcoming).

Kurtz, D. P., Kurtz, G., & Jarvis, S. (1991). Problems of Maltreated Runaway Youth. *Adolescence*, *26*, 543–555.

Loeber, R., & T. Dishion. (1983). Early Predictors of Male Delinquency: A Review. *Psychological Bulletin*, *94*, 68–99.

Loeber, R., & Stoughamer-Loeber, M. (1986). Familial Factors as Correlates and Predictors of Juvenile Conduct Problems and Delinquency. In: M. Tonry, & N. Morris (Eds), *Crime and Justice*, *7*. Chicago: University of Chicago Press.

Loeber, R., & LeBlanc, M. (1990). Toward a Developmental Criminology. In: M. Tonry, & N. Morris (Eds), *Crime and Justice*, *12*. Chicago: University of Chicago Press.

Moffitt, T. (1993). Adolescent Limited and Life-course Persistent Antisocial Behavior: A Developmental Taxonomy. *Psychological Review*, *100*, 674–701.

Moffitt, T., Caspi, A., Krueger, R., Magdol, L., Margolin, G., Silva, P., & Sydney, R. (1997). Do Partners Agree about Abuse in their Relationship? A Psychometric Evaluation of Interpartner Agreement. *Psychological Bulletin*, *9*, 47–56.

Malinosky-Rummell, R., & Hanson, D. (1993). Long-term Consequences of Childhood Physical Abuse. *Psychological Bulletin*, *114*, 68–79.

Paternoster, R., Dean, C., Piquero, A., Mazerolle, P., & Brame, R. (1997). Generality, Continuity, and Change in Offending. *Journal of Quantitative Criminology*, *13*, 231–267.

Patterson, G., & Yoerger, K. (1993). Developmental Models for Delinquent Behavior. In: M. Rutter & D. Hay (Eds), *Development Through Life*. Oxford, UK: Blackwell.

Piquero, A., Paternoster, R., Brame, R., Mazerolle, P., & Dean, C. (1999). Onset Age and Offending Specialization. *Journal of Research in Crime and Delinquency*, *36*, 235–274.

Rivera, B., & Widom, C. (1990). Childhood Victimization and Violent Offending. *Violence and Victims*, *5*, 19–35.

Singer, M., Miller, D., Gua, S., Flannery, D., Frierson, T., & Slovak, K. (1999). Contributors to Violent Behavior Among Elementary and Middle School Children. *Pediatrics*, *104*, 878–884.

Smith, C., & Thornberry, T. (1995). The Relationship Between Childhood Maltreatment and Adolescent Involvement in Delinquency. *Criminology*, *33*, 451–477

Spaccarelli, S. J., Coatsworth, D., & Bowden, S. (1995). Exposure to Serious Family violence Among Incarcerated Boys: Its Association with Violent Offending. *Violence and Victims*, *10*, 163–182.

Straus, M. (1979). Measuring Intrafamily Conflict and Violence: The Conflict Tactics Scales *Journal of Marriage and the Family*, *41*, 75–88.

Szinovacz, M. (1983). Using Couple Data as a Methodological Tool: The Case of Marital Violence. *Journal of Marriage and the Family*, *45*, 633–644.

Thornberry, T. (1994). Violent Families and Youth Violence. *Fact Sheet #21*. Washington, DC: Office of Juvenile Justice and Delinquency Prevention.

Tolan, P. H. (1987). Implications of Age of Onset for Delinquency Risk. *Journal of Abnormal Child Psychology*, *15*, 47–65.

Tolan P. H., & Lorion, R. (1988). Multivariate approaches to the identification of Delinquency Proneness in Adolescent Males. *American Journal of Community Psychology*, *16*, 547–561.

Tracy, P. E., & Kempf-Leonard, K. (1996). *Continuity and Discontinuity in Criminal Careers*. New York: Plenum.

U.S. Department of Health and Human Services, National Center of Child Abuse and Neglect. (1996). *Child Maltreatment 1994: Reports from the States to the National Center on Child Abuse and Neglect*. Washington, D.C.: U.S. Government Printing Office.

Widom, C. (1989a). Does Violence Beget Violence: A Critical Examination of the Literature. *Psychological Bulletin*, *106*, 3–28.

Widom, C. (1989b). Child Abuse, Neglect, and Violent Criminal Behavior. *Criminology 27*, 251–271.
Wolfe, D. A., Wekerle, C., Reitzel-Jaffe, D., & Lefebvre, L. (1998). Factors Associated with Abusive Relationships Among Maltreated and Nonmaltreated Youth. *Development and Psychopathology, 10*, 61–85.
Wolfgang, M., Figlio, R., & Sellin, T. (1972). *Delinquency in a Birth Cohort.* Chicago: University of Chicago.
Zingraff, M., Leiter, J., Myers, K., & Johnson, M. (1993). Child Maltreatment and Youthful Problem Behavior. *Criminology, 31*, 173–202.

APPENDIX

Measures of Exposure to Violence (Child Abuse and Observing Inter-Parental Violence) During Childhood

During an average year in respondent's childhood, how often did respondent's parents perpetrate the following acts towards the respondent (or each other):

Kicked, bit, hit with fist
Hit with something
Beat up
Burned/scalded
Choked
Threatened with knife/gun
Used knife or gun
Threaten to hit/throw at
Threw/smashed/hit/kicked something
Threw something at spouse/child
Pushed/grabbed/shoved
Slapped

Measures of Intimate Partner Violence in Adulthood

In the year prior to prison, how often the respondent perpetrated the following acts towards their partner:

Kicked, bit, hit with fist
Hit with something
Beat up
Burned/scalded
Choked
Threatened with knife/gun
Used knife or gun

Forced sex
Slapped

Early Delinquent Onset

At what age did respondent:

Take something from a store or business during business hours without paying for it?
Enter a store after business hours or a home in order to steal things?
Tried to get money from a person or business by threatening them in some manner?
Physically assaulted another person?

Violent Crime

Whether or not the respondent had ever:

Take anything of value from someone, using a weapon?
Tried to get money from a person or business by threatening them in some manner?
Kidnapped, or held anyone against their will?
Physically assaulted another person?
Attempted or had sexual intercourse or other sexual acts through the use of force?

Illicit Drug Use

Whether or not respondent reported using marijuana, cocaine, heroin, speed, downers, or LSD at least once per week for a month.

Alcohol Problems

Respondents were asked whether or not they experienced any of these alcohol-related problems:

Has your family objected because you were drinking too much?
Do you ever think you were an excessive drinker?
Have you ever drunk a fifth of liquor in one day?
Has there ever been a period of two weeks when every day you were drinking at least seven drinks?

Have friends, your doctor, your clergyman, or any other professional ever said you were drinking too much?

Have you ever wanted to stop drinking but couldn't?

Some people promise themselves not to drink before 5 o'clock or never to drink alone in order to control their drinking. Have you ever done anything like that?

Have you ever had job or school trouble because of drinking?

Did you ever lose a job (or get kicked out of school) because of drinking?

Have you ever gotten into trouble driving because of drinking?

Have you ever been arrested because of drinking, or for disturbing the peace while drinking?

Have you ever gotten into physical fights while drinking?

Have you ever gone on binges or benders where you kept drinking for a couple of days or more without sobering up?

Did you neglect some of your usual responsibilities while on binges/benders?

Have you ever had bad blackouts while drinking?

Do you ever need a drink when you get up in the morning?

Have you ever had the shakes after stopping or cutting down on drinking?

Have you ever had fits/seizures after stopping or cutting down on drinking?

Have you ever had the DTs when you quit drinking?

Have you ever continued to drink when you knew you had a serious illness that might be made worse by drinking?

Has there been a period in your life when you could not do your ordinary work well unless you had something to drink?

SONS IN PRISON AND THEIR MOTHERS: IS THERE A RELATIONSHIP BETWEEN CHILDHOOD HISTORIES OF PHYSICAL ABUSE?

Jaana Haapasalo

Evidence for the intergenerational cycle of violence hypothesis has been accumulated from various sources. Longitudinal and retrospective studies as well as studies using a 'follow-back' strategy have shown a relationship between violent and abusive childrearing practices and later violent adult criminality (Lewis, Mallouh & Webb, 1989; Luntz & Widom, 1994; McCord, 1996; Widom, 1989; Widom & White, 1997). Widom (1989) identified a sample of 908 substantiated cases of abused and/or neglected children and investigated their juvenile and adult court records. Compared to a matched non-maltreated control group, maltreated children had higher rates for adult criminality and violence. Lewis

***This project was supported by Grant No. 98-WT-VX-0007 awarded by the National Institute of Justice, Office of Justice Programs, U.S. Department of Justice. Points of view in this document are those of the author and do not necessarily represent the official position or policies of the U.S. Department of Justice.**

Author Note: Jaana Haapasalo, Ph.D., Department of Psychology, University of Jyväskylä, Finland. **I am indebted to Tiina Hämäläinen, M. A., Elina Pokela, M. A., Hanna Koivisto, M. A., and Marja Saarinen, M. A., for their help in data collection and Asko Tolvanen, M. A., for his statistical advice. The research was supported by a grant from the Academy of Finland.**

Families, Crime and Criminal Justice, Volume 2, pages 73–90.
2000 by Elsevier Science Inc.
ISBN: 0-7623-0737-4

and her colleagues (e.g., Lewis et al., 1988) used a retrospective design with clinical samples of delinquents and found high prevalences (around 60 to 80 %) of childhood maltreatment experiences among them. Equally high prevalences of childhood maltreatment have been found in criminal populations in Finland, Scandinavia, where the present study was carried out (Haapasalo & Hämäläinen, 1996; Hämäläinen & Haapasalo, 1996; Haapasalo & Petäjä, 1999).

Abused children, after they have become parents themselves, also appear to have an elevated risk for abusing their own children or behaving in a violent manner in adulthood (Egeland, 1993). For example, in a Finnish study on 50 mothers, the mothers' childhood abuse experiences predicted their abuse of their own child (Haapasalo & Aaltonen, 1999). Moreover, the childhood experience of physical violence has been found to be related to living in a violent marital relationship (Cascardi, O'Leary, Lawrence & Schlee, 1995). Despite the convincing evidence, there are exceptions to the rule of abuse generating abuse. Most parents who come from abusive families do not abuse their own children (Kaufman & Zigler, 1993). It has been estimated that approximately one third of abused individuals later abuse their own children (Kaufman & Zigler, 1987; Oliver, 1993). Studies have generally compared abused and non-abused populations in trying to examine whether parental history of abuse is related to abuse of one's own children. Relatively few studies have investigated the associations between the history of abuse of parents and that of their children in clinical samples with a high base rate of maltreatment, such as criminal offenders. It is likely that the association between the experiences of parents and their offspring will be stronger than usual in groups where parents have raised children who later commit crimes. Furthermore, there is a lack of information concerning factors that may moderate the relationship between abuse of parents and that of their offspring, especially in high base rate populations. The present study is targeted at these issues.

The link between the abuse experiences of parents and their offspring is far from given, since there are factors that may moderate the effects of abuse. A moderator is a variable that affects the direction and/or strength of the predictor-criterion variable relationship (Baron & Kenny, 1986). For example, marital conflict, parental substance abuse, and a low socioeconomic status may act as moderators between childhood physical abuse and later adverse outcomes (Malinosky-Rummell & Hansen, 1993). Also, placement in foster care or other child protection measures aiming at the prevention of child maltreatment are potential moderators (Fein, 1991; Jellinek et al., 1994; Widom, 1991). In the presence of protective factors, the effects of abuse may diminish. We can assume that the whole parenting context in the abusive family-of-origin is important when considering the making or breaking of the cycle of abuse. Family and parenting factors that co-occur with child maltreatment should be taken into account in the prediction of maltreated children's future functioning (Belt & Abidin, 1996).

The cycle of abuse could be assumed to be especially visible in samples with a high risk for problematic parenting, such as offenders' families-of-origin. Power-assertive, authoritarian, and harsh childrearing practices and parental rejection and aggression have been found in offenders' childhood family environments (Eron, Huesmann & Zelli, 1991; Farrington, 1992; McCord, 1988; Patterson, Reid & Dishion, 1992). Lack of maternal warmth (McCord, 1996) and negative, unfair and inconsistent disciplinary practices on the part of mothers (Dishion, Patterson & Kavanagh, 1992) have also been associated with antisocial outcomes in the child.

The transmission of abuse and violence has rarely been studied within the same family by independently inquiring about both parents' and their children's abuse experiences at the same point of time in their lives. If this could be done, we would be able to look for an association between parental childhood abuse experiences and those of their children in samples where the risk of the intergenerational transmission of abuse may be greatest. When studying the transmission of punitive disciplinary practices across two generations, it has been found that mothers' – but not fathers' – childhood experiences of corporal punishment predict the striking and beating they deliver to their own children (Stattin, Janson, Klackenberg-Larsson & Magnusson, 1995). In the light of these findings, it appears that the role of the mother as a disciplinarian may be crucial in comparison with that of the father.

This study sought to test the association between maternal and offspring childhood experiences of physical abuse in a sample of criminal offenders and their mothers. Offenders were chosen as the target group because, according to the literature, they carry a heightened risk of abusive childhood treatment. It is important to know which factors may motivate or deter the cycle of abuse in criminal populations. Therefore, in examining the relationship between the childhood abuse experiences of mothers and sons, a number of family problems (marital conflict, maternal problems, economic stress, child protection measures) were also tested with respect to their mediator and moderator effects.

METHOD

Participants

Offenders. In Finland, the great majority of young offenders under the age of 21 years are placed in a youth prison. The number of young prisoners under age 21 fluctuates from one day to another. For example, on October 1, 1994 (the year when the data collection started), there were 138 young inmates between the ages of 15 and 20 years serving their sentence in Finland. Originally, 131 offenders were randomly selected from among all male inmates born in 1972 or later. (There were no official statistics on the exact population of young inmates serving their sentence at the outset of the study, but it should range from 140 to 150.) They came from one youth prison and four facilities for adults. Of these, 97 (74%) gave their informed consent and 89 (68%) participated. Those eight prisoners who initially agreed to participate but later dropped out of the study did so for various reasons (e.g., prison leave, escape, trial). The refusal rate varied from one facility to another (0 to 40%) and was highest in the three prisons where a prison guard was present at the information meeting. The offenders' age ranged from 16 to 22 years ($M = 20$ years).

The sample can be considered as representative of young offenders in Finland, since young offenders who go to prison in Finland constitute a very homogeneous group. Moreover, the majority of young Finnish inmates participated in the study. For a young offender, a relatively serious and long history of offenses is a prerequisite for a prison sentence. In this sample, the mean number of convictions was eight. Mean age at the first official offense was 16 years, range 15–20 years. (Criminal responsibility begins at age 15 in Finland.)

The offenders were highly alike with respect to education, vocational training, and school performance. The proportions of the offenders at the various levels of basic education were as follows: no basic education 20%, special education class 20%, grade school 54%, part of high school 5%, high school 1%. For vocational training, the proportions were: no training 63%, vocational course(s) 20%, vocational school 17%. The majority had been maladjusted in school (66%) and received temporary special education (69%). For 20%, poor school performance was reported, and 11% were dyslexic. Half of the offenders had been taken into custody by the child protection services before the age of 16. They had convictions for various types of property, violent, traffic, public order and minor offenses based on their lifetime criminal records. Twenty-three offenders were convicted of assault, 16 of aggravated assault, 13 of manslaughter or attempted manslaughter, and the rest of the participants had convictions for various non-violent offenses.

All the offenders gave their informed consent before the study began. No incentives were used. The offenders chose whether they were willing to give information about their parents to the investigators, including addresses. Consequently, 66% of the offenders gave permission for their parents to be contacted. Those offenders who agreed to permit contact differed from those who did not in that they either had parents or had permanent contact with them.

There were offenders who had no parental figures, which was reflected in the fact that only 67% of the participants were able to respond to questions about their major paternal father figures (father/stepfather) and 91% about their major maternal caregiver.

Mothers. The mothers were asked to participate in a letter informing them about the study. The letter was sent to 48 mothers, of whom 8 failed to respond and 9 actively declined to participate. Hence, 31 mothers agreed to the study and signed an informed consent document. The mothers ranged in age from 33 to 58 years ($M = 49$ years). It is difficult to indicate whether those mothers who participated differed from the non-participating mothers, because no interview or questionnaire data were gathered in respect of the latter. However, 11 out of the 17 non-participating mothers but only 8 out of the 31 participating mothers were living with the offender's father. It could be assumed that the mothers who were single or separated/divorced from the boy's father were able to make a more independent decision on whether to participate than the married or cohabiting mothers. Their relationship with their son could also have been different than that of the mothers from the two-parent families.

Measures

Abuse Forms Interviews. The mothers responded to an eight-part interview. First, the mothers were asked about their childhood maltreatment experiences. A section entitled The Abuse Forms examined retrospective self-reports of childhood maltreatment experiences. It was designed by the author on the basis of the maltreatment classification schedule by Barnett, Manly & Cicchetti (1993), Hart, Germain & Brassard's (1987) classification of psychological abuse, and Zuravin's (1991) types of neglect. The following instruction was given to the mothers: "Now I'm going to ask you detailed questions about things that you may have experienced at the hands of your parents in your family-of-origin. I'd like you to tell me how many times, under the age of 15 years, you've experienced any of the things I mention." The respondents were then handed a form with the following options: never, once, twice, 3–4 times, 5–9 times, 10 times or more. The specific forms of physical abuse included, for example, punching with a fist, hitting with an object, been thrown against something, and strangulation (a total of 20 distinct and concrete forms). The concurrent validity of the Abuse Forms appeared good in the light of high and significant positive correlations (in the range of 0.61 to 0.78) with the Minor and Severe violence scales of Straus's (1992) Conflict Tactics Scales.

The second part of interview included a modified Family-of-Origin instrument (see Brassard, Hart & Hardy 1993). In addition, the mothers were given two

17-item Likert-type scales, designed by the author and entitled the Descriptions of Parenting. Next, the mothers responded to the Psychosocial Stressors and the Pregnancy, Birth, and Preschool Questionnaire of the Diagnostic Interview for Children and Adolescents – Parent Version (DICA-R-P; Reich, Shayka & Taibleson, 1991b). They were also asked about their marital relations and problems, mental health problems and treatment, substance abuse, and criminality. Finally, they answered a section entitled Child-rearing attitudes and behaviors.

The offenders were interviewed using a four-part structured computerized interview that shared three sections with the mother's interview (Abuse Forms, Family-of-Origin, Descriptions of Parenting). In addition, they were given six sections of the DICA-R-Adolescent version (Reich, Shayka & Taibleson, 1991a). For the offenders, however, the interview data were not used in the childhood maltreatment assessments, because more objective file data were available (see below). The complete interview protocols can be requested from the author. Five research psychologists were used as interviewers to minimize the subjective bias. The interview took approximately 5 to 6 hours per participant.

Files. The offenders were asked to write down all the places they had visited as clients in childhood and adolescence. Also, their prison files were reviewed to elicit additional information about previous social and mental health services. During the mothers' interviews, information about even more places were obtained. A request for a copy of the records, together with the offenders' written permission, was then distributed to each of these places, mostly various official agencies and institutions. Eventually, the file information consisted of all the available early records collected from the child protection services, various clinics, hospitals and institutions, and social agencies of whom the offenders had been clients in childhood and adolescence. The information in the files was originally produced by the clinicians and professionals who were working with the offenders at the time. The files were thus authentic descriptions of the participant's childhood and adolescent years. No files regarding the mothers' early years were separately collected, but in the majority of cases the child protection files included information about the mothers' childhood as well. The files were available for 78 offenders. All the 31 participating mothers were mothers of these 78 offenders with files, except for two cases in which the offender had no history of social or mental health services and thus no files.

Questionnaires. The participants also completed modified versions of the Conflict Tactics Scales (Straus, 1992) and a short form of Schaefer's (1965) Parent Behavior Inventory (Schludermann & Schludermann, 1970) (for mothers, the items were modified to describe the parents' own behavior towards their son). The detailed information together with the reliability data for these modifications are available from the author. This study used the questionnaire data only to form the merged physical abuse variable for the mothers (see below).

Major variables

Because no files were collected for the mothers, their childhood physical experiences were condensed into a variable based on all the information gathered from the interview, offender file and questionnaire sources. This merged information was collected for all mothers of the offenders for whom the files were available (n = 78). The merged data revealed that 30% of the offenders' mothers had experienced physical abuse in childhood (*Mother physically abused*). Physical abuse had occurred if the parent was at least beaten with an object by her/his parents in childhood according to the maternal interview, the questionnaires, or the son's files. Only when a clear-cut report of childhood physical abuse was available was physical abuse denoted to be present. Examples of physical abuse reported in the material are as follows: "Mother was disciplined by father who used a carpet beater and smacked her," "Both of mother's parents punished her using a coathanger and a shoe, for example," "Mother was belted in childhood," "Mother was beaten with an electric cord and slapped around by her mother so that her legs and buttocks were bleeding," and "Mother was disciplined by her father who struck her with objects and smacked her."

The files were available for the majority of the offenders (n = 78). For this reason, the files were used to assess childhood physical abuse experiences among the offenders. *Physical abuse* (prevalence: 37%) experienced by the offender at the hands of his parents was a three-category file-based (ordinal) variable, consisting of the following categories: 0 = no mention of physical abuse in files, 1 = corporal punishment, and 2 = extreme, life-endangering physical abuse. There was no need to compute interrater reliabilities for the mothers' physical abuse experiences or the offenders' file-based physical abuse, as their assessment was unequivocal and based on clear documentation in the files. Because the documents were written by professionals, it is reasonable to assume that at least the most extreme cases of abuse were registered. On the other hand, underreporting of abuse is possible, if the abuse was not suspected and substantiated or the family members denied it in the absence of clear evidence.

The variables describing family problems were solely based on the file information. *Maternal problems* was a sum of three dichotomous (present/absent) variables (percentage of cases where files indicated the presence of a problem in parentheses): Maternal psychiatric problems (12%), Maternal alcohol abuse (34%), and Maternal criminality (4%). Only when there was a clear report of a problem in the files, was it attested to be present. *Marital conflict* was a sum of the following variables: Verbal and physical conflicts between parents (0 = no conflict, 56%; 1 = minor conflicts, 24%; 2 = constant quarelling, 4%; 3 = spousal abuse, 16%), Son witnessed parental violence (present/absent; 17%), and Parental divorce (present/absent; 62%). Again, only if the files included a clear verified report of conflicts, witnessing parental violence or a divorce, was the incident considered to be present.

Child protection services was a sum of 10 variables describing whether the boy had been taken into care by the child protection services under age 2, between ages 3 and 5, ages 6 and 9, ages 10 and 13, and between ages 14 and 16, and whether open-care child protection interventions (home visits, therapy, help with household chores, etc.) were given to the family during the respective age intervals. Of the total sample, 86% had been subjected to at least some of the child protection interventions. This information about the interventions can be regarded as substantiated and reliable, because child protection records are available for each case that has been processed by the child protection services.

Economic stress in the family (40%) was assessed as being present, if it was clearly mentioned in the files. Examples of the notes that warranted the presence of economic stress were as follows: "difficulty in getting along, because the mother is unemployed," "the mother says that she had to borrow money, because her unemployment benefit has not been paid yet," "the family has unpaid bills," "the financial situation is very poor," and "the family is financially burdened." In addition, economic stress was present if the social services records showed that the family lived on welfare for longer periods or that the agency paid for the family's rent, food, bills, and other necessities. Table 1 provides a summary of the major variables in the study.

Data analyses

The relationship between maternal childhood abuse experiences and the son's abuse experiences was also examined in the mediator-moderator paradigm using a structural equation model in which family factors were tested as mediators and moderators (LISREL; Jöreskog & Sörbom, 1989). In assessing the goodness of fit between the model and the data, three statistics were used. The goodness-of-fit (GFI) should preferably be over 0.90, the root-mean-square residual (RMR) below 0.10, and the chi square test have a p value over 0.05 (see e.g., Cole, 1987; Marsh, Balla & McDonald, 1988). Polychoric correlations (SPSSx PRELIS program) were computed between the major variables.

Table 1. Summary of Major Variables

Variable	Source	Type of variable
Mothers		
Mother physically abused	Files, interviews, questionnaires	Dichotomous
Maternal psychiatric problems	Files	Dichotomous
Maternal alcohol abuse	Files	Dichotomous
Maternal criminality	Files	Dichotomous
Maternal problems	Files (Maternal psychiatric problems + Maternal alcohol abuse + Maternal criminality)	Continuous
Offenders		
Physical abuse	Files	Ordinal
Verbal and physical conflicts between parents	Files	Ordinal
Son witnessed parental violence	Files	Dichotomous
Parental divorce	Files	Dichotomous
Marital Conflict	Files (Verbal/physical conflicts + Son witnessed parental violence + Parental divorce	Continuous
Child protection services	Files (a sum of 10 variables)	Continuous
Economic stress in the family	Files	Dichotomous

RESULTS

Mediator Effects Between the Mother's and Son's Childhood Abuse Experiences

To examine whether and how maternal childhood maltreatment was associated with the sons' childhood maltreatment, a LISREL model with the mother's childhood physical abuse as a predictor and the son's childhood physical abuse as a criterion variable was tested with: (1) Maternal problems, Marital conflict, Economic stress, and Child Protection Services as mediators and (2) the same family factors as the moderator variables. Our first theoretical model stated that the mother's childhood abuse experiences might be related to their son's childhood abuse experiences through a number of indirect mediating effects between the maltreatment variables and family factors. The second model assumed that the relationship between the mother's and son's abuse experiences might be different depending on the level of family problems. For example, the

Table 2. Intercorrelations of Maltreatment and Family Variables (LISREL Model)

Variables	1.	2.	3.	4.	5.	6.
1. Mother abused	•					
2. Son abused	0.45***	•				
3. Economic stress	0.07	–0.19	•			
4. Marital conflict	0.49***	0.33*	0.06	•		
5. Maternal problems	–0.01	–0.32*	0.33*	0.21	•	
6. Child protection services	0.26	–0.10	0.32*	0.36**	0.55**	•

Note: The variables are described in Table 1.
*$p < 0.05$ **$p < 0.01$ ***$p < 0.001$

relationship could be stronger or weaker if certain family factors were present or absent.

Before testing the hypothetical models, polychoric correlations were computed between the maltreatment and family variables included in the model (Table 2). The intergenerational abuse correlation was evident. The correlations revealed that the mother's childhood history of abuse and physical abuse experienced by the son correlated strongly with Marital conflict. The family problems mostly had strong intercorrelations. Marital problems correlated positively with Economic stress, and Child protection services were associated with Economic stress, Marital conflict, and Maternal problems. In the hope of better understanding the interrelationships between the variables, a LISREL model testing the mediator effects was constructed. Figure 1 presents the observed relationships between the maltreatment and family variables. The goodness-of-fit index (GFI; Jöreskog & Sörbom, 1986) of the model was 0.973 and the root-mean-square residual (RMS; Jöreskog & Sörbom, 1986) was 0.051, indicating a good fit between the model and the data. The chi-square goodness of fit $\chi^2(6)$ was 7.22 with a p value 0.301. There was a strong, direct relationship between the mother's experiences of abuse and those of her son, and none of the family factors were needed to mediate this relationship. There were, however, several significant relationships between the maltreatment and family variables and amongst the family variables. Economic stress was related both to Maternal problems and Marital conflict, which, in turn, increased the number of Child protection interventions. Maternal childhood history of abuse was strongly related to Marital conflict and, to some extent, to the Child protection interventions. Surprisingly, Maternal problems were negatively correlated with the offenders' physical abuse experiences but highly positively correlated with the number of Child protection interventions.

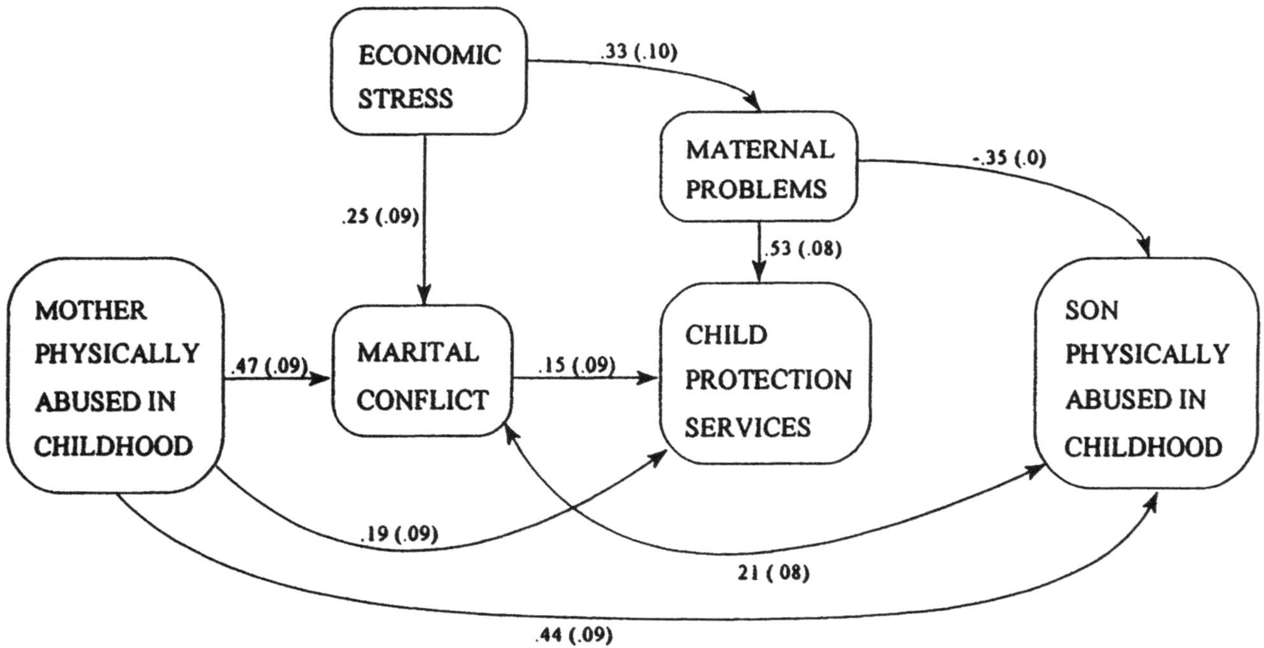

Figure 1. Significant standardized parameter estimates (standard errors in parentheses) for the LISREL model between mothers' and sons' childhood physical abuse experiences ($n = 78$).

Table 3. Correlations Between Mother's and Son's Childhood Abuse in the Groups of Tested Moderator Variables

Variable	Correlation	$X^2(1)$, p =
Maternal problems		
Low (n = 33)	0.43**	
High (n = 34)	0.58***	< 1, ns
Marital conflict		
Low (n = 30)	0.14	
High (n = 32)	0.72***	3.53, .060
Child protection services		
Low (n = 26)	0.22	
High (n = 41)	0.65***	3.43, .064
Economic stress		
Absent (n = 23)	−0.29	
Present (n = 34)	0.86***	18.79, .000

Note. The variables are described in Table 1. Ns vary due to missing information.
** $p < 0.01$ *** $p < 0.001$

Moderator Effects Between Abuse Experiences in Two Generations

Next, the moderator effects were tested to examine whether the relationship between the mother's and son's abuse experiences was different in the low and high groups of Economic stress, Maternal problems, Marital conflict, and Child protection services. The Economic stress variable was a dichotomous (present/absent) variable, but the continuous family variables were first dichotomized into high and low groups according to the mean value of the variable. To examine the moderator effects, the correlations between mother's childhood abuse and her son's childhood abuse were estimated separately for the family variable groups (present/absent or high/low) using a LISREL analysis. If $p < 0.05$ for the chi square test, the correlations were significantly different from each other in the present/absent or high/low groups.

The correlations with the chi square values are presented in Table 3. Economic stress was the only family factor that clearly moderated the relationship between the mother's and the son's abuse experiences: if economic stress was present in the offender's family-of-origin, maternal abuse experiences correlated positively and strongly with the son's abuse experiences, but not, if there was no documented economic stress. Marital conflict and Child protection services moderated the relationship to some degree but not significantly: the intergenerational relationship was stronger when much conflict or many child protection interventions were observed in the family. In contrast, the intergenerational correlation was equally strong in the low and high maternal problem groups.

DISCUSSION

In the present sample, a considerable proportion of the offenders and their parents were physically abused during their childhood. Some new results concerning the role of family problems in the intergenerational cycle of violence among criminal offenders were obtained. No mediator variables were needed between the mother's and son's childhood histories of physical abuse, but maternal problems and marital conflict appeared to account for the correlation between economic difficulties and child protection interventions. Economic stress, in turn, moderated the relationship between the mother's and son's childhood history of physical abuse. If economic stress was present in the offender's family-of-origin, the relationship was strong.

The study thus brought additional evidence for the often cited finding that offenders come from dysfunctional, violent families (Lewis et al., 1989). A strong positive relationship was found between maternal problems (alcoholism, criminality, psychiatric problems) and the number of child protection interventions in this specific sample. Widom (1991) observed that an alcoholic parent or a parent suffering from mental problems increased the likelihood of the child being placed outside the home by the child protection services. It can be assumed that those children whose mothers have a number of serious problems receive more attention from the authorities and may avoid the continuation of abuse. In contrast, those children whose mothers had fewer problems may have stayed in the dysfunctional and abusive environment for a longer period. This could explain the fact that the number of maternal problems was negatively related to the physical violence experienced by the son.

Demographic variables, such as low socioeconomic status, have often been associated with child physical abuse (Milner & Murphy, 1995). Family poverty has also been found to predict the more serious forms of antisocial behavior, including breaking and entering or beating someone innocent (Pagani, Boulerice, Vitaro & Tremblay, 1999). The present study supported the indirect or moderating effect of economic stress on the dysfunction of the family. The more economic stress there was in the family, the more there were both marital conflict and maternal problems. The presence of economic stress also acted as a moderator variable and intensified the association between maternal childhood abuse experiences and the offenders' childhood abuse. These findings are in accordance with previous findings indicating that economic pressures affect family processes and child outcomes through parental mental problems (e.g., depression and demoralization) and disruptions in nurturant parenting, thus increasing harsh and explosive disciplinary strategies (Conger & Elder, 1994). The observed interrelationships between contextual and family system variables suggest that no single-variable explanation of abuse is sufficient among criminal offenders either (see Kolko, 1996).

Maltreated children usually experience a variety of risk factors in their family-of-origin. To single out the contribution of each risk and their interactions in the intergenerational cycle of abuse would appear an extremely challenging task. One risk factor may strengthen or diminish the effect of another risk factor or they may have independent effects on the cycle of abuse. Also in this study, other factors than physical abuse and family problems may have influenced the offenders' development. In view of the fairly serious and early criminal behavior among the offenders, there are reasons to assume that even if they were not abused in childhood, other risk factors have been present.

The retrospective nature of the interview and questionnaire information and the possible underreporting of maltreatment in the file-based information may distort the rates of prevalence of maltreatment in the present sample. This is a common problem in interview- and file-based retrospective studies. In this study, the information about the mothers' childhood abuse was gathered from the merged interview, offender file and questionnaire data, whereas file data were used to assess the offenders' childhood abuse. The childhood abuse data concerning the mothers were thus more likely to be biased than that of the offenders. The setting of the study may have prompted negative memories rather than happy memories of family life. In contrast, the offenders' file data described the true and substantiated events of their childhood.

Retrospective self-reports of childhood maltreatment obtained in adulthood may be biased due to memory impairment, false memories or the avoidance of negative emotions related to childhood experiences (Belli & Loftus, 1996). Recent research has indicated that approximately 40% of those individuals for whom physical abuse was documented failed to report their experiences in adulthood (Widom & Shepard, 1996). Retrospective reports of abuse are influenced by the method used to elicit memories and the way people have defined their early experiences (Widom & Morris, 1997). There is also evidence of the preservation of salient and emotional events in the memory, which would argue for the reliability of retrospective childhood recollections (Brewin, Andrews & Gotlib, 1993; Loftus & Marburger, 1983). Emotional memories of traumatic childhood events are often well retained. However, lack of memory for trauma is also possible due to dissociation, forgetting and the creation of false memories (Brewin & Andrews, 1998; Melchert, 1998).

Physical abuse experienced by the offenders in childhood was used as an outcome variable in the present analyses, because it was externally validated file-based information. Those incidents of childhood physical abuse which were reported in the files were thus reliably documented cases of physical abuse directed at the offender in his childhood. The disadvantage of using file-based information is, of course, that not all incidents of physical abuse are likely to be found and registered in the file reports.

The re-victimization of abused individuals during their adult life, for example in violent marital relationships, is possible (Cascardi et al., 1995). In this study, the mother's childhood history of abuse was strongly related to parental (marital) conflict in adulthood. The existence of a relationship between the mother's childhood history of abuse and physical abuse experienced in childhood by her son does not necessarily indicate that the mother abused her son. It could as well result from the fact that a mother with a childhood history of abuse selected a violent partner who abused the son. Marital conflict also correlated with physical abuse directed at the son. In violent families, both aggressive parental conflicts and physical violence towards children may be present (Jouriles & Norwood, 1995; Straus & Donnelly, 1994). The present study focused on criminal sons within the cycle of abuse, but it would be interesting to examine whether the observed relationships would also apply to girls. Straus & Donnelly (1994) concluded that it was reasonable to think that the greater likelihood of boys being hit contributes to the greater violence of men. The cycle of abuse may be stronger among boys than girls. However, it appears likely that imprisoned girls come from at least equally as disturbed and abusive childhood family environments as boys who go to prison. Possibly exposure to a higher number of risk factors for antisocial behavior are needed for girls than for boys (see e.g., Loeber & Stouthamer-Loeber, 1998).

Research should be focused on the different generations within the same family and on finding reliable sources of abuse information. In intergenerational research, the comparability of abuse information concerning different generations may, however, constitute a problem. For older generations, the reliability of retrospective information is handicapped by the increase in memory distortion over time. Also, file-based information is likely to be qualitatively different and more difficult to trace than that on younger participants. Despite these pitfalls, more effort to involve several generations from the same family in the same study should be taken in trying to make the cycle of violence issue more intelligible.

One study is bound to be limited only to a few potentially important variables influencing the occurrence of childhood physical abuse, leaving many potential effects unaddressed. As Giles-Sims, Straus & Sugarman (1995) pointed out in the context of corporal punishment: "No single factor will explain even a majority of the variation in rates of spanking, but each piece of the puzzle contributes to our understanding.'" (p. 175). This also holds true for the problem of abuse transmission among criminal populations.

REFERENCES

Barnett, D., Manly, J. T., & Cicchetti, D. (1993). Defining child maltreatment: The interface between policy and research. In: D. Cicchetti, & S. L. Toth (Eds), *Child abuse, child development, and social policy*, 7–73. Norwood, NJ: Ablex.

Baron, R. M., & Kenny, D. A. (1986). The moderator-mediator variable distinction in social psychological research: Conceptual, strategic, and statistical considerations. *Journal of Personality and Social Psychology*, *51*, 1173–1182.

Belli, R. F., & Loftus, E. F. (1996). The pliability of autobiographical memory: Misinformation and the false memory problem. In: D. C. Rubin (Ed.), *Remembering our past. Studies in autobiographical memory*, pp. 157–179. Cambridge: Cambridge University Press.

Belt, W., & Abidin, R. R. (1996). The relation of childhood abuse and early parenting experiences to current marital quality in a nonclinical sample. *Child Abuse and Neglect*, *20*, 1019–1996.

Brassard, M. R., Hart, S. N., & Hardy, D. B. (1993). The psychological maltreatment rating scales. *Child Abuse and Neglect*, *17*, 715–729.

Brewin, C. R., & Andrews, B. (1998). Recovered memories of trauma: Phenomenology and cognitive mechanisms. *Clinical Psychology Review*, *18*, 949–970.

Brewin, C. R., Andrews, B., & Gotlib, I. H. (1993). Defining child maltreatment: The interface between policy and research. In: Cicchetti, & S. L. Toth (Eds), *Child abuse, child development, and social policy*, 7–73. Norwood, NJ: Ablex.

Cascardi, M., OLeary, K. D., Lawrence, E. E., & Schlee, K. A. (1995). Characteristics of women physically abused by their spouses and who seek treatment regarding marital conflict. *Journal of Consulting and Clinical Psychology*, *63*, 616–623.

Cole, D. A. (1987). Utility of confirmatory factor analysis in test validation research. *Journal of Consulting and Clinical Psychology*, *55*, 584–594.

Conger, R. D., & Elder, G. H. Jr. (1994). *Families in troubled times*. New York: Walter de Gruyter.

Dishion, T. J., G. R. Patterson, & K. A. Kavanagh. (1992). An experimental test of the coercion model: Linking theory, measurement and intervention. In: J. McCord and R. E. Tremblay (Eds), *Preventing antisocial behavior. Interventions from birth through adolescence*, 209–232. New York: Guilford Press.

Egeland, B. (1993). A history of abuse is a major risk factor for abusing the next generation. In: R. J. Gelles, & D. R. Loseke (Eds), *Current controversies on family violence*, 197–208 Newbury Park: Sage

Eron, L. D., Huesmann, L. R., & Zelli, A. (1991). The role of parental variables in the learning of aggression. In: D. J. Pepler & K. H. Rubin (Eds), *The development and treatment of childhood aggression*, 169–188. Hillsdale, NJ: Erlbaum.

Farrington, D. P. (1992). Explaining the beginning, progress, and ending of antisocial behavior from birth to adulthood. In. J. McCord (Ed.), *Facts, frameworks and forecasts. Advances in criminological theory*, *3*, 253–286. New Brunswick, NJ: Transaction.

Fein, E. (1991) Issues in foster family care: Where do we stand? *American Journal of Orthopsychiatry*, *61*, 578–583.

Giles-Sims, J., Straus, M. A., & Sugarman, D. B. (1995). Child, maternal, and family characteristics associated with spanking. *Family Relations, 44*, 170–176.

Haapasalo, J., & Aaltonen, T. (1999). Mothers abusive childhood predicts child abuse. *Child Abuse Review, 8*, 231–250.

Haapasalo, J., & Hämälainen, T. (1996). Childhood family problems and current psychiatric problems among young violent and property offenders. *Journal of the American Academy of Child and Adolescent Psychiatry, 35*, 1394–1401.

Haapasalo, J., & Petäjä, S. (1999). Mothers who killed or attempted to kill their child: Life circumstances, childhood abuse, and types of killing. *Violence and Victims, 14*, 1–21.

Hart, S. N., Germain, R. B., & Brassard, M. R. (1987). The challenge. To better understand and combat psychological maltreatment of children and youth. In: M. R. Brassard, R. Germain, & S. N. Hart (Eds), *Psychological maltreatment of children and youth*, 3–24. New York: Pergamon.

Jellinek, M. S., Little, M., Benedict, K., Murphy, J. J., Pagano, M., Poitrast, F., & Quinn, D. (1995). Placement outcomes of 206 severely maltreated children in the Boston juvenile court system: A 7.5-year follow-up study. *Child Abuse and Neglect, 19*, 1051–1064.

Jouriles, E. N., & Norwood, W. D. (1995). Physical aggression toward boys and girls in families characterized by the battering of women. *Journal of Family Psychology, 9*, 69–78.

Jöreskog, K. G., & Örbom, D. (1989). *LISREL 7: A guide to the program and application.* (2nd ed.). Chicago: SPSS Inc.

Kaufman, J., & Zigler, E. (1987). Do abused children become abusive parents? *American Journal of Orthopsychiatry, 57*, 186–192.

Kaufman, J., & Zigler, E. (1993). The intergenerational transmission of abuse is overstated. In: R. J. Gelles & D. R. Loseke (Eds), *Current controversies on family violence*, 209–221. Newbury Park: Sage.

Kolko, D. J. (1996). Child physical abuse. In: J. Briere, L. Berliner, J. A. Bulkley, C. Jenny, & T. Reid (Eds), *The APSAC handbook on child maltreatment.* Thousand Oaks: Sage.

Lewis, D. O. (1992). From abuse to violence: Psychobiological consequences of maltreatment. *Journal of the American Academy of Child and Adolescent Psychiatry, 31*, 383–391.

Lewis, D. O., Lovely, R., Yeager, C., Ferguson, G., Friedman, M., Sloane, G., Friedman, H., & Pincus, J. H. (1988). Intrinsic and environmental characteristics of juvenile murderers. *Journal of the American Academy of Child and Adolescent Psychiatry, 27*, 582–587.

Lewis, D. O., Mallouh, C., & Webb, V. (1989). Child abuse, delinquency, and violent criminality. In: D. Cicchetti & V. Carlson (Eds), *Child maltreatment Theory and research on the causes and consequences of child abuse and neglect*, 707–721. Cambridge: Cambridge University Press.

Lewis, D. O., Yeager, C. A., Lovely, R., Stein, A., & Cobham-Portorreal, C. S. (1994). A clinical follow-up of delinquent males: ignored vulnerabilities, unmet needs, and the perpetuation of violence. *Journal of the American Academy of Child and Adolescent Psychiatry, 33*, 518–528.

Loeber, R., & Stouthamer-Loeber, M. (1998). Development of juvenile aggression and violence. Some common misconceptions and controversies. *American Psychologist, 53*, 242–259.

Loftus, E. F., & Marburger, W. (1983). Since the eruption of Mt. St. Helens has anyone beaten you up? Improving the accuracy of retrospective reports with landmark events. *Memory & Cognition, 11*, 114–120.

Luntz, B. K., & Widom, C. S. (1994). Antisocial personality disorder in abused and neglected children grown up. *American Journal of Psychiatry, 151*, 670–674.

Malinosky-Rummell, R., & Hansen, D. J. (1993). Long-term consequences of childhood physical abuse. *Psychological Bulletin, 114*, 68–79.

Marsh, H. W., Balla, J. R., & McDonald, R. P. (1988). Goodness-of-fit indexes in confirmatory factor analyses: The effect of sample size. *Psychological Bulletin, 103*, 391–410.

McCord, J. (1996). Family as crucible for violence: Comment on Gorman-Smith et al., 1996. *Journal of Family Psychology, 10*, 147–152.

McCord, J. (1988). Parental behavior in the cycle of aggression. *Psychiatry, 51*, 14–53.

Melchert, T. P. (1998). Family of origin history, psychological distress, quality of childhood memory, and content of first and recovered childhood memories. *Child Abuse and Neglect, 22*, 1203–1216.

Milner, J. S., & Murphy, W. D. (1995). Assessment of child physical and sexual abuse offenders. *Family Relations, 44*, 478–488.

Oliver, J. E. (1993). Intergenerational transmission of child abuse: rates, research, and clinical implications. *American Journal of Psychiatry, 150*, 1315–1324.

Pagani, L., Boulerice, B., Vitaro, F., & Tremblay, R. E. (1999). Effects of poverty on academic failure and delinquency in boys: A change and process model approach. *Journal of Child Psychology and Psychiatry, 40*, 1209–1219.

Patterson, G. R., Reid, J. B., & Dishion, T. J. (1992). Antisocial boys. Eugene, OR: Castalia.

Reich, W., Shayka, J. J., & Taibleson, C. (1991a). *Diagnostic Interview for Children and Adolescents. Adolescent version.* Washington, DC: Washington University.

Reich, W., Shayka, J. J., & Taibleson, C. (1991b). *Diagnostic Interview for Children and Adolescents. Parent version (DICA-R-P).* Washington, DC: Washington University.

Schaefer, E. S. (1965). Children's reports of parental behavior: An inventory. *Child Development, 36*, 413–424.

Schludermann, E., & Schludermann, S. (1970). Replicability of factors in children's report of parent behavior (CRBI). *Journal of Psychology, 76*, 239–249.

Stattin, H., Janson, H., Klackenberg-Larsson, I., & Magnusson, D. (1995). Corporal punishment in everyday life: an intergenerational perspective. In: J. McCord (Ed.), *Coercion and punishment in long-term perspectives*, 315–347. Cambridge: Cambridge University Press.

Straus, M. A. (1992). Manual for the Conflict Tactics Scales (CTS). Durham, NH: University of New Hampshire.

Straus, M. A., & Donnelly, D. A. (1994). *Beating the devil out of them Corporal punishment in American families.* New York: Lexington Books.

Widom, C. S. (1989). The cycle of violence. *Science, 244*, 160–166.

Widom, C. S. (1991). The role of placement experiences in mediating the criminal consequences of early childhood victimization. *American Journal of Orthopsychiatry, 61*, 195–209.

Widom, C. S. & Morris, S. (1997). Accuracy of adult recollections of childhood victimization: Part 2. Childhood sexual abuse. *Psychological Assessment, 9*, 34–36.

Widom, C. S., & Shepard, R. L. (1996). Accuracy of adult recollections of childhood victimization: Part 1. Childhood physical abuse. *Psychological Assessment 8*, 412–421.

Widom, C. S., & White, H. R (1997). Problem behaviours in abused and neglected children grown up: prrevalence and co-occurrence of substance abuse, crime and violence. *Criminal Behaviour and Mental Health, 7*, 287–310.

Zuravin, S. J. (1991). Research definitions of child physical abuse and neglect: Current problems. In: R. H. Starr, Jr. & D. A. Wolfe (Eds), The effects of child abuse and neglect, 100–128. London: Guilford.

VIOLENCE IN FAMILIES: THE INTERSECTION OF RACE, POVERTY, AND COMMUNITY CONTEXT

Michael L. Benson, Greer L. Fox, Alfred DeMaris and Judy Van Wyk

INTRODUCTION

For over two decades, criminologists and family researchers have observed marked variation in rates of domestic violence among demographically and socio-economically defined populations. Domestic violence has been associated with race and poverty. Higher rates are observed among racial minorities and people of low socio-economic status(Magdol, Moffitt & Silva, 1998; Straus, Gelles & Steinmetz, 1980; Sugarman & Hotaling, 1989; Stets, 1991). There is suggestive evidence that domestic violence also varies geographically across neighborhoods. At the neighborhood level, domestic violence appears to be more common in disorganized and disadvantaged neighborhoods than in their more well off counterparts (Miles-Doan, 1998).

Race, poverty, and community disadvantage have long been linked in the United States. Blacks are more likely to be poor than whites, and the poor are more likely to live in disadvantaged communities than the rich. Poor black people are particularly likely to live in disadvantaged areas. Despite the passage

---

***This project was supported by Grant No. 98-WT-VX-0007 awarded by the National Institute of Justice, Office of Justice Programs, U.S. Department of Justice. Points of view in this document are those of the authors and do not necessarily represent the official position or policies of the U.S. Department of Justice.**

Families, Crime and Criminal Justice, Volume 2, pages 91–109.

ISBN: 0-7623-0737-4

of the Fair Housing Act and other civil rights legislation, the United States remains a highly racially segregated society (Massey, 1990; Massey & Denton, 1993). Residential segregation has the effect of isolating high levels of black poverty within black neighborhoods (Wilson, 1987). Particularly during the last quarter of the 20th century pockets of ghetto poverty grew in the United States, and the concentration of poverty affected racial minorities more than whites (Hagan, 1994).

The concentration of poverty in inner city ghettos suggests that the problem of domestic violence may be particularly severe in these areas. It is important to ascertain whether the effects of individual poverty and community disadvantage on domestic violence are more pronounced for racial minorities, or whether these variables operate independently of race. Although variation in rates of domestic violence across race, economic status, and neighborhood context have been observed for some time, there has been little research on their joint effects. Whether and how the convergence of race, poverty, and community disadvantage influences domestic violence deserves thorough investigation. In this chapter, we address this issue using a data set that merges wave 2 of the National Survey of Families and Households with tract level data from the 1990 U.S. Census. This data set permits us to investigate whether the intersection of race and poverty with domestic violence varies with community context.

DOMESTIC VIOLENCE AND CRIME

Domestic violence refers to physical violence between spouses and cohabitors, in which one or both of the partners physically attacks the other. Before turning our attention to the research questions that motivate this investigation, it is appropriate to consider briefly the reasons why a study of domestic violence belongs in a volume on families and crime.

First, domestic violence can be viewed as crime that occurs in families (Straus & Lincoln, 1985). The question of what counts as crime among family members is complex (Straus & Lincoln, 1985). In many cases, even though violence between intimates could be legally defined as criminal assault, neither the victim nor the perpetrator define the event as a criminal act. Yet, in many other cases, it is clear that at least the victim defines the assault as a crime by reporting it to the police. The proliferation of state and local laws mandating arrests in domestic violence cases clearly indicates that society agrees with this assessment and views domestic assault as part of the crime problem (Sherman & Cohn, 1989). Sherman & Cohn (1989) note that the growth of mandatory arrest laws for domestic violence may be associated with dramatic increases in national per capita arrest rates for simple assault. Thus, a second reason why it is appropriate to include domestic violence in this volume is that violence between intimates is an important pathway into the criminal justice system. When family members become ensnared in the justice system it can have profound effects on the family as a whole (see Sandifer & Kurth, this volume).

Third, domestic violence is implicated as a cause of crime among children and adolescents. Families in which spouses behave violently toward one another may increase the risk that children in the household will become involved in crime and delinquency (Loeber & Stouthamer-Loeber, 1986; Thornberry, 1994). Violent parents may neglect their children emotionally, or through their violent behavior they may inadvertently teach their children that violence is an appropriate way of dealing with others (Loeber & Stouthamer-Loeber, 1986). Thus, violence within families exacerbates the problem of crime in society at large.

Finally, domestic violence is important because it is a major form of victimization for women. Women are substantially more likely to be assaulted by intimates than strangers (Tjaden & Thoennes, 1998). Thus, domestic violence bears upon the family and crime nexus in several important ways: as a form of crime that occurs in families, as a means through which husbands, fathers, wives, and mothers become ensnared in the justice system, as a background cause of crimes that are committed outside of the family, and as the location of substantial victimization of women.

RACE, POVERTY, AND DOMESTIC VIOLENCE

Domestic violence has received substantial research attention over the past several decades. Although most of the scientific and public attention to this problem has focused on violence by males against their female partners, a growing body of research suggests that in many relationships, either the female or the male or both may engage in occasional outbursts of violence (Johnson, 1995; Giordano, Millhollin, Cernkovich, Pugh & Rudolph, 1999; Straus & Gelles, 1986). Johnson (1995) refers to these episodic outbursts of violence as 'common couple violence'. He distinguishes between this type of domestic violence and 'patriarchal terrorism', which involves systematic and extreme violence by males towards their female partners. Patriarchal terrorism is much more rare than common couple violence. Because of limitations of the data at our disposal, the analyses reported here focus on common couple violence. Although common couple violence may pose a less severe threat to the health and well-being of individuals, it is nevertheless important to understand its causes and correlates. Failure to recognize and to distinguish between the two forms of violence may lead to inappropriate and ineffective educational and therapeutic efforts (Johnson, 1995, 292).

Data on the individual and couple-level correlates of domestic violence come from multiple sources: women's shelters, hospitals, the courts, and general surveys of the public. The survey methodology is particularly good at tapping the common couple form of domestic violence, while data from shelters, hospitals, and the courts are more likely to capture women experiencing patriarchal terrorism (Johnson, 1995). Because the data used in this study come from a national survey and because we are focusing on common couple violence, we confine our review of previous findings to survey research.

Surveys of the general population show higher rates of domestic violence among blacks than whites. For example, in the first wave of the National Family Violence Survey higher rates of family violence were found among blacks (Straus et al., 1980). An early analysis of National Crime Victimization Survey (NCVS) conducted in the 1970s found black women were more likely than whites to be victimized by intimates (Gaquin, 1977–1978). These differences have persisted in the ensuing years. In the 1985 National Family Violence Survey married black women were 2.36 times as likely as white wives to experience severe partner violence (Hampton & Gelles, 1994). More recent analyses of NCVS data reveal higher rates of violence by intimates among blacks, especially black women compared to white women, in the 1990s (Greenfeld et al., 1998). Finally, analyses of the National Survey of Families and Households also show significant race differences in violence against women in relationships (Stets, 1991; Umberson, Anderson, Glick & Shapiro, 1998; Sorenson, Upchurch & Shen, 1996).

The pressure and stress created by the financial strains of poverty have been hypothesized to lead to intimate violence (Fagan & Browne, 1994; Brush, 1990). Evidence suggests that this hypothesis is correct as general population surveys find domestic violence to be more common among low income couples (Stets, 1991; Sorenson et al., 1996). In the NCVS, women in low-income households were found to experience a substantially higher rate of violence by intimates than women in households with larger incomes (Greenfeld et al., 1998). For example, in the early 1990s, women in households with incomes of less than $7,500 experienced intimate violence at a rate that was nearly five times higher than the rate for women in households with incomes of more than $75,000.

COMMUNITY CONTEXT AND DOMESTIC VIOLENCE

Compared to what we know about the influence of race and wealth on domestic violence, the effects of community context are neither well established nor well understood. Few studies have examined how intimate violence might be affected by the characteristics of neighborhood settings (Miles-Doan, 1998). Nonetheless, there are a number of reasons why community context may be related to variation in domestic violence. Economically disadvantaged neighborhoods with little residential stability tend to foster a sense of anonymity, and correspondingly reduced social control (Sampson, Morenoff & Earls, 1999; Stark, 1987). In such situations, neighbors may exhibit little willingness to become involved in domestic disputes by personally intervening or calling the police. Violently inclined spouses may therefore aggress against their partners with greater impunity, feeling that they have little to fear from outside influences. Hence, they may feel free to commit violence against their spouses and cohabiting partners without fear of social disapproval. Recent ethnographic research suggests that among some disadvantaged area residents, especially minority men, the use of violence in interpersonal relationships is viewed not only as normatively acceptable but as virtually required by local cultural codes (Anderson, 1990).

Social disorganization theory also suggests other reasons why intimate violence may be particularly pervasive in disadvantaged areas. Since residents of socially disorganized areas are more likely to have weak social bonds to their neighbors (Bursik, 1988; Kornhauser, 1978), spouses and cohabitors involved with potentially abusive partners are more likely to be isolated and thus at greater risk of violent victimization (Stets, 1991). To exit an abusive relationship requires resources. Individuals who want to leave abusive partners need money and friends. These resources are likely to be less accessible to those who live in poverty stricken and socially fragmented areas. In addition, couples in disadvantaged areas are likely to live in overcrowded households, which often leads to pathological consequences for family relationships (Gove, Hughes & Galle, 1979). Finally, official forms of social control are weaker in disorganized areas. Stark (1987) suggests that law enforcement is more lenient in these areas because the police view these neighborhoods as filled with morally disreputable people whose problems do not warrant quick intervention. Extensive social disorganization may create particularly pervasive and persistent forms of domestic violence.

There is some empirical evidence to support the theoretical link between community economic disadvantage and intimate violence. Studies of calls to the police indicate that calls related to domestic violence incidents tend to come disproportionately from poor neighborhoods (Sherman & Berk, 1984). A recent study investigated the effects of neighborhood resource deprivation, as indicated by the absolute poverty level and income inequality, on incidents of assault between intimates reported to the police (Miles-Doan, 1998). Miles-Doan found that neighborhoods suffering from great resource deprivation experienced dramatically higher rates of officially reported intimate violence. Although the NCVS is not designed to permit neighborhood level analyses, it too contains suggestive evidence of geographic variation in domestic violence. Higher rates female victimization by intimates are observed in urban than suburban or rural areas (Greenfeld et al., 1998).

Because race, poverty, and community context overlap so strongly in the United States and because all of these variables appear to be strongly related to domestic violence, it becomes important to investigate their joint effects on this social problem. We turn to this task now and begin by describing our data.

RESEARCH METHODS

Sample and Data

Data for this study come from wave 2 of the National Survey of Families and Households (NSFH) and from the 1990 U.S. Census. The NSFH is a nationally representative sample of households. Completed in 1988, the first wave of the NSFH included interviews with a probability sample of 13,007 adult respondents, representing 9,637 households. Face-to-face interviews were conducted with a randomly selected *primary respondent* from each household. To facilitate the collection of sensitive information, portions of the interview with the primary respondent were self-administered. The primary respondent's spouse or cohabiting partner, hereinafter referred to as the *secondary respondent*, was given a shorter self-administered questionnaire (Sweet, Bumpass & Call, 1988). In wave 2, completed in 1994, interviews were conducted with all surviving members of the original sample (n = 10,007) and with the current spouse or cohabiting partner of the primary respondent (n = 5,624). Interviews were also conducted with the spouse or partner of the primary respondent in cases where the relationship had ended (n = 789) (Sweet & Bumpass, 1996). This study is based on a sub-sample of households in which respondents were married or cohabiting at the time of wave 2 and who were either white, black, or Hispanic (n = 6,067).

The NSFH has a number of design strengths that make it an especially advantageous data set to use to enhance understanding of the interplay of race, economic distress and community context in regards to intimate violence. Because of the sample size, an adequate number of households with intimate violence and households of racial and ethnic sub-groups are available for analysis. Minority households and cohabiting couples were over-sampled. Detailed measures on a variety of economic and employment indicators are available. Finally, NSFH has recently made it possible to merge the survey data with aggregate census data at the tract level. Merging the individual level survey data with aggregate level tract data enables us to investigate the influence of community context on domestic violence more thoroughly than has been possible in previous research.

Variables

Intimate Violence. The NSFH asks respondents how often during the past year they became 'physically violent' in the course of arguments with the spouse or partner. Respondents are also asked how often arguments resulted in their spouse or partner becoming physically violent with them. The five category response set ranges from 'none' to 'four or more' times. To measure violence by men, we created a dichotomous variable in which 0 indicates that neither partner reported violence and 1 indicates that either the male reported becoming physically violent with the female, or the female reported that the male became physically violent with her, or both. In cases where the reports did not agree, we assigned a code of 1 – indicating wife assault – provided that at least one partner reported violence against the female partner. We created an identical measure of violence by the female member of a couple.

Race. We examine the distribution of violence in families among three racial groups: white, black, and Hispanic. Although individuals of five other racial and ethnic backgrounds are represented in the NSFH sample, the numbers available for analysis in these groups are very small. Altogether the five other groups constitute less than four percent of the couples. Hence, we restrict our analyses here to whites, blacks, and Hispanics.

Community Context. The NSFH can define the location of respondents to the census tract level and can append census tract level information provided by researchers to survey respondents. We abstracted data from the 1990 census for the appropriate census tracts and merged it with the NSFH survey data. Although census tracts are not coterminous with local neighborhoods, they provide more reasonable approximations of local conditions than counties or ZIP codes.

To measure community context, we created an index of neighborhood disadvantage modeled on the work of Sampson and colleagues (Sampson et al., 1999). This index is based on five census tract measures that reflect what Sampson et al., (1999) call *concentrated disadvantage.* It is defined by the percent of single parents, percent unemployed, percent non-white, percent of families on public assistance, and percent below the poverty line. After transforming the items to z-scores, we summed them and divided by the number of indicators to form the index of concentrated disadvantage, which has an alpha reliability of 0.92. The index ranges from –1.07 to 5.37, with high scores indicating greater neighborhood disadvantage. For the analyses presented here, we split the sample at the 80th percentile of the index, designating couples with index scores above 0.347 as living in areas of high concentrated disadvantage.

Poverty. To measure poverty, we used data on household size and the total household income and calculated the income to needs ratio, which is the household income divided by the poverty line for households of that size. This technique adjusts for the size of the household. We then subdivided the sample into three groups: poor, moderately well-off, and rich. Those classified as poor had incomes adjusted for household size that ranged from 1.9 times the poverty level downward. The moderately well-off have adjusted incomes between 2 and 3.9 times the poverty level, and the rich have incomes that is 4 or more times the poverty level. Table 1 provides descriptive information on the variables.

Analytic Strategy

The causes of violence in families are complex. Comprehensive theories of domestic violence eventually will have to take into account cultural, structural, contextual, personality, and situational factors (Panel on Research on Violence Against Women, 1996). In this paper, our aim is not to develop or test a comprehensive theory of violence in families. Our goals are more modest than that. We hope to specify the social location of domestic violence in regards to race, poverty, and community context considered separately and in combination.

Table 1. Descriptive Statistics

Independent Variables	Percent	N
Race		
Black	11.7	708
White	83.6	5,074
Hispanic	4 7	285
Poverty		
Low	34.7	2,036
Medium	47.7	2,803
High	17 6	1,032
Concentrated Disadvantage		
High	19.9	1,201
Low	80.1	4,845
Dependent Variables		
Male Domestic Violence	5.2	314
Female Domestic Violence	5.0	306

Hence, we rely on cross-tabular rather than advanced multivariate analytical techniques. We begin by analyzing rates of domestic violence within the categories of our independent variables. We then investigate how various combinations of the variables raise or lower these base rates.

RESULTS

Before examining how race, poverty, and community disadvantage are related to domestic violence, we first briefly report on how they are related to one another in the NSFH sample. As expected, we find that they intersect strongly, especially in the case of racial minorities. Fully 40 percent of the low income couples in the sample live in disadvantaged areas compared to only 10% of high income couples. Almost one-third (31.5%) of the black couples in the sample fall into the lowest income category in our measure of poverty and nearly half (47%) of Hispanics are found there, too. In contrast, only 14% of whites are in the lowest category. Finally, in the NSFH sample, nearly seven out of ten black couples (68.8%) live in disadvantaged communities and so do 61.1% of Hispanics. In contrast, only 10.7% of whites live in disadvantaged areas.

Zero-Order Effects of Race, Poverty, and Community Disadvantage

As expected, race, poverty, and community disadvantage are significantly related to the prevalence of both male and female domestic violence. As Table 2 shows, at 9.6%, blacks have the highest rate of male violence, one that is close to twice the size of the base rate of 5.2% for the sample as a whole. The rate for whites (4.7%) is slightly lower than the base rate, while the rate for Hispanics is slightly higher (5.8%). Female violence follows a similar pattern with blacks having a notably higher rate than Hispanics and whites.

Poverty is also clearly related to an increased risk of domestic violence. In couples with incomes close to the poverty rate, male violence occurs at a rate of 8.1%, which is nearly three times the rate at which it occurs among couples that are the most well-off financially. Female violence is also significantly higher among poor couples than among their counterparts who are more well off.

Finally, both male and female violence are much more likely to occur among couples that reside in the most disadvantaged communities. Males located in highly disadvantaged areas are more than twice as likely to engage in domestic violence as males who live in advantaged areas (9.2% to 4.4%, respectively). There is a nearly equal sized disparity between females in disadvantaged areas compared to those in advantaged areas.

Table 2. Rates of Male and Female Domestic Violence by Race, Economic Strain, and Community Disadvantage

Variable	Male Violence	Female Violence
Race		
Blacks	9.6% (676)	8.3% (678)
Hispanics	5.8% (275)	5.4% (276)
Whites	4.7% (4,971)	4.7% (4,969)
Economic Strain		
High	8.1% (972)	7.6% (972)
Medium	5.7% (2,741)	5.2% (2,742)
Low	3.0% (2,017)	3.6% (2,017)
Community Disadvantage		
High	9.2% (1,156)	8.1% (1,157)
Low	4.4% (4,745)	4.4% (4,745)
Total	5.2% (5,922)	5.0% (5,923)

Note: Ns in parentheses. All variables are significant at the .05 level or better.

First Order Effects

An important question regarding the impact of race, poverty, and community disadvantage on domestic violence is whether the effects of these variables are cumulative or not. For example, is the effect of poverty status on domestic violence the same regardless of the type of neighborhood in which you live? Or, does it vary depending on whether you live in a disadvantaged or non-disadvantaged community. Similar questions need to be answered regarding the combined effects of race and poverty and the effects of race and community disadvantage. In Tables 3, 4, and 5, we begin to address these questions.

Table 3 shows the relationship between poverty status and both female and male domestic violence in disadvantaged versus non-disadvantaged communities. For both female and male domestic violence, community disadvantage appears to influence the effect of poverty status. In disadvantaged communities the relationship between poverty and violence is not significant. However, in communities that score low on the index of community disadvantage, poverty status still exerts a significant effect on the likelihood of domestic violence by males and females. There are also pronounced differences in the likelihood of violence when poverty status and community disadvantage are considered together. Note, for example, that the prevalence of domestic violence is 2.8% for males who are well-off and who live in advantaged communities, while among poverty stricken males who live in disadvantaged communities the rate

Table 3. Rates of Male and Female Domestic Violence by Poverty as a Function of Community Disadvantage.

Female Domestic Violence

Community Disadvantage	Poverty High	Medium	Low	Total	
High	8.3 % (385) (χ^2 = 2.045	8.5% (528)	5.4% (202)	 (1,115)	7.9%
Low	7.0% (583) (χ^2 = 14.751***	4.4% (2,208)	3.3% (1,805)		4.3% (4,596)

Male Domestic Violence

Community Disadvantage	Poverty High	Medium	Low	Total
High	9.1% (385) (χ^2 = 3.865	10.1% (527)	5.4% (202)	8.9% (1,114)
Low	7.5% (583) (χ^2 = 25.996***	4.7% (2,208)	2.8% (1,805)	4.3% (4,596)

Note: Ns in parentheses.
$p < 0.10$
** $p < 0.05$
*** $p < 0.01$

is over three times higher at 9.1%. A similar pattern is notable for female violence. Taken together, individual poverty and community disadvantage dramatically increase the risk of domestic violence.

In Table 4 we examine how community context influences the relationship between race and domestic violence. In disadvantaged communities, the relationship between race and domestic violence is not significant. Although both male and female domestic violence rates are higher for blacks than whites and Hispanics, these differences are not statistically significant. In communities that are not disadvantaged, however, the effect of race is still significant for male violence and moderately so for female violence. Black males located in disadvantaged communities have the highest rate of 10.2%. Community disadvantage

Table 4. Rates of Female and Male Domestic Violence by Race as a Function of Community Disadvantage

Female Domestic Violence				
		Race		
Community Disadvantage	Black	White	Hispanic	Total
High (463)	8.6% (527)	8.3% (167)	6.0% (1,063)	8.1% (χ^2 = 1.221
Low	7.5% (212) (χ^2 = 5.137*	4 3% (4,424)	4.6% (109)	4.4% (4,745)

Male Domestic Violence				
		Race		
Community Disadvantage	Blacks	White	Hispanic	Total
High	10.2% (461) (χ^2 = 1.100	8.7% (528)	7.8% (167)	9.2% (1,156)
Low	8.5% (212) (χ^2 = 9.458***	4.2% (4,425)	2.8% (108)	4.4% (4,745)

Note. Ns in parentheses
*p < 0.10
** p < 0.05
*** p < 0 01

appears to have particularly strong effects on white and Hispanic male violence. Note that the rate for white males in disadvantaged communities is 8.7%. This rate more than doubles the rate for white males in advantaged communities (4.2%). For Hispanics, the rate increases threefold from 2.3% in advantaged communities to 7.8% in disadvantaged communities. Thus, the effects of race appear to interact with community context to raise the risk of victimization of females in all racial groups.

Finally, Table 5 shows the effects of poverty controlling for race. Within racial groups, both male and female domestic violence is significantly more likely among the poor than the well-to-do. The effect of poverty status appears particularly strong among blacks. Black males in the low category of our index of poverty status have a violence rate of 3.4%, while their counterparts in the high category have a rate that is almost four times higher (12.9%). For white

Table 5. Rates of Female and Male Domestic Violence by Poverty as a Function of Race.

Female Domestic Violence

Race	Poverty High	Poverty Medium	Poverty Low	Total
Black	12.3%	7.9%	2.7%	8.1%
	(203)	(305)	146)	(654)
	(χ^2 = 10.498***			
White	6.5%	4 9%	3.7%	4.6%
	(646)	(2,325)	(1,847)	(4,818)
	(χ^2 = 9.168***			
Hispanic	5.7%	5.4%	0.0%	5.0%
	(123)	(112)	(24)	(259
	(χ^2 = 1.412			

Male Domestic Violence

Race	Poverty High	Poverty Medium	Poverty Low	Total
Black	12.9%	9.9%	3.4%	9.4%
	(202)	(304)	(146)	(652)
	(χ^2 = 9.094**			
White	6.8%	5.2%	3.0%	4.6%
	(647)	(2,326)	(1,847)	(4,820)
	(χ^2 = 19.467***			
Hispanic	7.3%	4.5%	0.0%	5.4%
	(123)	(111)	(24)	(258)

Note. Ns in parentheses
p < 10
** p < .05
*** p < .01

males, those in the low poverty category have a rate of 3.0%, while those in the high poverty category have a rate that is more than double in size (6.8%). From the victims point of view, poor black females have the highest rate of victimization at 12.9%. This rate is more than four times higher than the victimization rate for well-to-do white females.

Second Order Effects: Race, Poverty and Community Disadvantage

In Tables 6a and 6b, we examine the effects of poverty on domestic violence, as a function of both race and community disadvantage. Working with three variables at the same time reduces the Ns available for analysis, particularly for Hispanics. The reduction in sample size makes it more difficult to discover significant effects and produces some instability in the pattern of relationships. Nevertheless, some important patterns are clearly apparent. For instance, consider the effect of poverty among blacks in communities of high disadvantage. Poor black males in disadvantaged communities have a rate of violence of 12.7%. However, black males in disadvantaged communities who are not poor have a rate of only 3.7%. This rate is lower than the overall rate for white males and close to the rate (2.8%) for well-to-do white males in communities with low disadvantage. The effect of poverty among blacks in communities that are not disadvantaged is not significant, but the pattern is the same as in disadvantaged communities with poor men having the highest rate. Small sample sizes may be responsible for the non-significant findings. In contrast, among white males, poverty seems to be a more important influence on violence in advantaged rather than disadvantaged communities. The violence rate for poor white men in advantaged communities is 7.1% compared to only 2.8% for well-to-do white men in these communities. The rate of 7.1% is higher than that observed for poor white men in disadvantaged communities.

As Table 6b shows, the patterns for female domestic violence largely mirror those observed for males in Table 6a. In disadvantaged communities, changes in poverty status are associated with dramatic differences in rates of female violence. Poor black females in disadvantaged areas have a violence rate of 12.7%, while females who are not poor but who reside in disadvantaged areas have a rate of only 2.5%. This is considerably lower than the base rate for white females of 4.7% (see Table 2).

DISCUSSION

The Social Location of Domestic Violence

Our results show that as with other forms of crime and violence, domestic violence is not spread evenly across geographic locations. Although violence occurs in well-to-do neighborhoods, it is much less likely there than elsewhere. Both male and female domestic violence is considerably more prevalent among couples located in disadvantaged communities. Indeed, male violence is twice as common in disadvantaged communities as it is in advantaged communities. Neither is the risk of domestic violence spread evenly across economic strata. Well-to-do men and women do engage in violence against their partners, but at rates that are roughly two to three times less than their poverty stricken counterparts. Well-off women, for example, have a violence rate of 3.6% compared

Table 6a. Rate of Male Domestic Violence by Poverty Status as a Function of Race and Community Disadvantage.

High Community Disadvantage				
	Poverty			
Race	High	Medium	Low	Total
Black	12.7%	10.2%	3.7%	9.9%
	(157)	(205)	(81)	(443)
	(χ^2 = 4.917*			
White	5.7%	10.8%	7.0%	8.5%
	(140)	(260)	(115)	(515)
	(χ^2 = 3.453			
Hispanic	8.0%	6.5%	0.0%	7.1%
	(88)	(62)	(6)	(156)
	(χ^2 = 0.599			

Low Community Disadvantage				
	Poverty			
Race	High	Medium	Low	Total
Black	14.0%	9.1%	3.1%	8.3%
	(43)	(99)	(64)	(206)
	(χ^2 = 4.160			
White	7.1%	4.5%	2.8%	4.1%
	(505)	(2,060)	(1,723)	(4,288)
	(χ^2 = 20.111***			
Hispanic	5.7%	2.0%	0.0%	2.9%
	(35)	(49)	(18)	(102)
	(χ^2 = 1.627			

Note: Ns in parentheses.
* $p < 0.10$
** $p < 0.05$
*** $p < 0.01$

to 7.6% for poor women. Even larger differentials are observed between rich and poor men. Thus, like patriarchal terrorism, common couple violence is concentrated among the poor.

More important than the simple zero-order effects of race, poverty, and community disadvantage on domestic violence, however, are their combined or higher order effects. An important example involves race and community disadvantage. Although blacks have higher rates of domestic violence than whites overall, a good part of the zero-order difference results from the association of

Table 6b. Rate of Female Domestic Violence by Poverty Status as a Function of Race and Community Disadvantage.

	High Community Disadvantage			
		Poverty		
Race	High	Medium	Low	Total
Black	12.7%	7.3%	2.5%	9.9%
	(158)	(206)	(81)	(445)
	χ^2 = 7.829**			
White	5.0%	10.4%	7.8%	8.5%
	(139)	(260)	(115)	(514)
	χ^2 = 3.436			
Hispanic	5.7%	4.8%	0.0%	7.1%
	(88)	(62)	(6)	(156)
	(χ^2 = 0.390			
	Low Community Disadvantage			
		Poverty		
Race	High	Medium	Low	Total
Black	11.6%	9.1%	3.1%	7.8%
	(43)	(99)	(64)	(206)
	χ^2 = 3.062			
White	6.7%	4.2%	3.4%	4.2%
	(505)	(2,059)	(1,723)	(4,287)
	χ^2 = 11.127***			
Hispanic	5.7%	6.0%	0.0%	4.9%
	(35)	(50)	(18)	(103)
	χ^2 = 1.116			

Note: Ns in parentheses
* $p < 0.10$
** $p < 0.05$
*** $p < 0.01$

race with residence in disadvantaged communities. In disadvantaged communities, the relationship between race and domestic violence is not significant for both males and females. We did find that race was still significant in communities that were not disadvantaged, with blacks having higher rates than whites. However, we believe this finding may be a methodological artifact.

Recall that to identify severely disadvantaged communities we dichotomized the measure of community disadvantage at the 80th percentile. This procedure lumps together four-fifths of the sample in the non-disadvantaged category. Of course, within this sub-sample, there is considerable variation in the types of communities in which respondents live. Some of the communities are very close to being severely disadvantaged, while others are extremely well off. Black respondents may be disproportionately represented in communities that are close to the cut off point in our measure. To investigate this possibility, we compared the means on the original index of concentrated disadvantage for blacks and whites. They are significantly different. The mean index score for blacks is significantly higher than the mean for whites. Even though these black respondents do not live in severely disadvantaged communities, they nevertheless live in relatively less advantaged communities than whites. This result suggests that the significant race effect observed in non-disadvantaged communities may be at least in part a methodological artifact. To further investigate the effects of race and community disadvantage on domestic violence, we used logistic regression. In this model, the index of disadvantage had a significant effect on male violence but race did not.

Our analyses also show that the disproportionate representation of blacks among the poverty stricken influences the effect of race on domestic violence. In couples with high incomes, there is virtually no difference between blacks and whites in domestic violence rates by either males or females. But blacks are much more likely to have low incomes than whites. In addition, for reasons that are not clear, among low income couples, blacks have notably higher rates of domestic violence than whites. One interpretation of the apparent interaction of race and poverty may be that for blacks poverty may appear to be a more permanent condition, one from which they are not likely to escape. Poor blacks may feel that they are destined to live in poverty in a society in which riches are flaunted everywhere, leading to feelings of resentment, frustration, hopelessness and alienation (Hagan, 1994). One potential outcome of these feelings is 'diffuse aggression' that may be expressed toward intimates or strangers (Blau & Blau, 1982). Because they do not suffer the additional stigma of being in a disadvantaged racial group, poor whites may be more inclined to view poverty as a temporary condition from which it is possible to exit. For whites, poverty may be less frustrating and alienating and less likely to promote aggression.

Relative deprivation may also be at work in our finding that rates of domestic violence are higher for poor black and white men living in communities of low rather than high disadvantage. Men who are poor but who live in close proximity to men who are well-off may feel their straitened circumstances more keenly than equally poor men in disadvantaged neighborhoods. For the poor man who lives in a nice community, proof of his lack of material success relative to his neighbors is readily available. It may fuel feelings of frustration and failure that are manifested as violence toward his intimate partner at home.

As we noted at the outset of this paper, domestic violence is a complex phenomenon. Without doubt, violence in families has multiple causes that operate at conceptually distinct levels – individual, couple, situational, and contextual. Individual differences in personality and temperament certainly play a key role, as well as do the unique interactional dynamics that couples develop over time. Also important are the situational stresses, such as loss of a job or other financial setbacks, which arise from time to time as couples go through life. The research reported here was not designed to parse these different levels of causation or to test a comprehensive theory of domestic violence. We have attempted instead to provide information on where domestic violence is most likely to occur by describing its social location in terms of race, economic status, and community context. More research is needed to determine exactly why the risk of domestic violence seems to vary so systematically with these variables.

REFERENCES

Anderson, E. (1990). *Streetwise: Race, Class and Change in an Urban Community.* Chicago: University of Chicago Press.

Blau, J., & Blau, P. (1982). The Cost of Inequality: Metropolitan Structure and Violent Crime. *American Sociological Review, 47*, 114–28.

Brush, L. D. (1990). Violent Acts and Injurious Outcomes: Methdological Issues in the National Survey of Families and Households. *Gender & Society, 4* 56–67.

Bursik, R. J. (1988). Social Disorganization and Theories of Crime and Delinquency: Problems and Prospects. *Criminology, 26*, 519–51.

Fagan, J., & Browne, A. (1994). Violence Between Spouses and Intimates: Physical Aggression Between Men and Women in Intimate Relationships. In: A. J. Reiss, Jr. & J. A. Roth (Eds), *Understanding and Preventing Violence, 3*, 115–292. Washington, D. C.: National Academy Press.

Gaquin, D. 1977–1978. Spouse Abuse: Data From the National Crime Survey. *Victimology, 2*, 634–43.

Giordano, P. C., Millhollin, T. J., Cernkovich, S. A., Pugh, M. D., & Rudolph, J. L. (1999). Delinquency, Identity, and Womens Involvement in Relationship Violence. *Criminology, 37*, 17–40.

Gove, W. R., Hughes, M., & Galle, O. R. (1979). Overcrowding in the Home: An Empirical Investigation of Its Possible Pathological Consequences. *American Sociological Review, 44*, 59–80.

Greenfeld, L. A., Rand, L. R., Craven, D., Klaus, P. A., Perkins, C., Warchol, G., Maston, C., & Fox, J. A. (1998). *Violence by Intimates: Analysis of Data on Crimes by Current or Former Spouses, Boyfriends, and Girlfriends.* Washington, D. C.: U.S. Department of Justice.

Hagan, J. (1994). *Crime and Disrepute.* Thousand Oaks, CA.: Pine Forge Press.

Hampton, R. L., & Gelles, R. J. (1994). Violence Toward Black Women in a Nationally Representative Sample of Black Families. *Journal of Comparative Family Studies, 25*, 105–19

Johnson, M. P. (1995). Patriarchal Terrorism and Common Couple Violence: Two Forms of Violence Against Women. *Journal of Marriage and the Family, 57*, 283–94.
Kornhauser, R. D. (1978). *Social Sources of Delinquency.* Chicago: University of Chicago Press.
Loeber, R. & Stouthamer-Loeber, M. (1986). Family Factors As Correlates and Predictors of Juvenile Conduct Problems and Delinquency. In: M. Tonry, & N. Morris (Eds), *Crime and Justice: An Annual Review of Research, 7*, 29–149 . Chicago: University of Chicago Press.
Magdol, L., Moffitt, T. E., Avshalom, C , & Silva, P. A. (1998). Developmental Antecedents of Partner Abuse: A Prospective – Longitudinal Study. *Journal of Abnormal Psychology, 107*, 375–89.
Massey, D. (1990). American Apartheid: Segregation and the Making of the Underclass. *American Journal of Sociology, 96*, 329–57.
Massey, D. & Denton, N. (1993). *American Apartheid. Segregation and the Making of the Underclass.* Cambridge, MA: Harvard University Press.
Miles-Doan, R.. (1998). Violence Between Spouses and Intimates: Does Neighborhood Context Matter. *Social Forces, 77*, 623–45.
Panel on Research on Violence Against Women. 1996. *Understanding Violence Against Women.* Washington, D. C.: National Academy Press.
Sampson, R. J., Morenoff, J. D., & Earls, F. (1999). Beyond Social Capital: Spatial Dynamics of Collective Efficacy for Children. *American Sociological Review, 64*, 633–60.
Sherman, L., & Cohn, E. G. (1989). The Impact of Research on Legal Policy: The Minneapolis Domestic Violence Experiment. *Law and Society Review, 23*, 117–44.
Sherman, L. W. & Berk, R. A (1984). *The Minneapolis Domestic Violence Experiment.* Washington, D C.: The Police Foundation.
Sorenson, S. B., Upchurch, D. M., & Shen, H. (1996). Violence and Injury in Marital Arguments: Risk Patterns and Gender Differences. *American Journal of Public Health, 86*, 35–40.
Stark, R. (1987). Deviant Places: A Theory of the Ecology of Crime. *Criminology, 25*, 893–909.
Stets, J. E. (1991). Cohabiting and Marital Aggression: The Role of Social Isolation. *Journal of Marriage and the Family, 53*, 669–80.
Straus, M. A., Gelles, R. A., & Steinmetz, S. K. (1980). *Behind Closed Doors: Violence in the American Family.* Garden City, N. Y.: Doubleday.
Straus, M. A., & Gelles, R. J. (1986). Societal Change and Change in Family Violence From 1975 to 1985 As Revealed by Two National Surveys. *Journal of Marriage and the Family, 48*, 465–79.
Straus, M. A., & Lincoln, A. J. (1985) *Crime and the Family.* C. C Thomas.
Sugarman, D. B., & Hotaling, G. T. (1989). Violent Men in Intimate Relationships: An Analysis of Risk Markers. *Journal of Applied Social Psychology, 19*, 1034–48.
Thornberry, T P. (1994). *Violent Families and Youth Violence.* Washington, D. C.: Office of Juvenile Justice and Delinquency Prevention, U.S. Department of Justice.
Tjaden, P., & Thoennes, N. (1998). *Prevalence, Incidence, and Consequences of Violence Against Women: Findings From the National Violence Against Women Survey.* Washington, D. C.: U S. Department of Justice.
Umberson, D., K. Anderson, J. Glick, & A. Shapiro. (1998). Domestic Violence, Personal Control, and Gender. *Journal of Marriage and the Family 60*, 442–52.
Wilson, W. J. (1987). *The Truly Disadvantaged.* Chicago: University of Chicago Press.

PART II

THE FAMILY'S ROLE IN THE PRODUCTION, MEDIATION AND PREVENTION OF CRIME

A GENERAL STRAIN THEORY APPROACH TO FAMILIES AND DELINQUENCY

Robert Agnew, Cesar Rebellon and Sherod Thaxton

Criminologists typically explain the effect of family factors on delinquency in terms of control and social learning theories. Control theorists focus on the extent to which parents directly control juvenile behavior by sanctioning delinquency, indirectly control behavior by establishing a close emotional bond with their children, and teach their children to control their own behavior. Social learning theorists focus on the extent to which parents teach juveniles to engage in conventional versus delinquent behavior; with such instruction occurring through the reinforcements and punishments they provide, the beliefs they teach, and the behaviors they model. Strain theorists have not had much to say about the impact of family factors on delinquency, except to argue that child abuse and harsh discipline are major sources of strain that contribute to delinquency (e.g. Agnew, 1985, 1992; Agnew & White, 1992; Brezina, 1998, 1999).

This paper draws on Agnew's (1992) General Strain Theory to more fully describe the relationship between family factors and delinquency. Drawing on the family research, it is first argued that parental strain contributes to those parenting practices that increase the likelihood of juvenile delinquency. The most important of these practices are poor supervision (i.e., lax, inconsistent, and/or harsh

***This project was supported by Grant No. 98-WT-VX-0007 awarded by the National Institute of Justice, Office of Justice Programs, U.S. Department of Justice. Points of view in this document are those of the authors and do not necessarily represent the official position or policies of the U.S. Department of Justice.**

Families, Crime and Criminal Justice, Volume 2, pages 113–138.

ISBN: 0-7623-0737-4

discipline) and the failure to establish a close emotional bond with children. It is next argued that these parenting practices contribute to delinquency *partly* because they increase the juvenile's level of strain. Further, certain family factors increase the likelihood that juveniles will respond to strain with delinquency.

The strain model that is presented helps integrate certain of the research on the family and delinquency, helps better explain why certain family factors impact delinquency, and draws attention to certain family factors that have been overlooked in the research on families and delinquency. This chapter does not present a formal test of the model, but supporting evidence is noted and two 'illustrative analyses' that provide tentative support for parts of the model are presented.

A Brief Overview of Control and Social Learning Explanations of the Family-Delinquency Link

While control theorists focus on a range of family factors, they attach special importance to the emotional bond between family members and the adequacy of parental supervision (e.g. Hirschi, 1969, 1995; Gottfredson & Hirschi, 1990; Sampson & Laub, 1993).

Delinquency is said to be less likely if there is a strong emotional bond between parents and the juvenile. This bond has a direct effect on delinquency since it provides juveniles with a 'stake in conformity'. In particular, juveniles who are bonded to their parents will be less likely to engage in delinquency for fear that their actions might hurt or jeopardize their ties to their parents. And this bond has an indirect effect on delinquency, since it increases the likelihood that parents will attempt to supervise their children, increases the effectiveness of their supervision (e.g. parental sanctions are more effective when children are bonded to parents), and increases the likelihood that juveniles will internalize parental beliefs.

Effective supervision involves the statement of clear rules prohibiting delinquency and actions that might contribute to delinquency, the monitoring of juvenile behavior to ensure that these rules are obeyed, and the consistent sanctioning of rule violations. Such supervision has a direct effect on delinquency, since juveniles are more likely to refrain from actions that result in consistent punishment. And supervision has an indirect effect on delinquency, since it increases the likelihood that juveniles will internalize prohibitions against delinquency and develop self-control.

Control theorists argue that 'poor parenting' is more likely when parents are low in social and self-control. In particular, poor parenting is more likely when parents: (a) are not attached to their children; (b) are not well monitored by those who might sanction poor parenting, perhaps because the parents are socially isolated; (c) have little to lose by engaging in poor parenting, perhaps because the parents do not have good reputations in the community or good jobs that would be jeopardized by poor parenting; and (d) have not been taught beliefs that encourage good parenting. Further, poor parenting is more likely when parents possess traits that reduce their motivation and ability to engage in good parenting; traits like impulsivity, self-centeredness, and irritability – which are components of low self-control. These arguments help explain why poor parenting is more common among such people as criminal parents, economically disadvantaged parents, and teenage mothers. These people are lower in social and/or self-control.

Social learning theorists also emphasize the importance of the bond between family members and effective supervision. These factors play a vital role in teaching juveniles to conform. They go on, however, to argue that parents sometimes teach their children to engage in delinquency. This instruction is seldom deliberate, but rather is usually the unintended result of parental actions like modeling aggressive behavior (e.g. spouse and child abuse), teaching beliefs conducive to delinquency (e.g. the importance of being tough), and inadvertently reinforcing aggressive/delinquent behavior. Patterson and his colleagues, for example, found that the following interactional pattern was common in families with antisocial or delinquent children: a parent would attempt to discipline their child, often through scolding and nagging; the child would react to these disciplinary attempts with aversive behaviors, like crying, yelling, and hitting; and the parent would often retreat or back down in response to such behaviors. The child's aversive behaviors, then, would be negatively reinforced. The parent, however, was usually unaware of the impact of his or her actions (Patterson et al., 1989, 1992).

Social learning theorists argue that parents learn to engage in poor parenting from others. Among other things, they were exposed to poor parenting as children and so model the behavior of their parents (e.g. Patterson et al., 1992); they learn values conducive to poor parenting at work (Kohn, 1977); and they develop interactional patterns with their children that result in the reinforcement of poor parenting (Patterson et al., 1992).

A STRAIN MODEL OF THE FAMILY-DELINQUENCY RELATIONSHIP

Strain theory has not had much to say about the effect of family factors on crime. The classic strain theories of Merton (1938), Cohen (1955), and Cloward & Ohlin (1960) argue that crime results from the inability to achieve monetary success/middle class status through legitimate channels. The frustration or strain that results from such goal blockage is said to increase the likelihood that individuals will strike out at others or attempt to achieve their goals through illegitimate channels like theft. Such theories state that the children of poor families are less able to achieve their goals through legitimate channels than the children of wealthier families. But beyond that, there is little discussion of family factors and delinquency. Such theories focus on the impact of class on delinquency, rather than the impact of family structure and processes.

General strain theory (GST), however, has broadened the scope of the classic strain theories (see Agnew, 1992, forthcoming – a). GST focuses on *three major types* of strain. First, strain occurs when people are unable to achieve their goals through legitimate channels. GST focuses on a broad range of goals, including monetary success, status goals – including masculine status, autonomy goals, and the desire to be treated in a just or fair manner. Second, strain occurs when people lose something they value – like material possessions and relationships with valued others. Third, strain occurs when people are presented with noxious or negative stimuli – like verbal insults and physical assaults. These types of strain make people feel bad: angry, frustrated, depressed, etc. These negative feelings create pressure for corrective action, and delinquency is one possible response. Delinquency may be a method for reducing strain; that is, for achieving goals, for protecting or retrieving positive stimuli, or for terminating or escaping from negative stimuli. For example, juveniles may steal the money they desire, assault the peers who harass them, or run away from the parents who abuse them. Delinquency may also be a method for seeking revenge or taking anger out on others. And delinquency may occur as individuals try to alleviate their negative emotions through illicit drug use.

GST recognizes that only some strained individuals turn to delinquency, and the theory pays special attention to those factors that condition the effect of strain on delinquency. At the most general level, strain is most likely to lead to delinquency when individuals are unable to cope through legitimate channels, the costs of delinquency are low, and they are motivated to engage in delinquency. These conditions are most likely to hold when individuals possess traits like impulsivity and irritability, have limited coping skills and resources, are low in conventional social support, are low in social and self-control, blame their strain on the deliberate acts of others, and have been taught to respond to strain with crime (for a fuller discussion, see Agnew, 1992, forthcoming – a).

GST is in a better position to examine the relationship between family factors and delinquency than the classic strain theories. A range of family factors may contribute to the types of strain described above. Also, family factors may condition the effect of strain on delinquency. As such, GST can shed additional light on the ways in which the family contributes to delinquency, supplementing the control and social learning approaches that now dominate the literature. Further, the types of strain described by GST may also contribute to poor parenting. In fact, GST is quite compatible with much of the research on the causes of poor parenting – as indicated below.

The discussion which follows examines: (a) the effect of *parental* strain on poor parenting, and (b) the effect of poor parenting – a major source of *juvenile strain* – on juvenile delinquency. Those factors that condition the effect of parental strain and of poor parenting are also discussed. Figure 1 illustrates the strain model that is presented.

Parental Strain As A Source of Poor Parenting

Data suggest that many of the types of strain or stress identified in GST are a major source of those parenting practices that contribute to delinquency. While delinquency researchers have focused on a range of parenting practices, most research has focused on the emotional bond between family members and parental supervision. Data suggest that delinquency is more likely when: (a) there is a weak or negative emotional bond between parents and children, and (b) parental supervision is lax, overly strict, inconsistent, and/or harsh (i.e., excessive given the infraction, physically or verbally abusive). Data also suggest that delinquency is more likely when marital conflict is high, parents fail to teach good social and problem-solving shills, parents fail to provide adequate social support, and parents teach beliefs conducive to delinquency – like the 'code of the street' (see Anderson, 1999). Other family factors, like family structure (e.g. whether both natural parents are present, family size, age of mother) and parental criminality are said to affect delinquency primarily through their effect on these processes. (For overviews of this research, see Agnew, forthcoming b; Loeber & Stouthamer-Loeber, 1986; Patterson et al., 1992; Rankin & Wells, 1990; Rutter et al., 1998; Sampson & Laub, 1993; Van Voorhis, 1988; Wright & Wright, 1995)

Researchers have recently begun to ask why some parents are more likely to engage in the above types of 'poor parenting', While control and social learning theories are frequently mentioned, the dominant answer is that parental strain or stress fosters poor parenting. In brief, researchers argue that a range of stressors increase the likelihood that parents will experience negative affect – like depression and irritability. This negative affect increases the likelihood of poor parenting. Depressed and angry parents are more likely to withdraw from their children, reject them, and treat them in a harsh manner. Whether stressed parents engage in poor parenting, however, is influenced by several factors with most research focusing on the parents' level of social support.

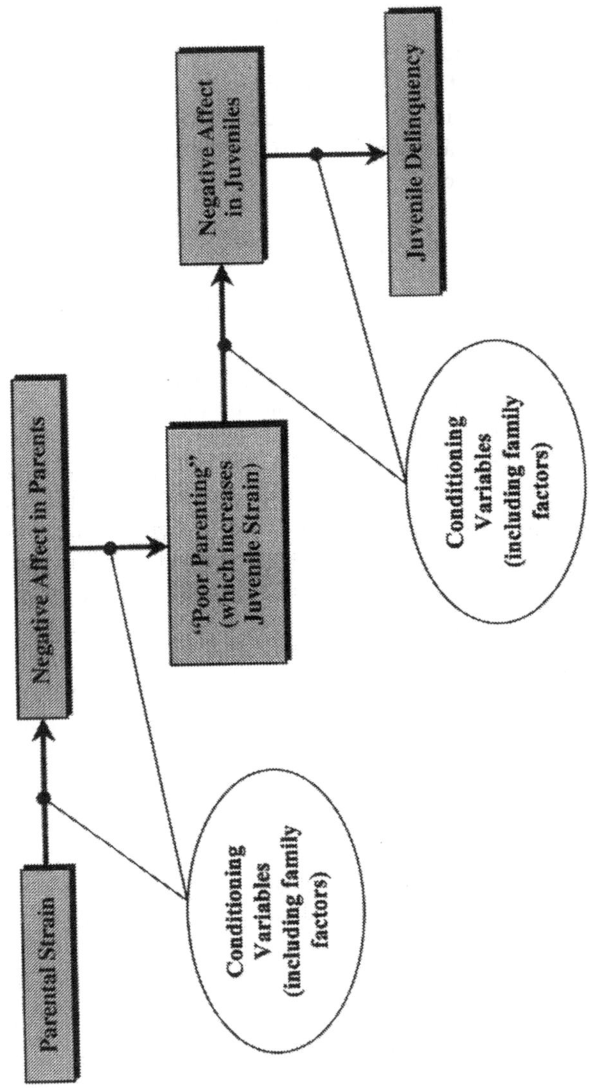

Figure 1. A General Strain Theory Model of the Family-Delinquency Relationship.

Most of the research on the causes of poor parenting, then, is quite compatible with GST. This is not surprising, because both this research and GST draw heavily on the stress literatures in sociology and psychology. GST, however, can contribute to this research by pointing to additional types of parental stress and additional factors that condition the impact of stress on poor parenting. Likewise, this family research can contribute to GST – as indicated below.

Types of Parental Strain

Most of the research on stress and poor parenting has focused on the impact of economic hardship (e.g. Belsky, 1990; Conger et al., 1992, 1994, 1995; Elder & Caspi, 1985; Elder et al., 1985; Larzelere & Patterson, 1990; Lempers et al., 1989; McLoyd et al., 1994; Simons et al., 1991, 1993a, 1993b; Voydanoff, 1990). This is not surprising, since much data indicate that economic hardship is a major stressor in itself and that it increases the likelihood of experiencing other stressors, including marital conflict, divorce, and a range of stressful life events. Family researchers have been quite sophisticated in their conceptualization and measurement of economic hardship (e.g. Conger et al., 1992; Fox & Chancey, 1998; Voydanoff, 1990). Economic hardship is not simply measured in terms of low income or a two or three item index of socioeconomic status. Rather, researchers typically examine such things as: (1) family per capita income; (2) unstable work history, which includes changing to a worse job, demotions, and being fired or laid off; (3) debt-to-asset ratio; and (4) increases or decreases in family income in the past year. Further, researchers recognize that these types of economic hardship do not impact all families in the same way. So a more direct measure of 'economic strain' is sometimes employed as well. For example, parents are asked about the extent to which the family has enough money for clothing, food, medical care, and bills. And they are asked about the changes they have had to make to cope with economic hardship, like moving, taking an additional job, canceling medical insurance, and obtaining government assistance. While economic hardship is correlated with economic strain, the correlation is far from perfect. Researchers typically find that economic hardship has a substantial effect on poor parenting, with much or all of this effect being mediated by economic strain (e.g. Conger et al., 1992, 1994; Simons et al., 1992, 1993b; Voydanoff, 1990; also see Elder & Caspi, 1985; McLoyd et al., 1994). Strain researchers in criminology can profit from this work, since they often employ rather simple measures of economic hardship and seldom measure 'economic strain' (although see Agnew et al., 1996).

Researchers have also examined other types of parental stressors (see Belsky, 1984, 1990 for overviews). Certain of these stressors focus on family structure and processes, like divorce/separation, large family size, teenage motherhood, and marital conflict (e.g. Biglan et al., 1990; Conger et al., 1984, 1992, 1993; Simons et al., 1993b, 1996). Other stressors focus on the characteristics of the child, including the antisocial or 'difficult' behavior of the child. Much data indicate that difficult children are a major source of strain for parents – with recent data also indicating that adolescent delinquency is a major source of strain as well (Ambert, 1999; Belsky, 1984, 1990; Biglan et al., 1990; Downey & Coyne, 1990; Elder & Caspi, 1985; Patterson & Forgatch, 1990; Rutter et al., 1998; Sampson & Laub, 1993). Finally, researchers sometimes employ more general inventories of stressful life events and daily life hassles. These inventories measure things like whether the parents have recently been fired or laid off, moved, experienced the death of friend, had an unwanted pregnancy, gotten robbed, or had an auto accident (e.g. Conger et al., 1995; Patterson & Forgatch, 1990; Patterson et al., 1989, 1992; Simons et al., 1993b, 1996; Straus & Gelles, 1990). Data suggest that these additional stressors also contribute to poor parenting. These stressors are correlated with most forms of poor parenting, even in longitudinal studies that attempt to control for relevant third variables. And – more generally – data indicate that poor parenting is most likely on those days that parents report high levels of stress (Patterson & Forgatch, 1990; Patterson et al., 1992).

GST points to other potentially important types of parental strain. While family researchers have devoted much attention to the failure to achieve economic goals, they might also consider other types of goal blockage. Strain researchers have devoted much attention to the failure to achieve status goals, with much recent research focusing on the inability of certain males to achieve masculine status through conventional channels. In this connection, certain accounts suggest that the failure to 'accomplish masculinity' through conventional channels may contribute to such problems as teenage motherhood, single parenthood, family violence, the neglect of parental responsibilities, and harsh parenting (see Anderson, 1999). In addition, researchers should examine the achievement of family related goals. Do parents feel that they are meeting their parental expectations or responsibilities. Likewise, do parents feel that their children are meeting their expectations. It has been suggested that the failure to satisfy such family-related expectations contributes to child abuse (Straus & Gelles, 1990). Finally, do parents feel that their partners and other family members are treating them in a just or fair manner. There has been some research on the factors that influence perceptions of fairness in the family, but little research on the impact of such perceptions on parenting practices. (see Barnett

& Shen, 1997; Gager, 1998; Rosenbluth et al., 1998; Ross & Van Willigen, 1996; Sanchez & Kane, 1996). GST also points to several additional types of strain involving the loss of positive stimuli/presentation of negative stimuli, including work in the secondary labor market, residence in high poverty communities, and – at the family level – the demands associated with nurturing others (see Agnew, 1992; forthcoming (a), (b); Broidy and Agnew, 1997; Voydanoff & Donnelly, 1988).

Effect of Strain on Negative Emotions

GST would argue that the above types of strain lead to poor parenting because they increase negative affect. Strain makes parents feel bad, and these bad feelings create pressure for corrective action. Parents may cope in a variety of ways, with poor parenting being one possible response. In particular, parents may respond to their depression and anger by withdrawing from their children – both emotionally and physically (Downey & Coyne, 1990). Effective parenting requires much time, effort, and patience; and these qualities are in short supply among parents who are worried and upset about the stressors in their life, like paying bills and spouse abuse. Withdrawal from parental responsibilities may be an attractive coping mechanism. As a result of such withdrawal, the emotional bond between parents and the child is weakened, parents do little to monitor or sanction misbehavior, and they engage in little 'positive parenting' – like providing instruction in social/problem solving skills. Parents may also respond to their depression and anger by treating their child in a hostile manner (see Conger et al., 1984, 1992, 1994; Downey & Coyne, 1990). Such parents are easily irritated and they often take out their anger on their children. As a consequence, parents may become excessively strict, their reaction to misbehavior may become excessive and harsh/abusive, there may be little effort to engage in effective problem-solving, and family conflict may increase. We might expect these reactions to be especially likely when the child is seen as a source of stress, as is the case with 'difficult' or antisocial children. It is easier to withdraw from or aggress against others when they are seen at fault. So GST argues that strain affects poor parenting by creating negative affect, with poor parenting being a response to or attempt to cope with such affect.

It is important to note, however, that the above types of 'strain' may lead to poor parenting for other reasons. For example, a control theorist might argue that economic hardship contributes to poor parenting because economically disadvantaged families are lower in control. They may be less likely to face sanctions if they engage in poor parenting, perhaps because they live in disorganized communities where deviance is more often ignored or tolerated.

They have a lower stake in conformity or less to lose by engaging in poor parenting (e.g. they are unemployed, they have low educational and occupational expectations for their children). And they may not hold beliefs that value good parenting. A social learning theorist might argue that economically disadvantaged people are more often taught to engage in poor parenting, like coercive discipline. Data, for example, indicate that blue-collar workers are more likely to hold values conducive to coercive discipline (see Conger et al., 1984; Kohn, 1977; Simons et al., 1993b). Control and social learning theorists might also argue that any association between the above strains and poor parenting is spurious: these strains and poor parenting may be associated because both are caused by low control or a learned disposition for crime/deviance. Low self-control, for example, may cause both economic hardship and poor parenting.

So it is important for a strain theorist to demonstrate that the above strains lead to poor parenting through their effect on negative emotions. Several studies have in fact tried to determine whether negative emotions mediate the effect of strain on poor parenting. Certain of these studies employ longitudinal data and attempt to control for relevant third variables. Most research has focused on the emotion of depression, although some studies also measure anger/irritability and anxiety. Data strongly suggest that measures of strain influence poor parenting at least partly through their effect on negative emotions. Stressors usually have a sizeable effect on negative emotions, and these emotions typically mediate some or all of the effect of stressors on poor parenting (Belsky, 1984; Conger et al., 1984, 1992, 1994, 1995; Elder & Caspi, 1985; Fox & Chancey, 1998; McLoyd et al., 1994; Patterson & Forgatch, 1990; Simons et al., 1993a, b, 1996; Voydanoff, 1990; Voydanoff & Donnelly, 1998).

Once more, strain researchers in criminology can benefit from this work. GST, in particular, focuses on the negative emotions of anger and frustration. The family research, however, demonstrates that it may be important to consider depression as well. While depression may not energize the individual for action, many forms of deviance involve withdrawal. Drug use, running away, and truancy may be examples in some cases. At the same time, the work on GST suggests that family researchers would do well to accord more attention to the emotions of anger (only considered in some studies) and frustration. While depression and anger/frustration usually occur together, that is not always the case (Downey & Coyne, 1990; Patterson & Forgatch, 1990). A consideration of these different emotional reactions may help explain different patterns of poor parenting; with withdrawal/neglect being more likely when depression dominates, harsh/abusive parenting when anger dominates, and inconsistent parenting when depression and anger occur together.

Conditioning Effects

GST argues that strain is more likely to lead to deviance in some cases than others. And family researchers have devoted some attention to the factors that influence the effect of stressors on poor parenting. Drawing on the larger stress literature, they have focused on the role of social support. Individuals may receive advice, emotional support, and assistance from others. This may reduce the impact of strain on negative affect and it may reduce the likelihood that parents respond to negative affect with poor parenting (see Belsky, 1981, 1984; Simons et al., 1993). Researchers have focused on both support from one's spouse and support from others, such as friends, relatives, and neighbors. It is argued that support from one's spouse is most critical given the centrality of this relationship (Belsky, 1984; Simons et al., 1993). Although the data are somewhat mixed, they suggest that support – especially spouse support – does reduce the negative effects of strain on parenting practices (Belsky, 1990; Elder & Caspi, 1985; McLoyd et al., 1994; Patterson et al., 1992; Simons et al., 1992, 1993a; Voydanoff, 1990).

GST – as well as the larger stress literature – points to still other conditioning variables. At most general level, three sets of factors may be important. First, do parents have the *ability* to cope with stress in a non-deviant manner. In particular, do they have the skills, resources, and support necessary for effective coping. Individual characteristics – like psychological health, temperament, and problem-solving skills – are important here. Some data, for example, suggest that irritable parents are more likely to respond to stress with poor parenting (see Belsky, 1990; Elder & Caspi, 1988; Patterson & Forgatch, 1990). The parents' financial resources are also important, since money can mitigate the impact of many stressors. And the parents' level of conventional social support is of course important. Second, are the *costs* of poor parenting low? The costs are influenced by the parent's level of social and self-control (e.g. are they socially isolated, do they hold good jobs and have good reputations in the community, do they hold beliefs that condemn poor parenting, are they aware of and concerned about the consequences of poor parenting). Third, are the parents *motivated* to engage in poor parenting. Motivation is influenced by individual traits like impulsivity and irritability, which may create a predisposition for poor parenting. It is influenced by whether parents blame their child for their strain. And it is influenced by the parents' learning history: were they taught to cope with strain by withdrawing from or striking out at others (see Straus & Gelles, 1990).

So parental strain is most likely to lead to poor parenting when the *ability* to cope with strain in a legitimate manner is low, the *costs* of poor parenting are low, and the *motivation* for poor parenting is high. Unfortunately, parents with the highest levels of strain are also the most likely to respond to strain with poor parenting. Strains like economic hardship reduce the ability to cope, the costs of poor parenting, and the motivation for good parenting (e.g. high levels of strain may emotionally overwhelm and physically exhaust parents, making it difficult to engage in good parenting). Likewise, these latter factors increase the likelihood of strain. Individuals with limited coping abilities, for example, often evoke negative reactions from others in particular situations and sort themselves into environments where negative treatment is likely – like bad marriages and bad jobs. Further, traits like irritability may be genetically based to some extent, so irritable parents are more likely to have irritable children (Moffitt, 1993).

SUMMARY

GST is quite compatible with the research on the causes of poor parenting. Much data suggest that parental strain contributes to negative emotions and that negative emotions at least partially mediate the effect of parental strain on poor parenting. The effect of parental strain, however, is conditioned by social support and certain other factors. GST, as indicated, can both contribute to and benefit from this research on the causes of poor parenting.

POOR PARENTING AS A SOURCE OF JUVENILE STRAIN

While strain/stress theory is commonly used to explain the causes of poor parenting, it is seldom used to explain the effect of poor parenting on delinquency – with the limited exception of harsh/abusive parenting. The effect of poor parenting on delinquency is typically explained in terms of control and social learning theories, as indicated above. This section argues that poor parenting is a major source of *juvenile* strain and that such strain is an important cause of both negative emotions and delinquency. Juveniles, in particular, may attempt to cope with poor parenting by escaping from their parents (e.g. running away, curfew violation, stealing to reduce financial dependence on parents), seeking revenge or rebelling against their parents (e.g. Brezina, 1999), taking out their anger on related targets, and/or managing their negative affect through illicit drug use. This argument applies not only to harsh/abusive parenting, but to the other dimensions of poor parenting as well – including poor supervision and weak/negative emotional bonding between parent and child.

Poor Parental Supervision as a Source of Juvenile Strain

At a minimum, effective supervision involves setting clear rules, monitoring behavior, and consistently sanctioning rule violations in an appropriate manner. Such actions are strongly related to delinquency and this relationship is typically explained in terms of control theory. Effective supervision, however, is also related to delinquency because it substantially reduces juvenile strain.

Parents supervise children for at least two reasons. The first – emphasized by control theorists – is to prevent misbehavior. The second has not received much attention from criminologists, but it is perhaps more important than the first. As a quick visit to any playground or discussion with any parent will reveal, *parents want to protect their children from harm (i.e., strain).* So the rules that parents set for their children not only forbid delinquent behavior, they also forbid behavior that might result in harm. Parents, for example, typically forbid their children from interacting with people who may harm them (e.g. strangers, delinquent peers), going to places where they might be harmed (e.g. certain sections of the city), and engaging in behaviors that might prove harmful (e.g. risky play activity). And one of the reasons that parents forbid delinquent behavior is because they fear that it might directly harm their child (e.g. drug use, fighting) or might provoke others into harming their child (e.g. peers who seek revenge, police who arrest).

Certain research has described the efforts of parents to protect their children from harm, including the extensive efforts sometimes employed by parents in high risk environments. Anderson (1999), for example, describes how a mother in a high-crime neighborhood frequently kept her daughter at home and enrolled her in a private school, despite the financial difficulties involved. This was done, in part, to protect the daughter from harm. Garbarino et al., (1992) describe how a grandmother established a neighborhood watch for her granddaughter, so she would be protected on the route home from school. Neighbors would notify the grandmother if the granddaughter did not pass by at the expected time and a search party would then be established. Such supervisory efforts seek to *prevent* harm. They are distinct from efforts to provide assistance *once harm has been experienced* (such efforts are typically referred to as social support).

Poor supervision, then, may increase the likelihood that juveniles experience a range of harms or strains. In particular, juveniles who are not effectively supervised may be more likely to interact with others who harm them, be in situations where the likelihood of harm is high, engage in behaviors that harm themselves, and act in ways that evoke negative or harmful reactions from others. Criminologists have not devoted much attention to this argument. But if it is correct, we would expect to find a negative association between measures of effective supervision and many measures of strain (some types of strain – like financial hardship and the death of family members – cannot be prevented through parental supervision). Further, we would expect the relationship between supervision and delinquency to be reduced when we control for strain. Ideally, we should conduct a longitudinal analysis since adolescent strain may influence parental supervision. We should also control for third variables that might affect both supervision and strain, such as economic hardship. And we should broaden our conception of supervision to include parental actions that seek to prevent *both* delinquency and juvenile harm.

Poor supervision may not only contribute to strain, but may also be a source of strain in and of itself. *Lax and inconsistent supervision* may promote feelings of normlessness; children may be uncertain how they should behave and how others will behave in certain settings. Research suggests that such uncertainty is a source of stress or strain, with children – particularly in high-risk environments – having a need for structure and control (see Garbarino et al., 1992). *Inconsistent discipline* may also be a source of strain because people like to be treated in a fair or just manner. Inconsistent discipline is quite likely to be experienced as unfair, since the same behavior may sometimes result in punishment and sometimes be ignored. In this area, data indicate that inconsistent discipline contributes to both psychological distress (depression) and delinquency (Lempers et al., 1989). Related to this, data indicate that the non-contingent use of punishment or the application of aversive stimuli promotes negative affect and aggression. Further, discipline that is seen as unjust or unfair is associated with delinquency (Berkowitz, 1993; Loeber & Stouthamer-Loeber, 1986; Rohner et al., 1996).

Excessively strict supervision may also be stressful, since it is likely to be seen as unfair and may interfere with the autonomy goals of many adolescents. Such discipline may involve rules that severely limit behavior, excessive monitoring, and sanctions that are excessive given the infraction. Limited data suggest that excessively strict discipline is associated with higher levels of delinquency (Rankin & Wells, 1990). It is difficult to explain this association in terms of control theory, but a strain explanation readily suggests itself. Future research should determine whether such discipline affects delinquency partly through its effect on negative emotions.

Finally, *harsh discipline (e.g. screaming, threatening, ridiculing, nagging, hitting), abusive behavior, and parental neglect* are important sources of strain. Most of the family research drawing on strain theory focuses on these types of behavior. This is understandable, since harsh discipline and abusive behavior clearly involve the presentation of negative stimuli. And parental neglect clearly involves the loss of positive stimuli and/or the inability to achieve positively-valued goals. Much data suggest that these types of parental behavior contribute to delinquency – with their effect on delinquency being partly mediated by negative emotions like anger (Agnew, 1985; Brezina, 1998; McLoyd et al., 1994; Straus & Yodanis, 1996; Turner & Finkelhor, 1996).

Weak/Negative Emotional Bonds Between Parent and Child as a Source of Juvenile Strain

Control theorists argue that a weak emotional bond between parent and child promotes delinquency for the reasons indicated above. A weak emotional bond may also be a source of strain, particularly for juveniles who desire good relations with parents. In this area, some data suggest that weak bonding is related to negative emotions like anger and depression and that these emotions partly mediate the relationship between weak bonding and delinquency (Agnew, 1993; Conger et al., 1992; Lempers et al., 1989). GST, however, would go on to argue that a negative emotional bond between parent and child contributes to delinquency over and above the effect of a weak bond. Juveniles who dislike or hate their parents probably do so because they believe their parents treat them in a negative manner. So a negative emotional bond indexes or reflects a range of negative parental treatment. But beyond that, it is stressful to live with someone you dislike or hate. If this argument is correct, we would expect juveniles who dislike or hate their parents to be higher in delinquency than those who are simply weakly attached to their parents – and this relationship should hold even after other measures of poor parenting are controlled.

Surprisingly, few studies have examined this argument (although see Nye, 1958, 71). Most measures of emotional bonding simply determine whether juveniles *do or do not like their* parents; they do not distinguish the absence of positive affect from negative affect. That is, they do not distinguish juveniles who say they *do not like their parents* from those who say they *dislike or hate their parents*. And the few studies that do contain measures making this distinction typically do not compare juveniles with weak parental bonds to juveniles with negative bonds. This failure to distinguish the absence of positive affect from negative affect likely reflects the influence of control theory. Drawing on control theory, it should not make a difference whether a juvenile is weakly bonded or negatively bonded to parents. In either case, the juvenile has little to lose by engaging in delinquency. Strain theory, however, predicts that juveniles who dislike or hate their parents will be more delinquent than those who simply do not like or care about their parents. While both groups are low in control, the juveniles who dislike/hate their parents are also high in strain. This argument is explored in the 'illustrative' analysis below.

Illustrative Analysis: The Effect of Negative Bonding on Delinquency

The 1994 Monitoring the Future survey is one of the few data sets that allows us to distinguish weak parental bonds from negative parental bonds. We use data from the 'Form 1' subsample, which consists of a nationally representative sample of 2,691 high school seniors (see Bachman et al., 1994 for complete information on the survey). The survey contains a single-item measure of parental bonding – with respondents being asked "how satisfied are you with how you get along with your parents?" Response categories are: completely dissatisfied, quite dissatisfied, somewhat dissatisfied, neutral, somewhat satisfied, quite satisfied, and completely satisfied. For the purposes of analysis, respondents were divided into three groups: dissatisfied with parents (16%), neutral (16%), and satisfied with parents (68%). The survey also contains measures of drug and alcohol use; no measures of other types of delinquency are available in the subsample we used. We measured the past-year frequency of marijuana/hashish use, other illicit drug use (cocaine, amphetamines, LSD, other psychedelics, heroin, other narcotics, barbiturates, tranquilizers, and quaaludes), and drunkenness. Response categories for each measure are: 0, 1–2, 3–5, 6–9, 10–19, 20–29, 30–39, and 40+. We also examine the respondent's *anticipated* frequency of marijuana/hashish use in the next five years (response categories range from "I will definitely not" use to "I definitely will" use).

Each measure of drug and alcohol use was regressed on the measure of parental satisfaction. Parental satisfaction was treated as a dummy variable, with the group being 'neutral' on the satisfaction measure taken as the reference category. This allows us to determine whether dissatisfied respondents are higher in drug/alcohol use than neutral respondents. We controlled for race (white=0; black=1), sex (female=0, male=1), and parents' education. This analysis is not designed to provide a comprehensive test of the above argument regarding negative bonding. It is done largely for illustrative purposes: that is, to suggest how one might go about testing this argument and to explore whether such a test is worth pursuing. A complete test would employ better measures of bonding and delinquency, would control for more variables – including measures of parental supervision and abuse/neglect, would examine a broader sample of juveniles, and would employ longitudinal data. Unfortunately, no data sets allowing for such a test are publicly available.

Regression results are shown in Table 1. As can be seen, the dissatisfaction measure is significantly related to all drug measures at the 0.05 level or better. It is related to the drunkenness measure at the 0.10 level. Seniors who report that they are dissatisfied with their parents are more likely to use drugs and (possibly) get drunk than seniors who report their relations with parents are

Table 1. Juvenile Alcohol and Drug Use Regressed on Satisfaction with Parents and Control Variables (Standardized Coefficients Shown with Unstandardized Coefficients in Parentheses)

	Endogenous Variables			
	(1) Marijuana/ Hashish	(2) Other Illicit Drugs	(3) Drunkenness	(4) Anticipated Drug Use
Sex (Male)[a]	0.111**	–0.004 n.s.	0.078**	0.090**
	(0.232)	(–0.002)	(0.493)	(0.142)
Race (Black)[b]	–0.031 n.s.	–0.074**	–0.140**	–0.020 n.s.
	(–0.096)	(–0.055)	(–0.715)	(0.046)
Social Class	0.048**	–0.044*	–0.016 n.s.	0.024 n.s.
	(0.045)	(–0.009)	(0.025)	(0 017)
Dissatisfied[c]	0.090**	0.052*	0.048*	0.058*
	(0.256)	(0.035)	(0.225)	(0.125)
Satisfied[c]	–0.043 n.s.	–0.078*	–0.069*	–0.081**
	(–0.097)	(–0.042)	(–0.259)	(–0.139)
Constant	(0.299)**	(0.136)**	(2.255)**	(1.392)**
R^2	0.027	0.018	0.052	0.022

*$p < 0.10$; *$p < 0.05$; **$p < 0.01$; (one-tailed)
[a]Female is the reference category.
[b]White/Caucasian is the reference category.
[c]'Neutral' Relationship with Parents is the reference category.

'neutral.' These data – although very tentative – suggest that negative bonding may contribute to delinquency (at least drug/alcohol use) over and above weak or neutral bonding.

Other Family Factors as a Source of Juvenile Strain

The above discussion has focused on supervision and bonding given their central role in explanations of delinquency. Other family factors, however, are also related to delinquency and this relationship can be explained partly in terms of strain theory. Marital conflict is related to delinquency is many studies, especially when the conflict involves screaming, insulting, threatening, expressions of contempt, and hitting (e.g. Buehler et al., 1998; Loeber & Stouthamer-Loeber, 1986). Data suggest that such conflict can be a potent source of stress for juveniles (see McNeal & Amato, 1998), and it is reasonable to suppose that such conflict affects delinquency partly for reasons related to strain theory.

There is some evidence that delinquency is more likely when parents *fail to teach effective social and problem-solving skills* (Agnew, forthcoming – b). This too can be partly explained in terms of strain theory. Individuals who lack such skills are more likely to experience strain; they evoke negative reactions from others in particular situations and they sort themselves into environments where the likelihood of negative treatment is high – like delinquent peers groups (Patterson, 1986; Patterson et al., 1989).

Still other family factors can be mentioned, although many have not been the subject of much research. Most notably, there has not been any research on whether the failure of children to satisfy parental expectations is an important source of *juvenile* strain and delinquency. Finally, other family factors – like parental divorce, teenage motherhood, and parental criminality – affect juvenile strain through their impact on the family processes listed above.

The Effect of Family Factors on the Reaction to Strain

A range of family factors influence whether juveniles cope with strain in a legitimate or delinquent manner. In particular, parents have a major influence on the ability of juveniles to cope in a legitimate manner, the costs of delinquent coping, and the motivation for conventional versus delinquent coping.

Regarding ability, most attention has focused on the extent to which parents provide their children with social support. Parents can cope on behalf of their children or provide them with assistance in coping. Cullen (1994) provides an excellent overview of the key role that social support plays in preventing crime. And certain accounts discuss the different ways in which parents help their children cope with stressful events (e.g. Fox et al., 1996; Garbarino et al., 1992). Limited data suggest that parental social support can reduce the effect of stressors on adolescent distress and delinquency – although more research is needed here (Bowen & Chapman, 1996; Harris & Marmer, 1996; also see Turner & Finkelhor, 1996).

In addition to social support, parents can equip their children with the skills and knowledge to cope on their own behalf. They may provide direct instruction in legitimate coping, model legitimate coping, reinforce legitimate coping, and foster the development of those traits that enhance legitimate coping (e.g. self-efficacy, intelligence). Unfortunately, there has not been much research on the extent to which such parental behaviors condition the impact of juvenile strain on delinquency.

Parents can also affect the juvenile's reaction to strain by influencing the costs of delinquent coping. In this area, we would expect the effect of juvenile strain on delinquency to be conditioned by the juvenile's attachment to parents and level of supervision (although very strict supervision may function as a source of strain, as indicated earlier). There has not been much research in this area, although Luster & Small (1997) found that sexual abuse was less likely to lead to problem behaviors in adolescents who had high levels of parental monitoring and support. And the illustrative analysis below examines whether parental attachment conditions the effect of a particularly severe type of juvenile strain – criminal victimization – on juvenile delinquency.

Finally, parents can influence the juvenile's motivation for conventional versus delinquent coping through their effect on the juvenile's traits (e.g. irritability, impulsivity, beliefs) and through the models they present and the reinforcements/punishments they provide. In this area, Anderson (1999) and others argue that parents in high risk environments like poor inner-city communities sometimes teach their children to respond to strain with violence. In particular, some parents instruct their children in the 'code of the street', – which states that one should respond to certain insults and provocations with force. Individuals who do not respond with force may be victimized on a regular basis by others. These parents, then, are trying to protect their children from harm in an environment where victimization is common and the police cannot be relied upon for protection. The parents may not embrace or like the code of the street, but they believe that it is sometimes necessary to live by it.

Illustrative Analysis: Does Parental Attachment Condition the Effect of Strain on Delinquency?

The 1976 National Youth Survey contains 1725 juveniles who appear to be representative of the total 11- through 17-year-old youth population in the United States (Elliott et al., 1985: 92). We use this survey to examine whether parental attachment conditions the effect of criminal victimization on juvenile delinquency. We would expect such a conditioning effect since juveniles high in attachment have more to lose by engaging in delinquency. Further, it is reasonable to expect that such juveniles receive more social support from their parents and are more likely to internalize parental prohibitions against delinquency. As such, they should be less likely to respond to strain with delinquency.

Parental attachment is measured by a four-item scale, with high scorers stating that (a) they get along with their parents, (b) they engage in many activities with their parents, (c) they can count on their parents for comfort during difficult times, and (d) they can talk openly with their parents. Juvenile victimization is measured with a nine-item scale, with respondents reporting the number of times they have had things directly taken from them, had a car or bike stolen from them, had things taken from their car, had their property damaged by others, had things stolen from them in public places, been abused by a parent, been sexually assaulted, been attacked with a weapon, and been beaten up by others in the prior year. Delinquency is measured by a ten-item scale focusing on violence, since criminal victimization is most strongly related to this type of delinquency. Respondents report the number of times they have carried hidden weapons, attacked someone, been in gang fights, hit teachers, hit parents, hit other students, sexually assaulted someone, used force on a teacher to get something, used force on a student, and used force on someone else in the prior year.

Violence was first regressed on parental attachment, juvenile victimization, and three control measures: sex, race, and social class as measured by the Hollingshead Index. Regression results are shown in column 1 of Table 2. As expected, parental attachment has a negative effect on violence and juvenile victimization has a positive effect. We then created a product term between parental attachment and juvenile victimization. This product or interaction term allows us to determine whether the effect of juvenile victimization on violence varies by level of parental attachment. In order to reduce multicolinearity, all variables were standardized before creating the product term (see Aiken & West, 1991, for a full description of this procedure; diagnostic analyses suggest multicolinearity was not a problem). Violence was then regressed on the product term and the standardized values of the other variables. Regression results are shown in column 2 of Table 2. These results are roughly equivalent to standardized regression results given our standardization procedure. We find that the interaction term between parental attachment and juvenile victimization is significant at the 0.05 level. As predicted, the effect of victimization on violence becomes *smaller* as parental attachment increases. So parental attachment appears to reduce the impact of at least this type of strain on violence. These results, however, should also be viewed as tentative. Among other things, the cross-sectional nature of the data is a problem. An ideal analysis should use longitudinal data to examine the effect of prior victimization on subsequent delinquency or to examine the contemporaneous effects of victimization and delinquency on one another.

SUMMARY

General Strain Theory, then, has much to say about the relationship between family factors and crime. As indicated in Figure 1, parental strain may lead to negative affect and poor parenting – although family factors and other variables condition the impact of parental strain on negative affect/poor parenting. Poor

Table 2. Juvenile Violence Regressed on Parent-Child Attachment, Victimization, Attachment/Victimization Interaction, and Controls

(Unstandardized Coefficients Shown in Parentheses)

	Endogenous Variables	
	(1) Violence Score	(2) Violence Z-score[d]
Sex (Male)[a]	0.175** (0.818)	0.162**
Race (Black)[b]	0.022 (0.154)	0.024
Race (Hispanic)[c]	–0.003 –(0.041)	–0.002
Social Class	0.082** (0.012)	0.076**
Parent-Child Attachment	–0 128** –(0.091)	–0.123**
Child Victimization	0.396** (0.133)	0.672**
Attachment* Victimization		–0.074*
Constant	11.386**	–0.003
R^2	0 248	0.250

*$p < 0.05$; *$p < 0.01$; (one-tailed)
[a]Female is the reference category.
[b]Caucasian/Other is the reference category.
[c]Additive Predictors Standardized prior to Analysis

parenting, in turn, is a major source of juvenile strain. Poor parenting, in particular, increases the likelihood of negative affect and juvenile delinquency. Poor parenting includes not only harsh and abusive parenting, but poor discipline and weak/negative emotional bonding between parent and child. The effect of poor parenting on negative affect and delinquency is conditioned by a range of family factors and other variables.

Certain of these arguments have much support, particularly those regarding the effect of parental strain on negative affect and poor parenting. Some data also suggest that harsh/abusive parenting increases negative affect and delinquency. Other arguments, however, have not been the subject of much investigation. In particular, we have little idea whether the supervision and bonding variables listed above are sources of strain/negative emotions, and whether strain/negative emotions partly mediate their effect on delinquency.

Also, there has been little research on those family-related factors that may condition the effect of strain on delinquency.

It should be noted that the above discussion has focused on the direct effect of juvenile strain on delinquency. But juvenile strain may also affect delinquency indirectly by reducing social control and fostering the social learning of crime. Most notably, juveniles who experience the types of family strain described above may reduce their attachment to parents, spend less time with parents (making effective supervision more difficult), become less responsive to parental sanctions, and become less likely to internalize parental beliefs. They may also become more likely to adopt beliefs favoring delinquency and to associate with delinquent peers (see Agnew, 1995). Paternoster & Mazerolle (1994) find empirical support for certain of these arguments.

It should also be noted that the above discussion is at a rather general level. There is some evidence that there are gender, age, and race/ethnic differences in the types of strain experienced and in the reaction to strain. For example, it is commonly argued that economic stressors have a greater effect on males and relational stressors (e.g. family conflict) have a greater effect on females. It is also argued that males are more likely to respond to stressors with externalizing behaviors like aggression, while females are more likely to respond with internalizing behaviors. The data are somewhat mixed regarding gender, age, and race/ethnic differences in these areas, but the further exploration of such differences will likely be a major area of research in the future (e.g. Broidy & Agnew, 1997; Fox & Chancey, 1998; Ross and Van Willigen, 1996; Voydanoff, 1990).

In sum, GST has much to say about the relationship between family factors and delinquency, and provides a useful supplement to the control and social learning explanations that now dominate the research on the family determinants of delinquency.

REFERENCES

Agnew, R. (1985). A Revised Strain Theory of Delinquency. *Social Forces*, *64*, 151–67.

Agnew, R. (1992). Foundation for a General Strain Theory of Crime and Delinquency. *Criminology*, *30*, 47–87.

Agnew, R. (1993). Why Do They Do IT: An Examination of the Intervening Mechanisms Between Social Control Variables and Delinquency. *Journal of Research in Crime and Delinquency*, *30*, 245–66.

Agnew, R. (1995). The Contribution of Social-Psychological Strain Theory to the Explanation of Crime and Delinquency. In: F. Adler & W. S. Laufer (Eds), *The Legacy of Anomie Theory, Advances in Criminological Theory*, *6*, 113–137. New Brunswick, NJ: Transaction Publishers.

Agnew, R. Forthcoming, a. An Overview of General Strain Theory. In: Ray Paternoster (Ed.), *Essays in Criminological Theory*. Los Angeles: Roxbury.

Agnew, R. Forthcoming, b. Juvenile Delinquency: Causes and Control. Los Angeles: Roxbury.

Agnew, R., Cullen, F. T., Burton V. S. Jr., Evans, T. D., & Dunaway, R. G. (1996). A New Test of Classic Strain Theory. *Justice Quarterly, 13*, 681–704.

Agnew, R., & Raskin White, H. (1992). An Empirical Test of General Strain Theory. *Criminology, 30*, 475–499.

Aiken, L. S., & West, S. G. (1991). *Multiple Regression: Testing and Interpreting Interactions.* Newbury Park, CA: Sage.

Bachman, J. G., Johnston, L. D., & OMalley, P. M. (1998). *Monitoring the Future: A Continuing Study of the Lifestyles and Values of Youth, 1994* [computer file]. Ann Arbor, Michigan: Inter-university Consortium for Political and Social Research.

Barnett, R. C., & Shen, Y.-C. (1997). Gender, High- and Low-Schedule-Control Housework Tasks, and Psychological Distress *Journal Of Family Issues, 18*, 403–428.

Belsky, J. (1981). Early Human Experience: A Family Perspective. *Developmental Psychology, 17*, 3–23.

Belsky, J. (1984). The Determinants of Parenting: A Process Model. *Child Development, 55*, 83–96.

Belsky, J. (1990). Parental and Nonparental Child Care and Childrens Socioemotional Development: A Decade in Review. *Journal of Marriage and the Family, 52*, 885–903.

Berkowitz, L. (1993). *Aggression. Its Causes, Consequences, and Control.* Philadelphia: Temple University Press.

Biglan, A., Lewin, L., & Hops, H. (1990). A Contextual Approach to the Problem of Aversive Practices in Families. In: Gerald R. Patterson (Ed.), *Depression and Aggression in Family Interaction*, 103–129. Hillsdale, NJ: Lawrence Erlbaum Associates.

Bowen, G. L., & Chapman, M. V. (1996). Poverty, Neighborhood Danger, Social Support, and the Individual Adaptation Among At-Risk Youth in Urban Areas. *Journal of Family Issues, 17*, 641–666.

Brezina, T. (1998). Adolescent Maltreatment and Delinquency: The Question of Intervening Processes. *Journal of Research in Crime and Delinquency, 35*, 71–99.

Brezina, T. (1999). Teenage Violence Toward Parents as an Adaptation to Family Strain. *Youth and Society, 30*, 416–444.

Broidy, L., & Agnew, R. (1997). Gender and Crime: A General Strain Theory Perspective. *Journal of Research in Crime and Delinquency, 23*, 207–239.

Buehler, C., Krisnakumar, A., Stone, G., Anthony, C., Pemberton, S., Gerard, J., & Barber, B. K. (1998). Interparental Conflict Styles and Youth Problem Behaviors: A Two-Sample Replication Study. *Journal of Marriage and the Family, 60*, 119–32.

Cloward, R. A., & Ohlin, L. E. (1960). *Delinquency and Opportunity.* New York: Free Press.

Cohen, A. K. (1955). *Delinquent Boys.* New York: Free Press.

Conger, R. D., Conger, K. J., Elder, G H. Jr , Lorenz, F. O., Simons, R. L., & Whitbeck, L. B. (1992). A Family Process Model of Economic Hardship and Adjustment of Early Adolescent Boys. *Child Development, 63*, 526–541.

Conger, R. D., Ge, X , Elder, G. H. Jr., Lorenz, F. O., & Simons, R. L. (1994). Economic Stress, Coercive Family Process, and Developmental Problems of Adolescents. *Child Development, 65*, 541–561.

Conger, R. D., McCarty, J., Yang, R. K., Lahey, B. B., & Kropp, J. P. (1984). Perception of Child, Child-Rearing Values, and Emotional Distress as Mediating Links between Environmental Stressors and Observed Maternal Behavior. *Child Development, 55*, 2234–2247.

Conger, R. D., Patterson, G. R., & Ge, X (1995). It Takes Two to Replicate: A Mediational Model for the Impact of Parents Stress on Adolescent Adjustment. *Child Development, 66*, 80–97.

Cullen, F. T. (1994). Social Support as an Organizing Concept for Criminology. *Justice Quarterly, 11*, 527–59.

Downey, G., & Coyne, J. C. (1990). Children of Depressed Parents: An Integrative Review. *Psychological Bulletin, 108*, 50–76.

Elder, G. H. Jr., & Caspi, A (1988). Economic Stress in Lives: Developmental Perspectives. *Journal of Social Issues, 44*, 25–45.

Elder, G. H., Jr., Van Nguyen, T., & Caspi, A. (1985). Linking Family Hardship to Childrens Lives. *Child Development, 56*, 361–375.

Elliott, D. S., Huizinga, D., & Ageton, S (1985). Explaining Delinquency and Drug Use. Beverly Hills, CA: Sage.

Fox, G. L., & Chancey, D. (1998). Sources of Economic Distress: Individual and Family Outcomes. *Journal of Family Issues, 19*, 725–49.

Fox, G. L., Von Bargen, J., & Jester, M. (1996). Managing Murder: Parents as Mediators of Childrens Experience. *Journal of Family Issues, 17*, 732–57.

Gager, C. T. (1998). The Role of Valued Outcomes, Justifications, and Comparison Referents in Perceptions of Fairness Among Dual-Earner Couples. *Journal of Family Issues, 19*, 622–648.

Garbarino, J., Dubrow, N., Kostelny, K., & Pardo, C. (1992). *Children in Danger*. San Francisco: Jossey-Bass.

Gottfredson, M. R., & Hirschi, T (1990). A General Strain Theory of Crime. Stanford, CA: Stanford University Press.

Harris, K M., & Marmer, J. K (1996). Poverty, Paternal Involvement, and Adolescent Well-Being. *Journal of Family Issues, 17*, 614–640.

Hirschi, T. (1969). *Causes of Delinquency*. Berkeley, CA: University of California Press

Hirschi, T. (1995). The Family. In: James Q. Wilson & Joan Petersilia (Eds), *Crime*, 121–140. San Francisco: ICS Press.

Kohn, M. L. (1977). Class and Conformity. Chicago: University of Chicago Press.

Larzelere, R. E., & Patterson, G. R. (1990). Parental Management: Mediator of the Effect of Socioeconomic Status on Early Delinquency. *Criminology, 28*, 301–324.

Lempers, J D., Clark-Lempers, D., & Simons, R. L. (1989). Economic Hardship, Parenting, and Distress in Adolescents. *Child Development, 60*, 25–39.

Loeber, R., & Stouthamer-Loeber, M. (1986). Family Factors as Correlates and Predictors of Juvenile Conduct Problems and Delinquency. In: M. Tonry, & N. Morris (Eds), *Crime and Justice, 7*, 29–149. Chicago: University of Chicago Press.

Luster, T., & Small, S. (1997). Sexual Abuse History and Problems in Adolescence: Exploring the Effects of Moderating Variables. *Journal of Marriage and the Family 59*, 131–42.

McLoyd, V C., Jayaratne, T. E., Ceballo, R., & Borquez, J. (1994). Unemployment and Work Interruption Among African American Single Mothers: Effects on Parenting and Adolescent Socioemotional Functioning. *Child Development, 65*, 562–589.

McNeal, C., & Amato, P. R. (1998). Parents Marital Violence. *Journal of Family Issues, 19*, 123–139.

Merton, R K. (1938). Social Structure and Anomie. *American Sociological Review, 3*, 672–82.

Moffitt, T. E. (1993). Life-Course Persistent and Adolescent-Limited Antisocial Behavior: A Developmental Taxonomy. *Psychological Review, 100*, 674–701.

Nye, I. F. (1958). *Family Relationships and Delinquent Behavior*. New York: Wiley.

Paternoster, R., & Mazerolle, P (1994). General Strain Theory and Delinquency: A Replication and Extension. *Journal of Research in Crime and Delinquency, 31*, 235–263.

Patterson, G. R. (1986). Performance Models for Antisocial Boys. *American Psychologist, 41*, 432–444.

Patterson, G. R., DeBaryshe, B. D., & Ramsey, E. (1989). A Developmental Perspective on Antisocial Behavior. *American Psychologist, 44*, 329–335.

Patterson, G. R., & Forgatch, M. S. (1990). Initiation and Maintenance of Process Disrupting Single-Mother Families. In: G. R. Patterson (Ed.), *Depression and Aggression in Family Interaction*, 209–245. Hillsdale, NJ: Lawrence Erlbaum Associates.

Patterson, G. R., Reid, J. B., & Dishion, T. J. (1992). *Antisocial Boys*. Eugene, OR: Castalia.

Rankin, J. H., & Wells, L. E. (1990). The Effect of Parental Attachments and Direct Controls on Delinquency. *Journal of Research in Crime and Delinquency, 27*, 140–165.

Rohner, R. P., Bourque, S. L., & Elordi, C. A. (1996). Childrens Perceptions of Corporal Punishment, Caretaker Acceptance, and Psychological Adjustment in a Poor, Biracial Southern Community. *Journal of Marriage and the Family, 58*, 842–52.

Rosenbluth, S. C., Steil, J. M., & Whitcomb, J. H. (1998). Marital Equality: What Does It Mean? *Journal of Family Issues, 19*, 227–244.

Ross, C. E., & Van Willigen, M. (1996). Gender, Parenthood, and Anger. *Journal of Marriage and the Family, 58*, 527–84.

Rutter, M., Giller, H., & Hagell, A.. (1998). *Antisocial Behavior by Young People*. Cambridge, England: Cambridge University Press.

Sampson, R. J., & Laub, J. H. (1993). *Crime in the Making*. Cambridge, MA: Harvard University Press.

Sanchez, L., & Kane, E. W. (1996). Womens and Mens Constructions of Perceptions of Housework Fairness. *Journal of Family Issues, 17*, 358–387.

Simons, R. L., Beaman, J., Conger, R. D., & Chao, W. (1993b). Stress, Support, and Antisocial Behavior Trait as Determinants of Emotional Well-Being and Parenting Practices Among Single Mothers. *Journal of Marriage and the Family, 55*, 385–398.

Simons, R. L., Lorenz, F. O., Conger, R. D., & Wu, C.-I. (1992). Support from Spouse as Mediator and Moderator of the Disruptive Influence of Economic Strain on Parenting. *Child Development, 63*, 1282–1301.

Simons, R. L., Lorenz, F. O., Wu, C.-I., & Conger, R. D (1993a). Social Network and Marital Support as Mediators and Moderators of the Impact of Stress and Depression on Parental Behavior. *Developmental Psychology, 29*, 368–381.

Simons, R. L., & Associates. (1996). *Understanding Differences Between Divorced and Intact Families*. Thousand Oaks, CA: Sage.

Straus, M. A., & Gelles, R. J. (1990). *Physical Violence in American Families*. New Brunswick, NJ: Transaction Publishers.

Straus, M. A., & Yodanis, C. L. (1996). Corporal Punishment in Adolescence and Physical Assaults on Spouses in Later Life: What Accounts for the Link? *Journal of Marriage and the Family, 58*, 825–41.

Turner, H. A., & Finkelhor, D. (1996) Corporal Punishment as a Stressor Among Youth. *Journal of Marriage and the Family, 58*, 155–66.

Van Voorhis, P., Cullen, F. T., Matthers, R. A., & Garner, C. C (1988). The Impact of Family Structure and Quality on Delinquency: A Comparative Assessment of Structural and Functional Factors. *Criminology, 26*, 235–261.

Voydanoff, P. (1990). Economic Distress and Family Relations: A Review of the Eighties *Journal of Marriage and the Family, 52*, 1099–1115.

Voydanoff, P., & Donnelly, B. W. (1988). Economic Distress, Family Coping, and Quality of Family Life. In: P. Voydanoff & L. C. Majka (Eds), Families and Economic Distress, (97–116). Newbury Park, CA: Sage.

Voydanoff, P., & Donnelly, B. W. (1998). Parents Risk and Protective Factors as Predictors of Parental Well-Being and Behavior. *Journal of Marriage and the Family, 60*, 344–355

Wright, K. N., & Wright, K. E. (1995). *Family Life, Delinquency, and Crime: A Policymakers Guide*. Washington, D.C.: Office of Juvenile Justice and Delinquency Prevention.

PARENTAL SUPPORT AND JUVENILE DELINQUENCY

John Paul Wright, Francis T. Cullen and John D. Wooldredge

Over the past decade, the family has enjoyed a revival in criminological interest. In part, the 'rediscovery' of the family has been based on the expanding body of empirical literature linking family conditions and parental behavior to delinquency (Loeber & Stouthamer-Loeber, 1989). As such, an increasing number of theories have identified the family as being central to crime causation across the life-course (Gottfredson & Hirschi, 1990; Hagan, 1989; Sampson & Laub, 1993).

Although exceptions exist (see Agnew, 1992), criminologists have largely interpreted the impact of parents and families on crime within some variant of control theory. Building on Hirschi's (1969) social bond theory, a continuing perspective is the importance of how children's attachments to parents creates 'indirect' or 'relational' controls (Hagan, 1989) – that is, the psychological presence of parents even when they are not present that restrains behaviors. More recently, however, there has been a growing body of evidence highlighting the importance of 'direct parental controls,' which are the active steps that parents take to supervise their children, and, if necessary, to punish wayward conduct.

The ascension of direct controls over other parenting behaviors can be seen, for example, in 'self-control' and 'power-control' theory. Drawing mainly from the work of Patterson (1980), Gottfredson & Hirschi (1990, p. 97) argue that certain 'minimum conditions' must be met for children to be adequately socialized. "To teach the child self control," they argue, "parents must monitor their

Families, Crime and Criminal Justice, Volume 2, pages 139–161.
2000 by Elsevier Science Inc.
ISBN: 0-7623-0737-4

child's behavior, recognize deviant behavior, and punish inappropriate behavior." Parental actions that restrict opportunities for deviance or that teach children self-restraint are also theoretically central to Hagan's (1989) power-control theory. Similar to how Gottfredson and Hirschi view the role of direct controls in limiting crime, Hagan maintains that 'instrumental' controls exercised by parents reduce the 'risk taking preferences' of adolescents.

Given the central place direct controls occupies in family theory, it is interesting to note that the empirical evidence has produced mixed and inconsistent results, sometimes producing positive linear results, sometimes negative linear results, and periodically nonlinear results (Seydlitz, 1993). Wells & Rankin (1988), for example, found evidence that the relationship is curvilinear, with low and high levels of control associated with high levels of delinquency. Barnes & Farrell (1992; see also Barnes, 1984) reported parental control efforts to be linearly and positively associated with adolescent drinking and delinquency. Contrary to these findings, Cernkovich & Giordano (1987) found higher rates of self-reported delinquency to be associated with monotonic decreases in control. Finally, in a review of the extant literature, Rollins & Thomas (1979) concluded that the weight of the evidence on the effects of direct parental controls showed that high levels of direct controls were contrary to a number of outcome variables, such as depression, aggression, and academic achievement.

Since the evidence on the effects of direct parental controls on delinquency is inconsistent, the continued and almost exclusive focus on the control perspective is misguided. Even when the analysis moves away from parenting behavior and onto family structural factors, however, control remains the perspective of choice. Single-parents, it is argued, supervise less (Matsuada & Heimer, 1987), mobility decreases the effects of informal controls (Sampson & Groves, 1989), larger families translate to less overall control (Hirschi, 1969), and fewer economic resources demand longer hours at work thus reducing the oversight of parents (LeFlore, 1988). While these explanations may have some empirical backing and are theoretically consistent with a control theory perspective, they have generally been left unchallenged, with other potentially important theoretical constructs overlooked.

Within the criminological literature, the dominant concern with control has obscured the potential salient effects of *supportive* behaviors by parents. While most existing control theories do not say that parental efforts to love, nurture, and assist children are unimportant, they either do not mention them or relegate them to a passing mention. There are some exceptions to this general trend. For instance, Patterson's (1986) focus on 'positive' parenting and Wilson & Herrnstein's argument that parenting should not only be 'restrictive' but also 'warm' stand in contrast to other control theories.

The neglect of parental support in criminology becomes starkly apparent in comparison to theory and research within the field of child development, where the lack of supportive parenting is linked to a range of problem behaviors (Rollins & Thomas, 1979). Furthermore, in the field of mental illness, social support is a major paradigm which has resulted in numerous theoretical pieces and in hundreds of research articles (Thoits, 1995; Vaux, 1988).

THE CONCEPT OF PARENTAL SUPPORT

Definitions of the term support generally take the form of "an interpersonal transaction involving one or more of the following: (1) emotional concern (liking, love, empathy, (2) instrumental aid (goods and services), (3) information (about the environment), or (4) appraisal (information relevant to self-evaluation" (House, 1981, p. 39). Support is even more broadly defined by Lin (1986, p. 18) as "the perceived or actual instrumental and/or expressive provisions supplied by the community, social networks, and confiding partners." Moreover, Thoits (1982, p. 147) defined support as "the degree to which a person's basic social needs are gratified through interaction with others." Thoits identified some of these needs as including affection, belonging, identity, and security – needs of children we note commonly filled by parents (see also Vaux, 1988). The common link to all of these definitions is that support requires someone, in this case a parent, doing something helpful for someone else, such as their child. It is the action of providing help and assistance, we argue, that distinguishes the term from either direct control or attachment.

A criticism common to support research is that support is defined too broadly and is far too inclusive to be of theoretical value. Much the same could be said of the term control, where the restraint perspective is related to any relationship that might somehow limit behavior. Similar to control theory, however, support theorists have argued that support is a meta-construct composed of several dimensions, such as supportive network resources, supportive behavior, or support appraisals.

Research On Parental Support

To date, research on the effects of parental support comes mainly from the child development literature. Findings generated from diverse samples, methodologies, and analyses show that parental supports reduce delinquent involvement as well as a host of other problems, such as depression. The evidence in favor of the social support perspective appears widespread. "One cannot spend much time in a review of the child socialization research," note Rollins & Thomas (1979, p. 351), "without being struck with the ubiquitous nature of the effects of this variable" – support. Supportive parenting, according to their review, has been associated with positive child outcomes across time, method, stage of development, and culture. Finding such consistency or 'invariance' of effects is relatively infrequent in social science research.

Supportive parent-child transactions have been found to promote prosocial and altruistic behavior (Radke-Yarrow & Zahn-Waxler, 1986; Zhan-Waxler & Radke-Yarrow, 1982), to insulate children in high-risk environments (Barnes & Farrell, 1992; Cohen & Wills, 1985; Werner & Smith, 1982; Farrell, Barnes & Banerjee, 1995; Jessor, 1993; Smith, Lizotte, Thornberry & Krohn, 1995; Wright et al., 1995), and to reduce delinquency (Hepburn, 1977; Johnson, Su, Gerstein, Shin & Hoffman, 1995; Wright et al., 1995). In this latter regard, Andry (1960) found that compared to delinquent youths, non-delinquent youths were significantly more likely to participate in 'outings' with their parents. Families of delinquents were also less concerned for the welfare of their children than families of non-delinquents in the Gluecks' (1950) study. Parents of delinquents have also been found to display less supportive communications (Alexander, 1973), to share fewer evenings out with their children (Slocum & Stone, 1963), and to be less confiding (Dentler & Monroe, 1961). Zhang (1996) reports further that parental forgiveness, as opposed to a stigmatizing reaction to youthful misconduct, reduces delinquency.

Various theories of family functioning also maintain that parents who are responsive and accepting of their offspring generally produce children who are better adjusted and more socially competent (Ainsworth, Casper, Nozyce & Cunningham, 1990; Bowlby, 1969; Maccoby, 1992). Rothbaum & Weisz's (1994) meta-analyses of parental caregiving and child externalizing behavior lends empirical credence to this proposition. In a study of empirical reports from 1945 to 1992 of parent-child relationships and child behavior problems, Rothbaum & Weisz (1994) found that children of accepting, responsive parents experienced fewer externalizing problems. These findings were particularly strong for studies employing experimental designs.

Additional empirical evidence can be drawn from Loeber & Stouthamer-Loeber's (1986) comprehensive meta-analysis of family correlates of delinquency: factors indicating a lack of parental support clearly increase delinquent involvement. Loeber and Stouthamer-Loeber conclude that delinquency is related inversely to "child-parent involvement, such as the amount of intimate communication, confiding, sharing of activities, and seeking help" (1986, p. 42). Similarly, their analysis indicates that measures of parental rejection of children, such as "rejection, not warm, lack of love, lack of affection, less affectionate," were "consistently related to delinquency and aggression" (1986, p. 54; also see Sampson & Laub, 1993, p. 119). These 'support' elements, moreover, were among the most powerful family factors related to delinquency; their effects exceeded those of parental criminality, marital discord, parental absence, parental health, and family size (1993, p. 120–23).

Potential Effects of Parental Supports

Based on previous research, there are three interrelated reasons why parental support should reduce delinquency: First, support contributes to healthy human development. Research shows that early supportive relationships are central to the moral, psychological, and behavioral development of young children, the absence of which can have diverse and long-lasting behavioral, emotional, and cognitive effects (Harlow, 1958; Hartup, 1986; Radke-Yarrow & Sherman, 1990).[4]

Second, direct parental controls should be more effective when applied within the context of a supportive relationship between the parents and children. Support theorists stipulate that the effectiveness of direct parental controls is conditioned by contemporaneous levels of parental supports. Maccoby's insights seem particularly applicable:

> If the parent-child relationship is close and affectionate, parents can exercise what control is needed without having to apply heavy disciplinary pressure. It is as if parents' responsiveness, affection, and obvious commitment to their children's welfare gave them the right to make demands and exercise control (1980, p. 394).

Consistent with Braithwaite's (1989) work on shaming, supportive parents are more effective at exercising control. Nurturing parents have been found to be better able to reduce aggression in children (Sears, Whiting, Nowlis & Sears, 1953), to report that punishment is more effective (Eron, Walder & Lefkowitz, 1971; Sears, Maccoby & Levin, 1957), and to supervise more frequently (McCord, 1982). Moreover, Patterson (1980, 1982), argues that the effects of direct controls are likely to be varied and partially determined by the context in which parents control and punish their kids. Parental support thus seems to be a necessary precursor to effective control, and may function to reduce the need for much control (Cullen, 1994; Hay, 1994).

Third, support can mitigate the effects of criminogenic risks. Well known to support researchers is the inoculating and insulating effects produced by support. Thus, parental supports may function to mitigate the effects of certain criminogenic risk factors by (1) reducing the absolute number of risks, (2) providing social and psychological resources that adequately minimize risks and, (3) support early in life may promote individual resiliency (Cohen & Wills, 1985; Dunst, Trivette & Cross, 1986; Thoits, 1995).

RESEARCH STRATEGY

The existing empirical and theoretical research identifies family risk factors and direct controls as potentially major sources of delinquent involvement. To demonstrate the causal salience of support, it is necessary to determine if parental support will have significant effects when entered into an analysis that also includes direct controls and the risk factors. We undertake this analysis and hypothesize that independent of competing family factors and demographic controls, parental support will be inversely related to delinquency.

Our discussion above also suggests that social support should condition the effects of direct controls and the risk factors. We thus hypothesize that parental support will increase the effectiveness of control and mitigate the criminogenic effects of risk factors.

Further, in line with the preceding discussion, we attempt to explore why family structural risk factors may be related to delinquency. We have suggested that structural factors affect youths' behavior by shaping the quality of parent child relations. Theoretically, it is possible for structural disadvantage to shape delinquency by reducing social support, increasing relational risks, and/or by attenuating direct controls. We test this issue by developing a path model. In line with the emerging social support paradigm (Cullen, 1994), we hypothesize that family structural risk factors influence youths by diminishing parental support.

We must note one limitation to our analysis: The data set did not measure youths' attachment to parents. This omission is unfortunate, because attachment is perhaps the most prominent of Hirschi's (1969) social bonds and is most often seen as a measure of 'informal social control' – a concept our study is unable to assess. Although firm conclusions await systematic research, it is unclear whether including attachment in the analysis would have altered the results reported here. Reviews of research report that attachment tends to reduce delinquency, but this relationship is often weak and inconsistent across studies (Akers, 1994, 199; Mawhorr, 1992). Further, we would contend that parental attachment is best conceived as a byproduct of support; that is, it is the provision of supports to children that causes them to become attached to parents. In this reasoning, supports become the mechanism through which attachment – and thus informal social control – is initially established and then sustained.[5]

METHODS

Sample

The 1988 National Survey of Families and Households (hereinafter NSFH) serves as the data source for this project. The NSFH is a cross-sectional, multi-stage area probability sample of 13,079 individuals within 9,643 households. Data were collected through a structured interview of a randomly chosen parent, and if applicable, a secondary respondent (usually the other parent). Minorities and single-parent homes were double sampled to allow for comparison. The individual is the unit of analysis (Sweet, Bumpass & Call, 1988).

This study is based on an analytic subsample of 'focal children'. Within families a randomly chosen 'focal child' was selected by the interview staff. The parents were asked more detailed questions about this child than about other children in the family. Further, because questions concerning the behavior of the child depended on the age of the 'focal child', we restricted the age range of youths eligible for inclusion in the analytic sample from 12 to 18 years of age. The age restriction was imposed because prior research has identified the adolescent years as the most delinquent and because the delinquency measures found in the NSFH focus on activities of school-age adolescents. Data are weighted to account for disproportionate selection probabilities and are aligned with the age, race, and sex composition of the United States population in 1987. The sample remains a nationally representative cross-sectional probability sample of 1,775 focal children and their parents. The characteristics of the sample are shown in Table 1.

Dependent Variable

Parents were asked whether or not they were aware of a number of potential problem behaviors their adolescent children may have displayed. If the parents observed the behavior by their child, the variable received a score of 1; if unobserved, the variable was scored 0.

Parental reports are not typically used by criminologists. As with other measurement strategies, parental reports are not without limitations (parents, for example, may not be aware of certain instances of misbehavior, or may be unwilling to admit their child's delinquency to interviewers). Even so, research into the validity and reliability of parents as informants has found that parents reports of child conduct problems are (1) consistent (rho = 0.75) with child self-reports of conduct problems, (2) comparable to teacher reports (rho = 0.74) (Loeber, Green, Lahey & Stouthamer-Loeber, 1991), and (3) stable over time

Table 1. Descriptive Information for NSFH Analytic Sample (N = 1775)

	Mean	S.D
Age of Focal Child	15.2	1.98
Gender of Focal Child (Percent Male)	47.6	0.49
Race (Percent Non-White)	20.9	0.34
Marital Status (Percent Currently Intact)	82.2	0.38
Parental Conflict	9.7	6.35
Parental Deviance	0.04	0.21
Parental Rejection	2.01	1.03
Family Mobility	.66	1.25
Child's Temper	1.78	0.73
Child-Parent Conflict	4.69	2.77
Direct Parental Controls	2.96	1.59
Parental Support	0.01	6.29
Annual Household Income	34,858.24	34,906.50

(Edelbrock, Costello, Dulcan, Kalas & Conover, 1985). Moreover, parental informants have been used in past research on crime, delinquency, and family relationships (Jouriles, Barling & O'Leary, 1987; Palamara, Cullen & Gersten, 1986; Simcha-Fagan, Langer, Gersten & Eisenberg, 1975; Thomson, Hanson & McLanahan, 1994).

The variables that compose the delinquency measure represent several domains of behavior. Two variables tap into problem behaviors at school: parents being asked to meet with school officials because of the child's behavior, and whether the child has ever been suspended. Two variables measure officially recognized delinquency: the child ever being in trouble with the police, and whether the child has ever seen a doctor or therapist because of emotional or behavioral problems. Two other variables account for the presence or absence of alcohol or drug problems experienced by the child. Finally, one variable – 'mean to others' – measures whether the child has bullied, been cruel to, or mean to others. The individual items composing the delinquency measure were summed across all variables. The resulting index has a range of 0 to 9, and a reliability of a = 0.60. Moreover, factor analysis shows the index forms a unitary factor.[7]

Parental Support Scale

Borrowing from the work of Lin, Dean & Ensel (1986), Vaux (1988), and Barnes & Farrell (1992), we conceptualized parental support as parental behaviors toward the child that provide love and acceptance, and that demonstrate a willingness by parents to sacrifice for their child.

To measure parental support we developed an eleven-item scale that measures the degree of support parents provide to their child. Two items measure the degree in which parents provide expressive forms of support – that is, provide affective displays of support. Parents were asked how often they: (1) praise and (2) hug their child. Scores vary along a four-point Likert scale from 1 = never to 4 = very often. Four measures of supportive time parents spend with their children, such as leisure time, working and playing with each other, having private talks, and assisting in homework were also included. These variables were scored on a six-point Likert scale from 0 = never or rarely to 5 = almost every day. Two other measures of support were included that measure how many days in the last week they had eaten: (1) breakfast, and (2) dinner with at least one of their children.[8] Scores ranged from 0, representing no days, to seven, representing every day, for both breakfast and dinner. Finally, two items were also included that gauge the amount of 'especially enjoyable time' parents have spent with the child in the past month.

These items are standard support measures that have been employed numerous times in past research (Barnes et al., 1986; Barnes & Farrell, 1992; Hepburn, 1977; Johnson et al., 1995; Rollins & Thomas, 1979; Thomson et al., 1994; Thomson, McLanahan & Curtin, 1992). The resulting parental support scale was computed through standardizing all support variables and then summing the scores for each variable. The psychometric properties of the scale show it to be quite reliable (a = 0.88) and unitary.

Family Risk Measures

For our purposes we distinguish between family relational risk factors and family structural risk factors. Relational risks include variables that assess how parents: (1) act toward their children, (2) react toward their children, and (3) communicate, neglect, or invest in their offspring. Structural risks, on the other hand, are conceptualized as family variables that influence the family's position within the larger social structure. Taken together, the following risk factors represent most of the best known family correlates to crime (Loeber & Stouthamer-Loeber, 1986; Wright & Wright, 1994).

Relational Risks

Parental Conflict. To measure parental conflict, we included a series of questions that ask respondents about how often they argue about household tasks, money, spending time together, sex, having another child, in-laws, and the children. A seven-item additive scale was used with a response set ranging from 1 = never to 6 = almost every day (a = 0.74).

Child-Parent Conflict. Child-parent conflict was measured through two items. First, the main parental respondent was asked how often, in the past month, he or she had argued or had a lot of difficulty dealing with the focal child. Responses were scored on a six-point Likert scale ranging from 1 = never to 6 = almost every day. Second, respondents were asked how often, in the last 30 days, their husband, wife, or partner argued with or had a difficult time dealing with the focal child. Responses were scored in the same fashion, with the two items summed forming the final scale (a = 0.74).

Parental Deviance. We include a binary two-item measure of parental deviance. Both parents were asked who in the home had: (1) a problem drinking too much alcohol, and (2) a problem with drug use. If either of the respondents answered in the affirmative the parents received a score of 1, the absence of any report received a score of 0.

Parental Role Rejection. Recent reviews of families and crime found that parents' rejection of their child is a consistent and strong predictor of juvenile delinquency (Loeber & Stouthamer-Loeber, 1986; Wright & Wright, 1994). Although we do not have this variable in our data set, we were able to include a measure of parental role rejection. Although conceptually different from parental rejection, there appears to be a degree of overlap between the constructs. Parents who reject their roles as parents are likely to display more avoidance behaviors towards their children – that is, to marginalize their relationship with their children.

We measure parental role rejection through the use of a single-item measure. Using a five-point Likert scale, respondents were asked to what degree they agreed/disagreed with the following statement: "I often wish I could be free from the responsibility of being a parent." High scores on this variable represent greater role rejection.

Childs's Temperament. Children who lack self-control (impulsivity) and demonstrate little understanding or caring for others (empathy) have been shown to tax parenting resources, to engender conflict, and to limit the degree to which parents supervise, discipline, and presumably provide support (Patterson, 1980, 1982). Due to the child's difficult temperament, argue Snyder & Patterson (1987, 237) "the child is both a victim and an architect of his own environment," implying that tempermental children affect the quality of relationships within families.

To capture possible 'child effects', we include a single-item measure of the child's temperament. Parents were asked if the focal child loses his or her temper easily. The three-point response set includes 'often true', 'sometimes true', and 'not true'. The higher the score the more temperamental the child.

Family Structural Risk Factors

We include four variables that gauge the presence and extent of structural risks experienced by families. *Household income* is a composite measure of all economic resources held by the family (for example, wages, savings, bonds, governmental assistance) and represents the total amount of monetary resources available to the family. *Residential mobility* is measured as the number of times the family has relocated in the past five years. *Family size* is measured as the total number of children living within the household. Lastly, we include a dummy variable that assess whether the family is *intact* – that is, includes a mother and father – or not.

Direct Parental Control Scale

To conceptualize the direct control construct, we followed the lead of Rollins & Thomas (1979) and Barnes & Farrell (1992). Accordingly, direct controls are conceptualized as "parental behaviors toward the child which are intended to direct the child's behavior in a manner acceptable to the parent" (Barnes & Farrell, 1992, 764).

The direct control construct was measured through six-items. One item taps the extent to which focal children must inform their parents of their whereabouts. Parents were asked: "When (child) is away from home, is he or she supposed to let you know where his or she is?" Responses were measured through a four-point Likert scale ranging from 1 = hardly ever to 4 = all the time.

Two other questions were asked concerning time restrictions parents placed on their offspring. Respondents were asked: (1) by what time their child was to be home on school nights, and (2) by what time their child was to be home of Friday and Saturday nights. To reflect the age-graded nature of parental curfews, we computed the mean curfew time for each age and then ascertained whether the child's curfew was above or below the average. Scores at or below the average were scored 1 to reflect greater parental control.

Following the work of Snyder & Patterson (1987), Raine (1993), and Gottfredson & Hirschi (1990), we also included three measures detailing the extent to which parents regulate the completion of homework, chores, and the watching of television. First, respondents were asked if the child must do homework prior to watching television, playing, or going out after school. Second, parents were asked if they restrict the amount of television their child watches. Finally, parents indicated whether or not their child was required to complete chores before playing, watching, television or going out after school. Responses were coded 1 = no, 2 = sometimes, 3 = yes. The direct control scale was computed by standardizing all variables and then summing the values for all variables (a = 0.55).

RESULTS

Main and Interaction Effects

Table 2 shows the results of the regression equations predicting delinquency. Model 1, the main effects model, largely conforms to theoretical expectations. Eight of the nine family risk variables are positively and significantly related to delinquency. The largest effect is generated from the variable measuring the child's temper, followed by the effect produced by having an intact marriage.

The direct parental control scale, however, produced a significant and positive result. Further examination of the form of the relationship revealed that the direct control construct approached a curvilinear function. Including higher order terms within the models, however, failed to detect significant curvilinear effects. Thus higher levels of direct controls appear linearly related to delinquency.

Importantly, for the perspective advanced here, the parental support scale has a strong independent effect on delinquency, even though measures for family risks, direct parental controls, and family demographics are included in the model. The results indicate that, independent of other variables, parental supports are inversely and significantly related to delinquency. Moreover, the magnitude of the standardized partial coefficient is comparable to those produced by the majority of the family risk measures.

Model 2 tests for the effects of a support-by-risk interaction and therefore conforms to a slightly different standard. To simplify the analysis of interaction effects, we computed a risk scale by first standardizing all risk variables and then summing the variables to form single scale.

Tests for interactions are potentially problematic. Multiplicative terms are generally accompanied by unacceptable levels of multicollinearity with their shared covariates. To alleviate collinearity problems, we centered each main effect term prior to forming the multiplicative term (Tate, 1984). The results of the interactions are shown in Model 2. The interaction term measuring the effect of parental supports on risks did reach statistical significance, as did the three main effects. These results imply that the effects of elevated risks can be offset by increasing the number of parental supports.

Table 2. Standardized OLS Coefficients for the Regression of Child Delinquency on Family Risks, Direct Parental Controls, Parental Supports, and Interaction Terms

	Model	Model	Model
Variables	1	2	3
Parent Conflict	0.03 (.008)		
Parent Deviance	0.09 (.166)*		
Parent Rejection	0.08 (.033)*		
Family Mobility	0.10 (.027)*		
Household Size	0.06 (.026)*		
Child's Temper	0.26 (.064)*		
Child/Parent Conflict	0.19 (.014)*		
Nonintact Marriage	0.19 (.141)*		
Household Income	0.00 (.000)		
Direct Controls	0.05 (.025)*	0.07 (.026)*	0.05 (.024)*
Parental Support	–0.10 (.013)*	–0.19 (.012)*	–0.14 (.012)*
Standardized Risk		0.33 (.012)*	0.32 (.024)*
Support X Risk		–0.07 (.010)*	
Support X Direct Control			–0.04 (.027)*
Age of Child	0.03 (.02)	–0.01 (.021)	0.02 (.019)
Race of Child	–0.07 (.107)*	–0.05 (.107)*	–0.06 (.093)*
Gender of Child	–0.10 (.065)*	–0.10 (.069)*	–0.10 (.063)*
R2	0.25	0.16	0.15

*p < 0.05 (one-tailed tests)
Numbers in parentheses are standard errors

To assess further the direction of the interaction, we followed the suggestion of Stolzenberg (1978) and took the first order partial derivative of the dependant variable for each multiplicative and additive term and ascertained the conditional effects of the variables of interest. Thus we calculated the point estimate for the conditional effects of risk on delinquency controlling for levels of parental support.

The conditional effect of risk on delinquency, absent parental support, was found to equal 0.34. When support was set to its mean, the conditional effect of risk on delinquency equaled 0.19. And when support was set at its maximum value, the conditional effect of risk was reduced to 0.07. Increasing parental supports in light of risk appears to produce monotonic decreases in the conditional probabilities of delinquency.

To clarify further the effect parental support has on risk and its relationship to delinquency, we also calculated the conditional probabilities of risk and

Table 3 Conditional Probabilities of Risk on Delinquency Based on Varying Levels of Parental Supports

	Delinquency		
	High Risk %	Medium Risk %	Low Risk %
Parental Support			
High	0.52	0.39	0.20
Medium	0.54	0.44	0.23
Low	0.70	0.53	0.29
Prevalence	0.68	0.46	0.25
% Sample	10	51	39

delinquency controlling for variations in the numbers of supports. The results, shown in Table 3, depict two fairly clear patterns. First, the differences in the prevalence of delinquency across levels of risk are substantial, ranging from 25 percent in the low-risk group to 70 percent in the high-risk group. The second noticeable feature, and the one we are most concerned with, is found between the contrasting of various levels of support on levels of risk. Keeping the level of risk constant, variations in the levels of support are constantly – that is, across all groups – associated with decreases in the conditional probabilities of delinquency. Thus it appears that there is substantial evidence in favor of our hypothesis that parental supports interact with risk to reduce delinquency.

Referring again to Table 2, model 3 provides the results of the multiplicative estimates of the control-by-support effects on delinquency. The results show that the interaction is statistically significant but relatively small in magnitude. To examine other conditional effects of support on control we also factor analyzed the individual items that composed both scales. The analysis extracted two factors with eigenvalues above 1.0 and after rotation the individual items appeared to load on distinct factors, although substantial overlap was noted. An examination of the scree plot, however, indicated that the largest break between eigenvalues occurred between the first and second factor, indicating that a single factor could account for the structure of the data. We included the factor variable into the regression equation. The results of that analysis again showed the latent construct of support and control reduced delinquency (beta = –0.07).

Structural Determinants of Support

The final phase of this analysis employed full information maximum likelihood (FIML) covariance structural modeling to assess the effects of social structural determinants on parental functioning and delinquency. We estimated the equations using AMOS, a program designed to test and assess variance-covariance models. The program is similar to the well-known LISREL 7 package.

The model includes four exogenous variables that measure social structural risk features of families and three family functioning endogenous variables that are linked to the structural elements. The exogenous variables measure the family's mobility and size, the household income of the family and whether or not the family is represented by an intact marriage. Endogenous variables include the parental support scale, direct control scale, and a scale that measures family relational risk factors – that is, child-to-parent conflict, the child's temper, parental deviance, and parental rejection.

The model assumes that social structural features occur causally prior to family management variables and that structural risks have both direct effects on delinquency and indirect effects on delinquency through family process variables. Moreover, we also included a path from the direct control scale to the parental support measure to allow direct controls to affect delinquency indirectly through parental supports. We should also note that the effects of gender, race, and age are held constant throughout the analyses but are not presented because they add to the clutter of the diagram. The results from our theoretical model are shown in Figure 1.

In accord with our theoretical expectations, social structural risk variables were found to covary and influence delinquency directly and indirectly through family process variables. Most notably, the majority of structural risks correlate with single-parent status. Households headed by a single parent face significantly more relocations and substantially reduced income levels. Moreover, three of the structural variables had direct effects on delinquency: household size, mobility, and single-parent status.

Contrary to control theory expectations, however, the results show that structural risks substantially limit the degree to which parents deliver support, not control, to their offspring. Single-parent households, family mobility, and household income were not related to levels of direct parental controls. Conversely, only household size significantly predicted levels of direct controls. In light of the available literature, it is noteworthy that single parents and intact households report similar levels of direct controls.

The prime variable determining levels of parental support was single-parent status, although household income and family size were significantly related. Although non-intact families are associated with similar levels of direct controls, and face reduced relational risks compared to intact families, they are much less likely to provide comparable levels of parental supports. In turn, parental supports appear to be the mechanism through which delinquency is reduced.

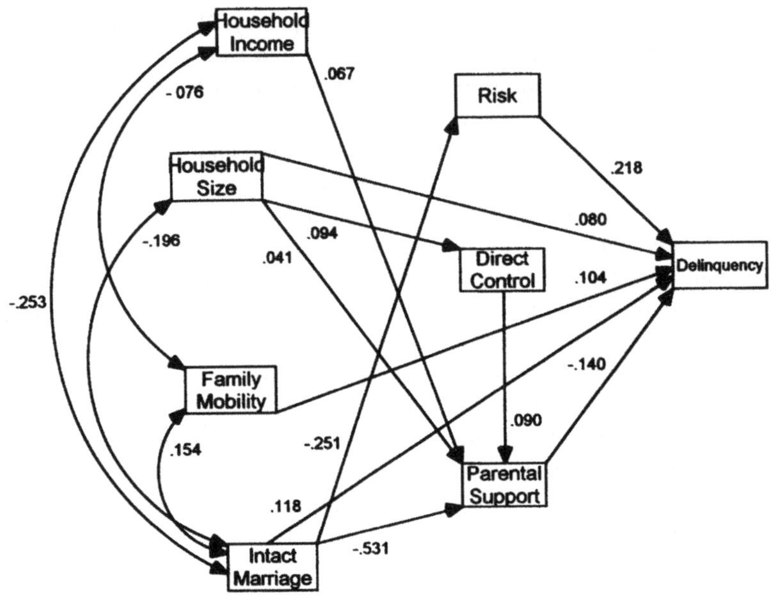

Figure 1. Path Model of Structural Effects on Family Process Variables.

The effect of direct parental controls is completely accounted for in this analysis. However, the path from direct controls to parental supports is significant, indicating that direct controls function through parental supports to decrease indirectly adolescent delinquency.

DISCUSSION

Families play an intricate and important role in delinquency. Until now though, most of the criminological treatment of families has centered around the various mechanisms of parental controls (Gottfredson & Hirschi, 1990; Sampson & Laub, 1993) or in linking risk to family functioning.

Various control models highlight the importance of 'doing to kids' – that is, punishing, monitoring, supervising, or setting rules – as opposed to 'doing for kids'. In contrast, we have argued that parental supports, or 'doing for kids', has remained an unexamined yet important variable in the etiology of delinquency.

Our analysis again highlights the inconsistent effects produced by direct parental controls. However, broadening the theoretical lens by which families are viewed to include parental supports resulted in a clearer specification of the effects and form of the direct control relationship to delinquency. In support of Maccoby's (1980) position, these data show that direct parental controls are effective when coupled with parental supports. In other words, parents who are supportive – that is, who share time in their child's life – are better able to exercise control. Conversely, parents removed from the daily happenings of their children are less successful in their control attempts and are likely to find that increasing control-related demands spurs greater misbehavior, not less.

These data also indicate that parental supports serve a salient ameliorative function in reducing the deleterious effects of risk. Supportive parents can mitigate to some extent the potent detrimental impact associated with risk variables. Moreover, the buffering effects of parental supports were found to be largely contingent on the overall level of risk: Children facing the highest level of risk were found to benefit the most from parental supports. It is telling that research on effective early and correctional interventions has also documented that those most at-risk are more likely to benefit from intervention (Trickett, Apfel, Rosenbaum & Zigler, 1982; Yoshikawa, 1994; Zigler & Hall, 1987).

Lastly, evidence garnered through our path model shows that structural risk variables reduce levels of parental supports, not controls as is generally hypothesized (Gottfredson & Hirschi, 1994). Although we could not fully account for the effects of structural risks, it is instructive that the variable most closely associated with contemporaneous reductions in parental supports was that of having a non-intact marriage.

These data show that children of single parents face an array of overlapping risk factors that predispose them toward a delinquent outcome. More importantly, however, single parents appear less able to offset the criminogenic effects associated with risk variables through increasing parental supports. Perhaps due to the dilution of parental resources in time and energy, single parents appear less capable of delivering support to their children. Similarly, the addition of another parent within the household is likely to increase the size of the support network available to children. Currently, we are not aware of any study that examines this issue.

Few perspectives are able to predict the intervening functions of key variables. Combined, these results indicate that parental supports produce substantive reductions in delinquency through multiple channels – that is, directly, by increasing the effects of direct controls and by reducing the effects of family risk variables. The consistency of the effects of parental supports clearly enhances our contention that including parental supports as a salient theoretical construct in delinquency theory is likely a profitable investment.

NOTES

1. Excellent summaries of support theory can be found in Vaux (1988) and Lin, Dean & Ensel (1986). The key element of support theory, at least for our purposes, is that it involves an active process of giving in time, emotion, and material.

2. Human and social capital perspectives emphasize informal social controls, such as attachment. Support theory, however, specifies the conditions by which informal controls are developed. Attachment then, according to support theory, is an outcome variable. Future research might benefit from testing this hypothesis.

3. Control theorists will argue that certain involvement variables, like parental rejection, matter only to the extent that they effect the supervision and subsequent punishment of the child (Gottfredson & Hirschi, 1990). However, this argument is misguided for two reasons: First, strong affective relationships have been shown to reduce the need for much control, thereby reducing its potential for behavioral modification (Hay, 1994). Second, the failure to distinguish supervision from involvement often distorts the fact that parents more involved in their children's life are also more likely to supervise and monitor their children, and are less rejecting and more attached.

4. Children are likely to suffer a number of emotional and behavioral handicaps across their life-span if not accorded appropriate levels of support early in life. Thus support theory directly addresses the issue of stability.

5. Our initial analysis of another national data set, which contains measures of support and attachment, confirms this causal link. In any case, it is noteworthy that much contemporary thinking about crime – popular and academic, including Hirschi in his move from social bond to self-control theory – emphasizes the importance of 'doing to kids' as opposed to 'doing for kids'.

6. We also note that these items measure observable behaviors that parents are likely to have knowledge of. Similarly, many of the dependant items were used in the reliability studies of Loeber et al., (1991). We suggest any bias to the use of parental reports in studying delinquency rests on the issue of the ease of collecting only child self-reports.

7. The moderate reliability is likely a function of the limited number of items composing the delinquency measure. Factor analysis, however, shows the scale forms a unitary factor. This is in line with much research on the generality of deviance and bolsters confidence in the validity of our dependant measure.

8. We see the meals item as a measure of parental support as opposed to a measure of direct control. First, involvement in planing meals and ensuring that kids attend them indicates a degree of parental investment. Second, family meals provide opportunities for parents and children to share sentiments and information. Finally, a factor analysis indicates that the meals items load on the support construct and not on the direct control construct.

REFERENCES

Agnew, R. (1992). Foundation for a General Strain Theory of Crime and Delinquency. *Criminology*, *30*, 47–87.

Akers, R. L. (1994). *Criminological Theories: Introduction and Evaluation*. Los Angeles: Roxbury.

Alexander, J. R. (1973). Defensive and Supportive Communications in Normal and Deviant Families. *Journal of Consulting and Clinical Psychology*, *40*, 223–31.

Anderson, E. (1990). *Street Wise: Race, Class, and Change in an Urban Community*. Chicago: University of Chicago Press.

Andry, R. G. (1960). *Delinquency and Parental Pathology*. London: Methuen.

Barnes, G. M. (1984). Adolescent Alcohol Abuse and Other Problem Behaviors: Their Relationships and Common Parental Influences. *Journal of Youth and Adolescence*, *13*, 329–48.

Barnes, G. M., & Farrell, M. P. (1992). Parental Support and Control as Predictors of Adolescent Drinking, Delinquency, and Related Problem Behaviors. *Journal of Marriage and the Family*, *54*, 763–776.

Braithwaite, J. (1989). *Crime, Shame and Reintegration*. Cambridge,UK: Cambridge University Press.

Cernkovich, S. A. & Giordano, P. C. (1987). Family Relationships and Delinquency. *Criminology*, *25*, 295–321.

Cohen, S., & Wills, T. A. (1985). Stress, Social Support, and the Buffering Hypothesis. *Psychological Bulletin*, *98*, 321–57.

Coleman, J. S. (1990). *Foundations of Social Theory*. Cambridge, Mass: Harvard University Press.

Cullen, F. T. (1994). Social Support as an Organizing Concept for Criminology: Presidential Address to the Academy of Criminal Justice Sciences. *Justice Quarterly*, *11*, 527–59.

Currie, E. (1985). *Confronting Crime: An American Challenge*. New York: Pantheon Books.

Dunst, C. J., & Trivette, C. M. (1988). Determinants of Parent and Child Interactive Behavior. In: K. Marfo (Ed.), *Parent-Child Interaction and Developmental Disability: Theory, Research and Intervention*, 3–31. New York: Praeger.

Dunst, C. J., & Trivette, C. M. (1990). Assessment of Social Support in Early Intervention Programs. In: S. J. Meisels and J. P. Shonkoff (Eds), *Handbook of Early Childhood Intervention*, (326–49). Cambridge, UK: Cambridge University Press.

Dunst, C. J., Trivette, C. M., & Cross, A. H. (1986). Mediating Influences of Social Support: Personal, Family, and Child Outcomes. *American Journal of Mental Deficiency*, *90*, 403–17.

Edelbrock, C. S., Costello, A. J., Dulcan, N. K., Kalas, R., & Conover, N. C. (1985). Age Differences in the Reliability of Psychiatric Interview of the Child. *Child Development*, *56*, 265–75.

Eron, L. D., Walder, L. O., & Lefkowitz, M. M. (1971). *Learning of Aggression in Children*. Boston: Little Brown.

Farrington, D. P. (1979). Longitudinal Research on Crime and Delinquency. In: N. Morris, & M. Tonry (Eds), *Crime and Justice: An Annual Review of Research* (Vol. 1). Chicago: Cambridge University Press.

Farrington, D. P. (1986). Stepping Stones to Adult Criminal Careers. In: D. Olweus, J. Block, & M. Radke-Yarrow (Eds), *Development of Antisocial and Prosocial Behavior*. New York: Academic Press.

Gottfredson, M. R., & Hirschi, T. (1990). *A General Theory of Crime*. California: Stanford University Press.

Glueck, S., & Glueck, E. (1950). *Unraveling Juvenile Delinquency*. Cambridge, MA: Harvard University Press.

Glueck, S., & Glueck, .E. (1968). *Delinquents and Nondelinquents in Perspective*. Cambridge, MA: Harvard University Press.

Hagan, J. (1989). *Structural Criminology*. New Brunswick: Rutgers University Press.

Hagan, J. (1993). Structural and Cultural Disinvestment and the New Ethnographies of Poverty and Crime. *Contemporary Sociology*, *22*, 327–32.

Harlow, H. F. (1958). The Nature of Love. *The American Psychologist*, *13*, 673–685.

Hartup, W. W. (1986). On Relationships and Development. In: W.W. Hartup & Z. Rubin (Eds), *Relationships and Development*, (1–26). New Jersey: Lawrence Erlbaum Associates.

Hepburn, J. M. (1977). Testing Alternative Models of Delinquency Causation. *Journal of Criminal Law and Criminology*, *67*, 450–60.

Hirschi, T. (1969). *Causes of Delinquency*. Berkeley: University of California Press.

Hirschi, T., & Gottfredson, M. R. (Eds). (1994). *The Generality of Deviance*. NJ: Transaction Publishers.

Jennings, K. D., Stagg, V., & Conners, R. E. (1991). Social Networks and Mothers Interactions with their Preschool Children. *Child Development*, *62*, 966–78.

Jessor, R. (1993). Successful Adolescent Development Among Youth in High-Risk Settings. *American Psychologist*, *48*, 117–26.

Johnson, D. L., & Breckenridge, J. N. (1982). The Houston Parent Child Development Center and the Primary Prevention of Behavior Problems in Young Children. *American Journal Of Community Psychology*, *10*, 305–316.

Johnson, R. A., Su, S. S., Gerstein, D. R., Shin, H., & Hoffmann, J. P. (1995). Parental Influences on Deviant Behavior in Early Adolescence: A Logistic Response Analysis of Age and Gender Differentiated Effects. *Journal of Quantitative Criminology*, *11*, 167–194.

Jouriles, E., Barling, J., & O'Leary. K. D. (1987). Predicting Child Behavior Problems in Maritally Violent Families. *Journal of Abnormal Child Psychology*, *15*, 165–73.

Kay, D. F. (1994). Prosocial Development. *Journal of Child Psychology and Psychiatry*, *35*, 29–74.

LeFlore, L. (1988). Delinquent Youths and Family. *Adolescence*, *23*, 629–42.

Loeber, R., & Dishion, T. (1983). Early Predictors of Male Delinquency, A Review. *Psychological Review*, *94*, 68–99.

Loeber, R., Green, S. M., Layey, B. B., & Stouthamer-Loeber, M. (1991). Differences and Similarities between Children,Mothers, and Teachers as Informants on Disruptive Child Behavior. *Journal of Abnormal Child Psychology*, *19*, 75–85.

Loeber, R., & Stouthamer-Loeber, M. (1986). Family Factors as Correlates and Predictors of Juvenile Conduct Problems and Delinquency. In: M. Tonry, & N. Morris (Eds), *Crime and Justice: An Annual Review of Research*, 29–149. Chicago: University of Chicago Press.

Luthar, S. S. (1991). Vulnerability and Resilience: A Study of High-Risk Youth. *Child Development*, *62*, 600–16.

Luthar, S. S. (1993). Annotation: Methodological and Conceptual Issues in Research on Childhood Resilience. *Journal of Child Psychology and Psychiatry*, *34*, 441–53.

Maccoby, E. E. (1980). *Social Development: Psychological Growth and the Parent-Child Relationship*. New York: Harcourt Brace Jovanovich.

Matsueda, R. L., & Heimer, K. (1987). Race, Family Structure, and Delinquency: A Test of Differential Association and Social Control Theories. *American Sociological Review*, *52*, 826–40.

Mawhorr, T. L. (1992). *Unraveling the Attachment-Delinquency Link*. Unpublished doctoral dissertation, Bowling Green State University.

McCord, J. (1979). Some Child-Rearing Antecedents of Criminal Behavior in Adult Men. *Journal of Personality and Social Psychology*, *37*, 1477–86.

McCord, J. (1982). A Longitudinal Study of the Link between Broken Homes and Criminality. In: J. Gunn, & D. P. Farrington (Eds), *Abnormal Offenders, Delinquency and the Criminal Justice System*. London: Wiley.

McLanahan, S., & Booth, K. (1989). Mother-Only Families: Problems, Prospects, and Politics *Journal of Marriage and Family*, *51*, 557–80.

McLeod, J. D., &Shanahan, M J. (1993). Poverty,Parenting, and Childrens Mental Health. *American Sociological Review*, *58*, 351–66

McLoyd, V. C. (1990). The Impact of Economic Hardship on Black Families and Children: Psychological distress, Parenting, and Socioemotional Development. *Child Development, 61*, 311–46

Nagin, D., & Patternoster, R. (1994). Personal Capital and Social Control: The Deterrence Implications of a Theory of Individual Differences in Criminal Offending. *Criminology, 32*, 581–606.

Nan, L., Dean, A., & Ensel, W. (1986). *Social Support, Life Events, and Depression.* FL: Academic Press.

Palamara, F., Cullen, F. T., & Gersten, J. C. (1986). The Effect of Police and Mental Health Intervention on Juvenile Deviance: Specifying Contingencies in the Impact of Formal Reaction. *Journal of Health and Social Behavior, 27*, 90–105

Patterson, G. R. (1980). Children Who Steal. In: T. Hirschi, & M. Gottfredson (Eds), *Understanding Crime: Current History and Research*, (73–90). Beverly Hills, CA: Sage Publications.

Patterson, G. R. (1982). *Coercive Family Process.* Eugene, OR: Castalia.

Patterson, G. R. (1986). Performance Models for Antisocial Boys. *American Psychologist, 44*, 105–111.

Patterson, G. R., & Stouthamer-Loeber, M. (1984). The Correlation of Family Management Practices and Delinquency. *Child Development, 55*, 1299–1307.

Radke-Yarrow, M., & Zahn-Waxler, C. (1986). The Role of Familial Factors in the Development of Prosocial Behavior, Research Findings and Questions. In: D. Olweus, J. Block, & M. Radke-Yarrow (Eds), *Development of Antisocial and Prosocial Behavior.* New York: Academic Press

Radke-Yarrow, M., & Sherman, T. (1990). Hard Growing: Children Who Survive. In: L. N. Robins and M. Rutter (Eds), *Straight and Devious Pathways from Childhood to Adulthood.* Cambridge, UK: Cambridge University Press.

Raine, A. (1993). *The Psychopathology of Crime: Criminal Behavior as a Clinical Disorder.* California: Academic Press Inc.

Rankin, J. H., & Wells, L. E. (1990). The Effects of Parental Attachments and Direct Controls on Delinquency. *Journal of Research in Crime and Delinquency, 27*, 140–65.

Reid, J. B., & Hendriks, A. F. C. J. (1973). A Preliminary Analysis of the Effectiveness of Direct Home Intervention for Treatment of Pre-delinquent Boys Who Steal. In: L. A. Hamerlynck, L. C. Handy, & E. J. Mash (Eds), *Behavior Therapy: Methodology, Concepts and Practice.* Champaign, Il.: Research Press.

Robins, L. (1966). *Deviant Children Grown Up: A Sociological and Psychiatric Study of Sociopathic Personality.* New York: Russell Sage Foundation.

Rollins, B. C., & Thomas, D. L. (1979). Parental Support, Power and Control Techniques in the Socialization of Children. In: Burr, W. R., R. Hill, F. I. Nye & I. L. Reiss (Eds), *Contemporary Theories About the Family*, Vol. 1, 317–64. New York: The Free Press.

Rutter, M. (1979). Protective Factors in Childrens Responses to Stress and Disadvantage. In: M. W. Kent & J. E. Rolf (Eds), *Primary Prevention of Psychopathology:* Vol. 3, 49–74. Social Competence in Children. New Hampshire: University Press of New England.

Sameroff, A. J., & Fiese, B. H. (1990). Transactional Regulation and Early Intervention. In: S. J. Meisels, & J. P. Shonkoff (Eds), *Handbook of Early Childhood Intervention*, 119–49. Cambridge, UK: Cambridge University Press.

Sampson, R. J., & Groves, W. B. (1989). Community Structure and Crime: Testing Social-Disorganization Theory. *American Journal of Sociology, 94*, 774–802

Sampson, R. J., & Laub, J. H. (1993). *Crime in the Making: Pathways and Turning Points Through Life.* MA: Harvard University Press.

Sears, R., J. Whiting, V. Nowlis, & P. Sears. (1953). Some Child-Rearing Antecedents of Aggression and Dependency in Young Children. *Genetic Psychology Monographs, 47*, 135–236.

Sears, R., Maccoby, E., & Levin, H. (1957). *Patterns of Child Rearing*. IL: Row, Peterson.

Seydlitz, R. (1993). Complexity in the Relationships Among Direct and Indirect Parental Controls and Delinquency. *Youth and Society, 24*, 243–75.

Shaw, D. S., & Vondra, J. I. (1993). Chronic Family Adversity and Infant Attachment Security. *Journal of Child Psychology and Psychiatry, 34*, 1205–15.

Shaw, D. S., Vondra, J. I., Hommerding, K. D., Keenan, K., & Dunn, M. (1994). Chronic Family Adversity and Early Child Behavior Problems: A Longitudinal Study of Low Income Families. *Journal of Child Psychology and Psychiatry, 35*, 1109–22.

Simcha-Fagan, O., Langner, T. S., Gersten, J. C., & Eisenberg, J. G. (1975). Violent and Antisocial Behavior: A Longitudinal Study of Urban Youth. Report no. OCD-CB-480. Washington, D.C.: Office of Child Development.

Snyder, J., & Patterson, G. R. (1987). Family Interaction and Delinquent Behavior. In: H. C Quay (Ed.), *Handbook of Juvenile Delinquency*, (216–43). New York: Wiley.

Steinberg, L., Lamborn, S. D., Dornbusch, S. M., & Darling, N. (1992). Impact of Parenting Practices on Adolescent Achievement: Authoritative Parenting, School Involvement, and Encouragement to Succeed. *Child Development, 63*, 1266–81.

Stolzenberg, R. M. (1978). Bringing the Boss Back In: Employer Size, Employer Schooling, and Socioeconomic Achievement. *American Sociological Review, 43*, 813–28.

Stouthamer-Loeber, M., & Loeber, R. (1986). Boys Who Lie. *Journal of Abnormal Child Psychology, 14*, 551–64.

Sullivan, M. L. (1989). *Getting Paid: Youth Crime and Work in the Inner City*. Ithaca: Cornell University Press.

Sweet, J., Bumpass, L., & Call, V. (1988). *The Design and Content of the National Survey of Families and Households*. Madison: Center For Demography and Ecology University of Wisconsin.

Tate, R. (1984). Limitations of Centering for Interactive Models. *Sociological Methods and Research, 13*, 251–71.

Thoits, P. A. (1995). Stress, Coping, and Social Support Processes: Where Are We? What Next? *Journal of Health and Social Behavior*, extra, 53–79.

Thomson, E., Thomas, T. L., & McLanahan, S S. (1994). Economic Resources vs. Parental Behaviors. *Social Forces, 73*, 221–42.

Trickett, P. K., Apfel, N. H., Rosenbaum, L.K., & Zigler, E. F. (1982). A Five Year Follow-up of Participants in the Yale Child Welfare Research Program. In: E. F. Zigler and E. W. Gordon (Eds), *Day Care: Scientific and Social Policy Issues*, (200–222). Boston: Auburn House.

Vaux, A. (1988). *Social Support: Theory, Research, and Intervention*. New York: Praeger.

Wells, L. E., & Rankin, J. H. (1988). Direct Parental Controls and Delinquency. *Criminology, 26*, 263–85.

Wells, L. E., & Rankin, J. H. (1991). Families and Delinquency: A Meta-Analysis of the Impact of Broken Homes. *Social Problems, 38*, 71–93.

Wilson, J. Q., & Herrnstein, R. J. (1985). *Crime and Human Nature: The Definitive Study of the Causes of Crime*. New York: Simon and Schuster.

Wright, K. N., & Wright, K. E. (1994) *Family Life, Delinquency, and Crime: A Policymakers Guide*. Washington D.C.: Office of Juvenile Justice and Delinquency Prevention.

Yoshikawa, H. (1994). Prevention as Cumulative Protection: Effects of Early Family Support and Education on Chronic Delinquency and Its Risks. *Psychological Bulletin, 115*, 28–54.

Zahn-Waxler, C., & Radke-Yarrow, M. (1982). The Development of Altruism: Alternative Research Strategies. In: N. Eisenberg (Ed.), *The Development of Prosocial Behavior*, (109–37). New York: Academic Press.

Zigler, E., & Hall, N. (1987). The Implication of Early Intervention Efforts for the Primary Prevention of Juvenile Delinquency. In: J. Q. Wilson & G. C. Loury (Eds), *From Children to Citizens, Volume III: Schools, Families and Delinquency Prevention*, (154–185). New York: Springer-Verlag.

NO TIME FOR INNOCENCE, NO PLACE FOR INNOCENTS: CHILDREN'S EXPOSURE TO EXTREME VIOLENCE

Greer Litton Fox

The focus of this chapter is on children who are exposed to extreme, traumatic violence.

Such children are second-hand victims in the background of high social dramas who absorb lessons about human nature and interpersonal interaction unwittingly but powerfully taught by the principals to such actions. This chapter will look closely at how the lives of children intersect with horrific violence and on the roles that family caregivers play in responding to children who have been exposed to extreme violence. Assessing the impacts on children may eventually allow for a more complete tally of the costs of the quintessentially American patterns of social violence, patterns that some have begun to accept as normal.

I became interested in the exposure of children to extreme violence while conducting research on the impact of homicide on family members (Fox, 1991). As part of that research, I spent several months in the inner offices of a local law enforcement agency, poring through their old homicide files, reading police reports, statements from witnesses, notes from detective work, and transcripts of police interrogations. I began to notice that children were present during some of the murders, that some of the interrogated witnesses were children, and that children were killers and victims as well. Of the 39 murder files I

Families, Crime and Criminal Justice, Volume 2, pages 163–181.

ISBN: 0-7623-0737-4

reviewed, 23 per cent of the murders – close to 1 in 4, were committed in the presence of a child.

The following four cases from those records illustrate some of the ways children were involved. One was a drug deal gone bad in a living room. The police records described a chaotic scene involving a two-year-old girl being handed off from one adult to another. Within seconds, the first adult lay dying of a gunshot wound, while the other shoved the child out through an open porch window, all the time screaming at the toddler to 'Get down! Get down!' A second case grew out of a seemingly mundane transfer of children from one custodial parent to another. This daylight killing occurred on a Sunday afternoon in the parking lot of the housing complex where a divorced father was returning his three sons to their mother after they had spent the weekend with him. The three boys watched as their father became engaged in an altercation with a friend of their mother's new boyfriend. Very quickly the argument escalated, and the father was shot to death in front of his sons. A third case involved a teenaged girl who was at home watching Saturday night television with her brother, mother, and grandmother. That evening their drunken, gun-wielding uncle, who had repeatedly subjected the family to his violent attacks and ravings, broke into the house again. Her teenaged brother grabbed a shotgun from the pantry and shot the uncle to death. A final example involved two brothers, aged 10 and 12, who lived in a downtown motel and who were together as they watched from the balcony of their motel room a midnight robbery/murder take place in the street below. Transcripts of the boys' interrogations as material witnesses by police, which were held at police headquarters during the wee hours of the night when most school-aged boys are at home in bed, were part of the file on this homicide.

These four cases illustrate several of the factors that are important in assessing the impact of witnessed traumatic violence on children. The age and developmental stage of the child, with age-differentiated cognitive and emotional skills and competencies, clearly bear on the impact of the trauma. The circumstances of the trauma, including the child's relationship with the principal persons involved, where the child was during the episode, and whether others were present also and how they reacted all impinge upon a child's encoding of the trauma. Finally, how the violence fits into the usual circumstances of the child's life – whether it is one of many similar traumas or stressful conditions, or a rare and unusual event – may affect a child's response.

The remainder of the chapter is structured as follows. First, traumatic or extreme violence is defined. This is followed by a brief discussion of three different conceptualizations of exposure to extreme violence. Then research is reviewed on the impact on children who witness extreme violence, using the three different conceptualizations of violence exposure. Finally, the role of family members as moderators of the aftermath of witnessing extreme violence is examined, and class-linked patterns in familial response are described.

TRAUMATIC OR EXTREME VIOLENCE

Traumatic or extreme violence refers to physical assaults by one or more persons on one or more others either with or without weapons in which the victim or victims are seriously wounded or injured, experience great pain, or are killed. Serious wounds and injuries include broken bones and bodily traumas that are accompanied by profuse bleeding. Extreme physical pain is evidenced by the victim's or victims' screams or cries. These events are horrific, grotesque, macabre, the stuff of nightmares.

Traumatic violence may be sudden and unexpected, or routine and commonplace. It is distinguished by the extremity of the physical or psychic pain it causes for its victims. It is often accompanied by destruction of material objects in the immediate setting, such as equipment and features of the landscape in outdoor settings or furniture, windows and dishes in indoor settings. In addition to the destruction of objects, traumatic violence can also involve the destruction of non-material elements, such as the shattering of a child's sense of time, the erosion of trust, the rendering of the familiar into the unrecognizable.

The impact of extreme violence for a child is heightened when the traumatic event is visited upon himself or herself or a loved one or is caused by someone known or trusted. Similarly, the impact is magnified when the event occurs in a setting, such as a home, restaurant, church or school, in which the child expected to be safe and protected.

RESEARCH ORIENTATIONS TO EXTREME VIOLENCE

At least three different definitions or conceptualizations of the nature of a child's exposure to extreme violence have been used to assess both the numbers of children involved in witnessing traumatic violence and the impact on children of doing so. Some researchers, most often those who come from a clinical disciplinary orientation, have focused on children exposed to single events of traumatic violence. Others have focused on extreme violence as a condition of a child's daily existence, an existence punctuated by discrete traumatic episodes that occur against a backdrop of commonplace violence. A third approach distinguishes children's exposure to episodes of extreme violence based upon their proximity to the violence and its principals.

Exposure to Traumatic Violence as a Discrete Event

Viewing traumatic violence as a discrete event or episode is exemplified by Terr's (1991) differentiation between acute traumas that occur a single time, such as the murder of a parent or a sniper attack in a school yard, and chronic repeated exposure to traumas that can be anticipated, such as physical or sexual abuse over an extended period during childhood. Terr's second type, the recurring trauma whose recurrence can be anticipated, shares similarity with the conceptualization of children's exposure to extreme violence as a condition of the social environment beyond the home.

Exposure to Traumatic Violence as a Routine Feature of a Child's Social Ecology

This view of traumatic violence emphasizes violence as a commonplace, everyday occurrence – a social backdrop against which the other aspects of a child's life are played out. This approach grew out of the study of children in war-ravaged areas of the world and was also applied fruitfully to an analysis of children's development in American environments, most especially those in impoverished, crime-ridden areas of our greater and lesser American cities (Dubrow & Garbarino, 1989; Garbarino, Kostelny et al., 1991; Horn & Trickett, 1998). The oxymoron, 'mundane, extreme environment', was coined to capture the reality of life for minority children in inner cities for whom violence at home was exacerbated by violence on the streets, at school, and in the neighborhood, and compounded by the hostility and rejection of the dominant, mainstream culture (Pierce, 1979; Peters & Massey, 1983).

Children share in the vulnerabilities of their parents. Poor children, more so than other segments of the American population living below the official poverty line, are likely to live in neighborhoods that are predominantly poor, that is, where more than 40 per cent of all residents are also poor (O'Hare, 1996). Violence and poverty are closely associated, as a long tradition of research on social disorganization attests (Hsieh & Pugh, 1993), suggesting that poor children living in poor neighborhoods are disproportionately exposed to high levels of community violence. In response to concerns about the potential impact on large numbers of children of exposure to community violence, The National Institutes of Mental Health initiated an intramural research program addressed to this issue (Martinez & Richters, 1993; Richters, 1993; Richters & Martinez, 1993), which has provided important impetus to research on the contextual factors which constrain the lives and well being of children.

Distal Violence as Indirect Exposure to Extreme Violence

A third useful concept in describing children's exposure to extreme violence is distal violence, that is, the exposure to traumatic events that are not witnessed directly but that nonetheless occur to persons known to the child. Multiple killings of students by other students in public schools across the nation – including recent, highly publicized incidents in the states of Oregon, Colorado, Arkansas, Georgia, Michigan, and Kentucky – provide examples of the exposure of large numbers of school-aged children to distal violence. Distal violence should be distinguished from media portrayals of fictitious violence, such as cartoons and action videos, and from televised newscasts of violent events. Distal violence shares similarities both with traumatic violence as directly experienced or witnessed and with traumatic violence that derives from daily exposure to an environment characterized by violence.

Differentiating among the three approaches to violence exposure can lead to fruitful questions, such as whether violence exposure is dose-specific. If one exposure leads to X amount of trauma, does exposure to 2 episodes generate 2X units of trauma? Does exposure to multiple episodes lead to near incapacitation due to trauma, or is there habituation to extreme violence over time, as has been found among children exposed to media violence (Slaby, 1994)? Is direct exposure to traumatic violence more traumatic than exposure to distal violence? Are children from homes in which domestic violence is endemic less affected or more so by exposure to community violence? Are there coping responses that are neighborhood-specific, such that violence-habituated neighborhoods informally teach their residents, including children, highly specific ways of responding to commonplace acts of extreme terror? And are these coping strategies similar to those employed by residents of less volatile neighborhoods, who may be less experienced in coping with tragedy and terror, but more competent in other ways of problem management? Partial responses to several of these questions are reviewed in the next sections.

CHILDREN EXPOSED TO EVENTS OF EXTREME TRAUMA

Only a few scholar/clinicians have looked closely at the reactions and responses of children of different ages who were present during such events of extreme violence as the murder of their mother, the beating and rape of their mother, parental homicide/suicide, or the murder of siblings (Eth, Silverstein & Pynoos, 1985; Pynoos & Eth, 1984; Zeanah & Burk, 1984). Terr identifies characteristic outcomes common to childhood trauma, which are hallmarks of post-traumatic stress disorder (Terr, 1991, 1988). These include: memories that are strongly visualized, recurring, and intrusive; behavioral re-enactments of the trauma, often repeated with high frequency; trauma-specific fears, including avoidance of transparently trauma-specific cues or reminders, such as settings, objects, or behaviors; an altered, more constrained, world view; a shattered sense of invulnerability or invincibility leading to hyper-vigilance and hyper-alertness; and a pervasive sense of distrust and wariness.

Terr further differentiates between children exposed to a single, discrete traumatic event, such as a parental murder, from those exposed to recurring and long-lasting victimization, such as ongoing sexual abuse. In contrast to children who are caught up in or simply present during single-event traumas, Terr (1991) suggests that children who are involved in recurring trauma utilize denial, psychic numbing, dissociation and other strategies to remove themselves emotionally from their setting. She notes further that such children are characterized by rage, often self-directed, and manifest extreme passivity.

Children present during single-event extreme traumas show recollections of the event as full, detailed, etched-in memories. These have been described as 'flash-bulb' memories by Pynoos and his colleagues (1984, 1987) in their work with similarly victimized children. In their work with children of murdered parents, Pynoos & Eth (1984) found that children had absorbed the traumatic event through all of their senses (cf. Burman & Allen-Meares, 1994; Zeanah & Burk, 1984). Children could recount the sounds of the violence – the shouting and screams, the sound of fists and knives and gunshots, the sight and smell of blood, the feel of the floorboards under the bed where they lay hidden from view. Children could easily identity what for them was the 'worst moment' and recount it in detail (Pynoos & Eth, 1984).

Terr indicates that traumatized children are also given to constant, repetitive cognitive reappraisals of the event. They go over and over the event in their minds, sometimes playing out different roles or different endings. Pynoos & Eth (1984) describe a similar pattern, especially notable among their middle school aged clients. That is, they describe a continuous replaying of the event, but with different outcomes, in which the children cast themselves in different roles, constantly questioning why events transpired as they did and rearranging them so that they could prevent or rescue the victim of the trauma (see also Burman & Allen-Meares 1994).

A consistent observation in the body of work on children exposed to extreme violence is the importance of the child's caretaker following the traumatic event. That person's actions, reactions to the traumatic event, and responses to the child can moderate or amplify the negative impacts of trauma, most especially the traumatic anxiety – or terror – experienced by children during and following extreme episodes (Burman & Allen-Meares, 1994; Freeman, Shaffer & Smith, 1996; Rigamer, 1986). The importance of this person is underscored when we keep in mind that children who were present when a parent is murdered or otherwise incapacitated will not only experience the trauma directly but will also experience a separation from familiar persons and settings and disruption of daily routine, transitions which in and of themselves are known to be upsetting to children. Small wonder that children's capacities to cope are often vastly exceeded, such that they must rely upon other adults to help them process these events.

CHILDREN IN EVERYDAY WORLDS OF EXTREME TRAUMA

A separate area of research concerns the impact on children of daily exposure to violence that comes from living in communities in which violence is commonplace (Garbarino, Kostelny & Dubrow, 1991; Dubrow & Garbarino, 1989; Horn & Trickett, 1998; Limber & Nation, 1998). An exemplary study conducted by Martinez & Richters (1993) compared a small sample of younger (first and second graders) and older (fifth and sixth graders) children and their mothers in what is described as a moderately violent section of Washington, D.C. Among the lessons to be learned from their work are that it is possible to measure reliably with self-report measures the extent of exposure to violence, direct violent victimization, and emotional and psychic distress among children as young as first and second-graders. Second, the amount of agreement between children and parents about the level of distress among children varied by the age and gender of the child, but in general there was little agreement between parents and their children's responses. Parents of boys in both age groups and of the older girls consistently underestimated their children's levels of anxiety and depression. For example, 49% of parents of older children reported that their child never worried about being safe, whereas none of the children said they never worried (Martinez & Richters, 1993, 29).

If exposure to violence through direct victimization or witnessing the victimization of others has an impact on children's emotional distress, we would expect significant bivariate correlations between these measures. The results of Martinez & Richters (1993) on this issue are striking. When parental assessments were examined, there was no relationship between the child's exposure to violence and the parent's assessment of the child's emotional distress; but, when children were the reporters, the relationships were strong and significant between their exposure to violent victimization of themselves or someone they knew and their self-reported levels of distress and depression. In short, these researchers found that parents, by and large, missed the mark as far as knowing what was happening in their children's lives and in their heads and hearts.

Finally, in additional analyses, the researchers learned that the mother's educational level moderated the impact of violence on children's distress. Children of women with little education experienced greater distress when exposed to violence than did children with more highly educated mothers. Although they were unable to discern what it is about education that accounts for this effect, nonetheless the study is provocative. It is also a first step that demands replication in additional sites and with larger samples to allow for multivariate analyses that can uncover potential moderators and mediators of the relationships described here.

A different community study focused on mothers of children in a Chicago housing project (Garbarino, Dubrow, Kostelny & Pardo, 1992). Events such as fights and bullying, random shootings throughout the neighborhood, and the risk of violent exchanges that accompanied drug trafficking characterized and disrupted daily activities in homes in the neighborhood. The researchers found that mothers were quite cognizant of the dangers faced by their children and drew upon a variety of coping strategies in an attempt to increase their children's safety. This study is corroborated by the research of Bell & Jenkins (1993), who also emphasize the importance of actions of adults in securing the safety of children who live in mundane, extreme environments. Clearly we need to know more about how children and their parents manage to survive and thrive in daunting environments. Much of the research, perhaps unavoidably, confounds several demographic variables, including race and social class, making it difficult to single out critical factors (Limber & Nation, 1998). Most of the U.S. neighborhoods that have been studied have been in large inner city areas, among people of lower income and education, many of whom are of minority status. It will be important to expand this work with additional comparison groups of neighborhoods to uncover the role of culture, education, and ideology, for example, in the responses that are made to children in danger. The work of Sampson and colleagues that uncovers systematic patterns in the distribution of collective efficacy of neighborhoods on behalf of children is exemplary in this regard (Sampson, Raudenbush & Earls, 1998; Sampson, Morenoff & Earls, 1999). Their work suggests that the collective motivation and ability of neighbors to act for children's wellbeing are distributed along the familiar structural axes of race and class that shape the spatial and social contours of advantage and disadvantage in this society.

DISTAL VIOLENCE AND CHILDREN

What evidence is there of the impact of distal violence on children, that is, violence that children are exposed to but do not actually participate in as witnesses or victims? Here I draw on my own work and that of my students on children's exposure to distal violence. In addition to discovering how children reacted to episodes of distal violence, we focused on the actions of the children's caretakers as mediators of children's experiences. In all instances in our studies these caretakers were the mothers of the children involved.

One study examined the aftermath of a neighborhood killing in which a young wife/mother bludgeoned her husband to death with their son's baseball bat, confessing to the crime only after three weeks of portraying herself to caring neighbors as a grieving widow (Fox, Von Bargen et al., 1996). The family was well known in the neighborhood, particularly to young children. Not only was the young son a member of a neighborhood play group, but the wife taught piano in their home and enrolled many neighborhood children as pupils. Moreover, several older children had been employed as babysitters for the son during piano lessons, and so were also quite familiar with the family members and their home, the setting of the murder. Parents of all the children enrolled as piano students were interviewed, as were neighbors; on-site observations and the use of documentary materials supplemented the interview material. The study is one of the first to allow an examination of the impact of an episode of known violence in the neighborhood on children from upper middle class homes.

The murder in effect took place in two phases. The first phase surrounded the discovery of the murder itself, while the second phase was initiated by their piano teacher's confession to the crime. The children had to process a double shock, the killing itself and then the betrayal of trust signified by the confession. This phasing provides a useful organizational scheme for discussing our findings about the impact of the episode on the children.

The amount of traumatic impact on children – as reported by their mothers some eight months after the event – surprised us, given the children's distance from the event itself and its principals, unlike those of studies reviewed earlier. Moreover, all 21 of the children involved in the study had access to two well-informed, well-educated parents with resources, unlike the children in many studies of violence in inner-city neighborhoods.

Following the discovery of the murder the children reactions, as reported by their mothers, mimicked PTSD symptoms. As a group the mothers reported their children showed fearfulness and anxiety; crying; acting upset; reversion to babyishness; changes in sleep routine, such as insisting upon sleeping with siblings or sleeping in the parents' room, and insisting on night lights; expressions of fear about being home alone, such as demanding a babysitter, not wanting to be out of sight of parents, not wanting to wait for the bus alone; avoidance of cues and reminders of the event, such as avoiding the piano, the street where the family lived, and refusing to play baseball; and emotional impact, such as expressions of sadness, shock, and disbelief.

We also asked mothers about their children's reactions to the confession which occurred some three weeks after the murder itself. While some mothers reported a cessation of fears, for many of the children there was continuation of a heightened level of fear, but now with a target – the piano teacher. Several reported their children developed new fears, both retrospective (for example, one child was haunted by the fact that "I hugged a murderer") and prospective (one mother reported her child's asking "will she come back for us?"). Mothers also reported their children showed a mix of emotions, such as anger, rage, along with extreme sadness and an overwhelming sense of loss. Several reported a continued avoidance of symbols and cues, reminders of the event or of the murderer herself.

It is notable that our findings diverge sharply from those of Martinez & Richters (1993) reviewed earlier. Mothers in their study underestimated both the degree of exposure of their children to violence and the impact on their children of such exposure. In addition to differences in the quantity or frequency of children's exposure to violence, there are substantial socioeconomic and educational differences between the mothers in the two studies. As elaborated in a later section, we believe these differences account for the difference between our studies and those of Martinez and Richters in terms of mothers' attentiveness to the violence exposure of their children.

One of the contributions of our study of an episode of distal violence to the literature on children's experiences with trauma is our exploration, albeit through their mothers, of the questions the children raised about the event. We were able to trace the cognitive processing of this event by the children. We came to see that for some children this event raised heretofore unthought and unknown fears and heightened their sense of vulnerability to harm. It is interesting to trace their questions across the phases of this murder. Immediately after the murder, their questions concerned personal and family safety – are we safe? Are we at risk? And then the children focused, as did their parents, on establishing matters of fact: Who did it? Will they come back? What happened here? After the confession, their questions were focused on only one thing – why? Why did the piano teacher do it? What could have made her do such a thing? How can we account for such a terrible thing? The children were clearly engaged in the construction of meaning, sense-making activity, in attempts to restore order and predictability to their worlds.

Even after the parents' attempts to answer and explain, some of the children continued to pepper their parents with questions as the children attempted to process their parents' explanations and apply them to their own lives. One mother reported that her daughter reacted to the mother's explanation that the piano teacher was enraged by her husband: "She killed out of anger. Could you have gotten that mad at us too?" Another mother reported that her child worried, "She sometimes was angry when my lesson wasn't good; would she have killed me too?" Some children feared the piano teacher's revenge: "Do you think she will come here? Will she come back to get us?" And finally, from a few came the question whether nice mom and dad are really any different from nice Mr. and Mrs. Piano Teacher. Clearly, only those children who felt free enough to give voice to their fears to their mothers are represented here. But, in essence, their questions – drawn from their experience with this single episode of distal violence — ask: "Can I trust the adults, on whom my life depends, to retain control over their rage, their emotions, and their marriage, which itself may not be what it seems to be? Am I safe even at home? Can I trust myself?" For more than a few of the children in this study, the murder has given shape and form, dimension and reality to thoughts that are no longer unthinkable.

Our findings suggest that the children in this case study showed initially many of the same kinds of post-traumatic responses seen among clinically assessed child witnesses of parental assault and murder ((Silvern, Karyl & Landis, 1995). The children in this study were universally frightened and disturbed, despite not having been physically present at the homicide. How long lasting were the effects? Our interviews were conducted 8 months after the confession, and by that time, by their mothers' accounts, most of the children's outer and inner lives had returned to normal. As one mother reported, "You go on. . . . You just move on."

It is here that the children in this neighborhood differ both from child witnesses of parental violent conflict, including murder, and children who live in violence-prone neighborhoods. For children in this study, the piano teacher murder was an aberration in their otherwise conventionally placid lives. Their own home lives were not changed by this episode. Indeed, their parents made themselves unusually available to their children during the weeks and months following the event. Unlike children living in violent settings, there was no pileup of stressors or cumulation of traumatic episodes, which can impede coping and recovery.

CLASS-LINKED RESPONSES TO CHILDREN AND DISTAL VIOLENCE

The difficulties of addressing violence in communities include working around or past strategies that residents have devised independently to ensure their own survival, but which may have counterproductive effects on the reduction of the overall amounts of violence in the neighborhood (Garbarino, Kostelny & Barry, 1998). Some of these self-protective strategies include withdrawal from the neighborhood and isolation of oneself and one's children from involvement with area co-residents (Brodsky, 1996). In a second project we found just such strategies being used by mothers and children living in a violence-ridden neighborhood. A rough analysis of class differences in the responses of mothers to their children's encounters with violence was made using data from the case study reviewed above and from research on responses of mothers and children to neighborhood violence endemic to their low-income housing development in the same city (Von Bargen, 1996). The class distinction between the mothers is based upon the neighborhoods in which they were living. In general, the levels of education and income of the women were consistent with residence in either the upper middle class neighborhood or the subsidized housing project. Both neighborhoods were interracial, but the middle class neighborhood was predominantly white and the lower income neighborhood was predominantly African American. Thus, this rough indicator of 'class' indexes several factors and, at the same time, masks unmeasured individual variations among women in both neighborhoods.

As shown in Fig. 1, mothers in our studies differed in their overall assessment of the degree of threat or harm posed to their children, in their assessment of the neighborhood as a resource, in their responses to their child and to the violent incident, and in their approach to civil authorities. The first two dimensions suggest that the middle class women reacted to the violent episode proactively, seeking collective solutions with other neighbors to a problem defined as a threat to safety shared in common with others. At the individual level, this reflects a mastery orientation to coping with stressor events (Boss, 1988). At the neighborhood or social level, this reaction is consistent with the concept of collective efficacy found disproportionately among neighborhoods high in social class advantage (Sampson et al., 1999).

In contrast, women in the low income housing project in the local disadvantaged neighborhood in the same city sought refuge within their own apartments from the violence in the streets outside their homes and withdrew from their neighbors as a safety strategy. Some did so, reporting they could not trust that their neighbors were not part of the problem of the commonplace, drug-related violence on the streets surrounding their home (Von Bargen, 1966). Their coping style was a passive one, constrained by circumstance to the individual level mechanisms within their power, such as withdrawal and retreat, hiding out in their apartments with the shades drawn, waiting for things to settle down, and forcing their children to do the same.

Dimensions of Response	Middle-Class Mothers	Poverty-Level Mothers
1. Overall response of mothers to violent episode	Proactive; took external action to define, contain violence; saw violence as threat to safety of neighborhood	Withdrew within home as sanctuary; state of siege. Personal interests defined apart from neighborhood.
2. Mothers' perception of Impact on child	Saw many behavioral indicators of distress – sleep disorders, fearfulness, regression, avoidance, anger, sadness, worry	Child seen as stoic, 'cool'. Few see fearfulness in child. Child is 'handling it,' not bothered by it
3. Response to child, attributions, interpretations of cues	Talk, talk, talk – talk is interpreted as processing, 'working it through.' Silence means pain, child is still suffering; silence is to be avoided. Mother asks questions, makes child talk. Humor, jokes signify healing.	Advises against, discourages discussion as ineffective ('Ain't nothing I can do anything about'), not relevant. Mother views child's silence as evidence child has gotten over, no longer thinks about a specific episode
4. Assessment of personal safety	In general, personal safety is taken as a given in this world; immediate safety of one's own or child's life is not a salient concern	Parent teaches personal safety skills – mind your own business, don't let anyone know your business, keep to yourself, stick up for yourself, don't rely on others.
5. Perceptions of neighbors	Neighbors viewed as resource for advice, assistance, support, co-caretakers of children; mothers call on neighbors for assistance, information, support, joint action in keeping children safe	Neighbors not viewed as coping resource; often neighbors viewed as potential threat to safety of family; mothers avoided contact with, discussion with neighbors
6. Assessment of safety assurance agencies (police, criminal justice system)	Skepticism about effectiveness; cynicism about weak 'justice', i.e. getting away with murder, along with belief in legitimacy of system	Reliance upon external agencies is dangerous; cooperation exposes family to retaliation from neighbors. Criminal justice alternatives are not racially fair and impartial, so are not viewed as legitimate
7. Nature of children's exposure to extreme violence	Extreme violence is episodic, unusual event. Direct victimization or witnessing victimization of others is extremely rare. Distal violence is rare.	Events of extreme violence are commonplace. Direct victimization or witnessing victimization of others is not uncommon. Distal violence is routine in lives of children.

Figure 1. Comparison of Middle-class and Poverty-level Parental Management Strategies for Children's Exposure to Neighborhood Violence.

Even within their own homes, the mothers felt vulnerable. Several recounted bullets slamming through the thin walls of their public housing apartments. One mother reported responding to a knock on the front door in the afternoon after her children had returned from school, only to be met by a man who pushed her aside and raced through her apartment and out the back door while being chased by a group of men who also ran headlong through her apartment. That neighborliness can be an unwise luxury was also exemplified by a mother who told of the following incident that occurred in the morning before her child went to school. After listening to the screams of her next door neighbor, the mother allowed the woman into her apartment to take shelter there from her violent husband. Moments later, the husband beat down their front door to continue his attack on his wife in the presence of the mother and her child. With the exception of a few women who had kinship ties with others in the same neighborhood, most of the mothers avoided contact with other residents and coped with their children alone (Von Bargen, 1996).

Mothers in both neighborhoods were asked how violent episodes had affected their children. The middle-class mothers saw plentiful evidence of impact. By contrast, the lower-income mothers saw little in their children's behavior that they defined as indicative of a child's having problems, consistent with the findings of Martinez & Richters (1993). Indeed, the mothers took pride in their children's stoicism and perceived ability to handle themselves on the streets while going to and from school.

When asked how they responded to their children following traumatic violent episodes, the middle class mothers used open communication and discussions, relying upon cognitively oriented verbal cues to guide their responses to their child. The lower income mothers discouraged discussion about violent events that had taken place in the neighborhood. Talking with their children was generally seen as ineffective and unneeded, and at worst, some mothers feared that careless talk might create vulnerabilities for their children if family discussions were repeated to non-family members. The middle class mothers interpreted a child's reluctance to talk about the violent episode as problematic and indicative that their child might be obsessively disturbed about the incident. So these mothers then probed for reactions and responses from their child. By contrast, the lower income mothers interpreted a child's silence as evidence that the child had moved on and was no longer bothered by an incident, if indeed the child had ever been disturbed by a specific incident.

Mothers also differed in terms of their general orientation to a sense of personal and family safety and in terms of the specific security strategies they taught to their children. For the middle class mothers safety was a concern only in a general abstract sense. Although some of the middle class mothers gave voice to concerns about 'where the world is headed' and about their perceptions of increasingly high levels of crime and violence, nonetheless they took for granted that the daily risks to their children and themselves from extreme violence were quite small. Few of them were aware of specific violence avoidance or coping strategies they had taught to their children, other than those the children learned at school (don't talk to strangers, look both ways before crossing the street, and the like). The residents of the low income neighborhood, on the other hand, were actively engaged in instructing and reminding their children daily about personal safety. Their instructions for safety were grounded in individual self reliance: mind your own business, don't talk about our family to outsiders, keep to yourself, stick up for yourself, come straight home and stay home. These were the self-same behaviors the mothers themselves relied upon to protect themselves and their children from their neighbors and the neighborhood.

The sixth panel in Fig. 1 suggests that cooperation with the police is problematic for residents of violence-torn neighborhoods. Cooperating with police investigations of crimes in the neighborhood runs the risk of retaliation from neighbors, a fear heightened not only by the low levels of trust among neighbors but also by perceptions of non-responsiveness and racial or class bias on the part of the police. The middle class mothers, by contrast, while giving voice to skepticism about organizational effectiveness, nonetheless took for granted that agencies would be responsive to their calls for assistance and would work on their behalf. Further, a few of the women suggested that if they were dissatisfied, they would use their social and political resources to ensure that their complaints were not only heard but responded to.

What is evident in this analysis is that the nature of exposure of children to extreme violence differs dramatically across the two types of neighborhoods. For children in the middle class neighborhood, exposure to traumatic violence of any kind was unusual. What we have labeled distal violence, that is, extreme violence that occurs to someone known to the child, was rare. Further, among these children direct victimization or witnessing the victimization of others was exceedingly rare. By contrast, violent events were commonplace in the daily lives of the children in the lower-income neighborhood. Direct victimization or witnessing the victimization of others was not uncommon, and distal violence was routine in the lives of children living here. This suggests that what is distal violence in the lives of middle class children (eg, the unwitnessed murder of a known person) is a likely part of the routine aspects of violence in lower class neighborhoods. Children in these neighborhoods probably routinely experience many more episodes of distal violence.

In sum, this comparative analysis shows that parents in both neighborhoods appeared to respond to extreme violence – and must be understood as doing so – in ways that they felt were feasible, effective, and rational in protecting their children, given their contexts. Drawing on the work of Sampson, Raudenbush & Earls (1998), we would argue that the distribution of a sense of security, and the routinized strategies used to engineer it, follow larger structural patterns of advantage and disadvantage that transcend the characteristics, values and beliefs of individuals. Such dynamics of distribution can easily overwhelm the best efforts of well-intentioned mothers living in the least advantaged of neighborhoods to protect their children from the effects of extreme violence.

CONCLUSION

The focus of this chapter, the exposure of children to extreme violence, is framed by a larger set of issues about children and violence. In general terms, children, just as adults, can play three different roles in episodes of violence. They can be perpetrators of or participants in violent acts; they can be targets of the violence of others; and they can be witnesses or passive observers of violent interactions. A more complete assessment of how children intersect with violence would address the following secondary questions about children in each of the roles of perpetrator, target, and witness to violence. First, how many children are involved? What are the consequences of such involvement to the children themselves, their families, and to the communities and larger social systems in which children are located? What is the nature of the ensuring harm and by what pathways is the harm transmitted? How many children fill multiple roles, that is, how many are both targets of parental abuse and witnesses of violence between their parents? How many are witnesses to both family and the neighborhood violence? How many replicate the patterns of violent interpersonal interactions they grow up with to become perpetrators in their own homes? Finally, can strategic points for action be identified? Are there moderators or buffers of the impact of violence? Are there protective influences in the child's natural environment? What are the potential points of intervention? Can strategic points for prevention be identified?

This chapter has looked only at the involvement of children as witnesses to episodes of extreme violence and has considered only the questions of the nature of harm, the consequences of a child's mere presence during acts of interpersonal violence. The importance of the subsequent actions of the child's caregivers as buffers of the experience has been suggested. At the same time, the structural constraints or advantages that accrue to families in different socioeconomic strata have also been suggested as factors that enter into the picture both directly and indirectly.

The importance of informal sources of assistance to children who are exposed to extreme violence cannot be overemphasized. Although it is clear that children respond to and benefit from intensive clinical and therapeutic interventions (Jaffe, Wolfe et al., 1990; Kolko, 1998; Silvern, Karyl et al., 1995), those mental health services resources are largely unavailable to most children. This is not solely because formal sources of accessible, cost-effective, and high quality mental health assistance to the families of children exposed to extreme trauma are unavailable in most communities, but also because those who come into contact with such children may be unaware of or predisposed not to notice children's distress (Payton & Krocker-Tuscan, 1988).

Those who come into first contact (usually the police) or regular contact (usually teachers and immediate caregivers) with children exposed to extreme trauma are in the best position, and for many children, the only position to offer assistance to children (Eth et al., 1985). Adults in these roles would benefit from specialized educational modules about the impact of traumatic violence on children, so that they can provide immediate and appropriate first-line help to those children. Secondarily, better informed first-line responders may be able to link children more effectively to the kinds of mental health services that are suited to their special needs. Ironically, these suggestions echo similar recommendations found in the literature a quarter century ago (Zeanah & Burk, 1984)

Research on the impact on children of marital aggression, hostility, and violent conflict suggests that many children are affected deeply by these 'mild' forms of overt and covert interpersonal conflicts and take away long-lasting lessons about intimacy and interaction from such experiences (Margolin, 1998). Children who are confronted with the much more intense episodes of traumatic violence, such as have been the focus of this article, are in even greater need of assistance in understanding and processing the situations that have been visited upon them. The fact that their numbers continue to grow and their needs for assistance continue to be unrecognized and unmet poses a challenge to those who value the well being of all our children.

REFERENCES

Arroyo, W., & Spencer E. (1995). Assessment Following Violence–Witnessing Trauma. In: E. Peled, P. G. Jaffe, & J. L. Edleson (Eds), *Ending the Cycle of Violence: Community Responses to Children of Battered Women*, (27–42). Thousand Oaks, CA: Sage.

Bell, C. C., & Jenkins., E. J. (1993). Community Violence and Children on Chicagos South Side. *Psychiatry: Interpersonal and Biological Processes 56*(1), 46–54

CCC. (1991). Traumatic Stress and Children. *Journal of Health Care for the Poor and Underserved 2*(1), 175–85.

Boss, P.. (1988). Family Stress Management. Newbury Park, CA: Sage.

Brodsky, A. E. (1996). AResilient Single Mothers in Risky Neighborhoods: Negative Psychological Sense of Community. *Journal of Community Psychology, 24*, 347–363.

Burman, S., & Allen-Meares, P. (1939). Neglected Victims of Murder: Childrens Witness to Parental Homicide. *Social Work, 1*, 28–34.

Dubrow, N., & Garbarino, J. (1989). Living in the War Zone: Mothers and Young Children in a Public Housing Development. *Child Welfare, 48*(1), 3–20.

Eth, S., Siverstein, S., & Pynoos, R. S. (1985). Mental Health Consultation to a Preschool Follwing the Murder of a Mother and Child. *Hospital and Community Psychiatry, 36*(1), 73–76.

Figley, C. R. (1983). Catastrophes: An Overview of Family Reactions. In: C. R. Figley & H. I. McCubbin (Eds), *Stress and the Family: Volume 2: Coping With Catastrophe, 2*, 3–20. New York, New York: Bruner/Mazel.

Finkelhor, D., & Dzuiba-Leatherman, J. (1984). Children As Victims of Violence: A National Survey. *Pediatrics, 94*(4), 413–20.

Fox, G. L. (1991). Families and the Criminal Justice System: The Aftermath of Homicide. *Academy of Criminal Justice Sciences* annual meeting. Nashville, TN.

Fox, G. L., Von Bargen, J., & Jester, M. (1996). Managing Murder: Parents As Mediators of Childrens Experiences. *Journal of Family Issues 17*(6), 732–57.

Freeman, L. N., Shaffer, D., & Smith, H. (1996). Neglected Victims of Homicide: The Needs of Young Siblings of Murder Victims. *American Journal of Orthopsychiatry 66*(3), 337–45.

Garbarino, J., & Kostelny, K. (1992). Child Maltreatment As a Community Problem *Child Abuse & Neglect, 16*, 455–64.

Garbarino, J., Kostelny, K., & Barry, F. (1998). Neighborhood-Based Programs. In: P. K. Trickett & C. J. Schellenbach (Eds), *Violence Against Children in the Family and the Community*, (287–314). Washington, D. C.: American Psychological Association.

Garbarino, J., Kostelny, K., & Dubrow, N. (1991). What Children Can Tell Us About Living in Danger. *American Psychologist, 46*, 376–83.

Horn, J. L., &Trickett, P. K. (1998). Community Violence and Child Development: A Review of Research. In: P. K. Trickett & C. J. Schellenbach (Eds), *Violence Against Children in the Family and the Community*, (103–38). Washington, D. C.: American Psychological Association.

Hsieh, C. C., & Pugh, M. D. (1993). Poverty, Income Inequality, and Violent Crime: A Meta-Analysis of Recent Aggregate Studies. *Criminal Justice Review, 18*(1), 182–202.

Limber, S. P., & Nation, M. A. (1998). Violence Within the Neighborhood and Community. In: P. K. Trickett & C. J. Schellenbach (Eds), *Violence Against Children in the Family and the Community*, (171–94). Washington, D. C.: American Psychological Association.

Malmquist, C. P. (1986). Children Who Witness Parental Murder: Posttraumatic Aspects. *Journal of the American Academy of Child Psychiatry, 25*(3), 320–325.

Margolin, G. (1998). Effects of Domestic Violence on Children In: P. K. Trickett & C. J. Schellenbach (Eds), *Violence Against Children in the Community and the Family*, (57–101). Washington, D.C.: American Psychological Association.

Martinez, P., & Richters, J. E. (1993). The NIMH Community Violence Project: II. Childrens Distres Symptoms Associated With Violence Exposure. *Psychiatry, 56*(1), 22–35.

O'Hare, W. P. (1996). A New Look at Poverty in America. *Population Bulletin, 51*(2), 1–41.

Payton, J. B., & Krocker-Tuscan, K. (1988). Childrens Reactions to Loss of a Parent Through Violence. *Journal of the American Academcy of Child and Adolescent Psychiatry, 27*(5), 563–66.

Peters, M. F. (1981). Parenting in Black Families With Young Children: A Historical Perspective. In: H. Pipes (Ed.), *McAdoo Black Families*, (211–24). Beverly Hills, CA: Sage.

Peters, M. F , & Massey, G. (1983). Mundane Extreme Environmental Stress in Family Stress Theories: The Case of Black Famılies in White America. *Marriage and Family Review, 6*, 193–218.

Pierce, C. M. (1979). The Mundane Extreme Environment and Its Effect on Learning. In: S. G. Brainerd (Ed.), *Learning Disabilities: Issues and Recommendations for Research*, 111–19. Washington, D. C.: National Institute of Education.

Pruett, K. D. (1979). Home Treatment for Two Infants Who Witnessed Their Mothers Murder. *Journal of the American Academcy of Chıld Psychiatry, 18*(4), 647–57.

Pynoos, R. S., & Eth, S. (1984). The Child As Witness to Homicide. *Journal of Social Issues 40*(2), 87–108.

Pynoos, R. S., Frederick, C , Nader, K , Arroyo, W., Steinberg, A., Eth, S., Nunez, F., & Fairbanks. L. (1987) Lıfe Threat and Posttraumatic Stress ın School-Age Children. *Archives of General Psychiatry, 44*(1), 1057–63.

Richters, J. E. (1993). Community Violence and Chıldrens Development: Toward a Research Agenda for the 1990s. *Psychiatry, 56*(1), 3–6.

Richters, J. E., & Martinez, P. (1993). The NIMH Community Violence Project: I. Children As Vıctims of and Witnesses to Violence. *Psychiatry, 56*(1), 7–21.

Rigamer, E. F. (1986). Psychological Management of Children in a National Crisis. *American Academy of Chıld Psychıatry, 25*(3), 364–69.

Sampson, R. J., Morenoff, J. D., & Earls, F. (1999). Beyond Social Capital: Spatial Dynamics of Collective Efficacy for Chıldren. *American Sociological Review, 64*(5), 633–60.

Sampson, R. J., Raudenbush, S. W., & Earls, F. (1997). Neighborhoods and Violent Crime: A Multilevel Study of Collective Efficacy. *Science, 277*, 918–24.

Silvern, L., Karyl, J., & Landis, T. Y. (1995). Individual Psychotherapy for the Traumatized Children of Abused Women. In: E. Peled, P. G. Jaffe & J. L. Edleson (Eds), *Endıng the Cycle of Vıolence. Community Responses to Chıldren of Battered Women*, (43–76). Thousand Oaks, CA: Sage.

Terr, L. C. (1991). Childhood Traumas: An Outlıne and Overview. *American Journal of Psychiatry, 148*(1), 10–19.

Terr, L. C. (1988). What Happens to Early Memories of Trauma: A Study of 20 Children Under Age 5 at the Time of Documented Traumatic Events. *Journal of the American Academy of Child Psychiatry, 27*(1), 96–104.

Von Bargen, J. (1996). Bullets Aınt Got No Eyes: The Experience of Neighborhood Violence for 10 African American Mothers and Their Children. University of Tennessee, Knoxville, Tennessee.

Zeanah, C. H., & Burk, G. S. January 1984. A Young Child Who Witnessed Her Mothers Murder: Therapeutic and Legal Considerations. *American Journal of Psychotherapy, 38*(1), 132–45.

NEIGHBORHOOD CHARACTERISTICS AND SUPPORTIVE PARENTING AMONG SINGLE MOTHERS

Gary L. Bowen, Natasha K. Bowen and Patricia G. Cook

From a contextual perspective, families operate in a larger social and cultural setting that frames and informs family processes and outcomes. This setting has potential implications for the interaction among family members, the nature of demands faced by the family system and its members, and the availability of resources for meeting the needs of the family and its members. In recent years, behavioral and social scientists have shown greater interest in examining the role of contextual variables in accounting for interactional relationships and processes in families (cf. Barber, 1997; Bowen & Pittman, 1995; Brooks-Gunn, Duncan & Aber, 1997a, 1997b; Duck, 1993; Luster & Okagaki, 1993; Moen, Elder & Luscher, 1995). Researchers have shown a particular interest in the influence of risk and protective factors in neighborhoods on parenting and child

*This project was supported by Grant No. 98-WT-VX-0007 awarded by the National Institute of Justice, Office of Justice Programs, U.S. Department of Justice. Points of view in this document are those of the authors and do not necessarily represent the official position or policies of the U.S. Department of Justice.

Author Note. This article was prepared with grant support from the Knight Foundation in Miami, Florida, for development and testing of the School Success Profile. Gary L. Bowen and Jack M. Richman are co-principal investigators.

Families, Crime and Criminal Justice, Volume 2, pages 183–206.

ISBN: 0-7623-0737-4

development (e.g. Bowen & Chapman, 1996; Brooks-Gunn, 1995; Garbarino & Kostelny, 1993; Sampson, 1997; Steinberg et al., 1995).

Based on a nationally representative sample of 2,099 middle and high school students, this chapter examines the influence of neighborhood characteristics on the parenting practices of single mothers with adolescents in the middle school years (6th–8th grade). Although there are consistent developmental themes across adolescence, students in early adolescence face a simultaneous and sudden onslaught of biological, cognitive, psychological, and social changes (Forehand & Wierson, 1993; Simmons et al., 1987). Concurrent transitions, new developmental tasks, an intense peer orientation, and less control over their physical settings are likely to render middle school students more sensitive than older students to situations and conditions in their neighborhoods.

The study is distinctive in examining the relationship between neighborhood characteristics and supportive parenting from the perspective of the adolescent child rather than from the parent's perspective. Adolescents are considered to have unique viewpoints toward the family system and its embeddedness in the larger social context. As a consequence, a measure of supportive parenting is developed assessing the warmth, involvement, and democratic communication components of an authoritative parenting style (Baumrind, 1978) and evaluated from the point of view of the adolescent respondent.

Two neighborhood characteristics are examined as potentially accounting for variation in the parent-child relationship reported by adolescents: neighborhood threats and collective efficacy. Neighborhood threats involve the extent to which adolescents describe their neighborhoods as characterized by events and situations, such as crime and violence and negative peer group behavior, that place them at potential risk of harm or negative outcomes. While the presence of neighborhood threats is viewed as a risk factor that decreases the probability of supportive parenting, the presence of collective efficacy is considered a protective factor that increases the probability of supportive parenting. Sampson, Raudenbush & Earls (1997) defined collective efficacy as "social cohesion among neighbors combined with their willingness to intervene on behalf of the common good" (p. 918). In the present study, collective efficacy is defined more narrowly as youth perceptions that neighborhood adults are interested in youth and willing to help youth and other adults in times of need.

Examination of neighborhood features and their effects on families and youth is consistent with an ecological perspective (Bronfenbrenner, 1979; Garbarino, 1992), social disorganization theory (Sampson, 1997; Shaw & McKay, 1969), and with growing evidence of the significance of the neighborhood microsystem as an influence on relational processes in families and youth functioning (Bowen & Bowen, 1999; Bowen, Richman & Bowen, 2000). Ecological theory suggests that characteristics of the extra-familial settings in which youth function, such as neighborhood, influence their developmental outcomes. Social disorganization theory stresses the potential deleterious effects of negative neighborhood features – specifically, an absence of neighborhood social organization around common positive goals of residents (Sampson et al., 1997). Because of the central role of parents in the socialization of youth (Bronfenbrenner, 1986) and the capability of parents to mediate the effects on children of negative conditions and events (Garbarino & Kostelny, 1993), it is especially important to understand the impact of neighborhoods on parenting performance. Implications of the results are discussed for targeting neighborhood characteristics in community-level interventions that support effective parenting and promote positive youth outcomes.

SINGLE MOTHER FAMILIES

The present investigation focuses on single mother families: a subgroup of families who are among those most economically disadvantaged in the U.S. and who, as a consequence, are likely to live in neighborhoods that potentially have fewer resources and higher levels of social peril. The economic plight of single mothers has been well chronicled in the literature (Bowen, Desimone & McKay, 1995). More than one-half of single mother families live in poverty, and their average income is less than one-half of the average for two-parent families (McLanahan & Garfinkel, 1995).

The economic peril faced by single mothers conspires to create other problems such as greater residential instability that disrupts social networks for both mothers and children (McLanahan & Booth, 1989). Because of their lack of financial resources, single mothers are more likely than their married counterparts to reside in more economically marginal neighborhoods (McLanahan & Garfinkel, 1989). Such neighborhoods are often plagued with other problems such as high rates of unemployment, high rates of crime and violence and juvenile delinquency, and poor access to community resources (cf. Chow, 1998; Coulton, 1997; Sampson et al., 1997). Although the literature suggests that single mothers maintain active social relationships with kin, friends, and neighbors (McLanahan & Booth, 1989), single mothers may withdraw from involvements in neighborhoods that present a high level of social peril for themselves and their children (Brodsky, 1996; Furstenberg, 1993; Garbarino & Kostelny, 1993). Brodsky (1996) describes this process of withdrawal by single mothers as a "negative psychological sense of community" (p. 347).

In some situations, other adults may be present in the household, potentially reducing single mothers' social isolation and providing them with instrumental and expressive support (Hunter et al., 1998). In Coleman's (1988) discussion of social capital in the family, both the presence of adults in the home and the amount of attention they give to children are significant. Additional adults in the household may function as resources to single mothers that increase the probability that adolescents will perceive their mothers as supportive parents.

The economic and social challenges that many single mothers experience create special demands and challenges for effective parenting (Brodsky, 1999; Garbarino, 1995). Research suggests that single mothers provide less supervision and regulation of their children's behavior than parents in married households (McLanahan & Booth, 1989). In a study of recently divorced mothers with school age sons (kindergarten through grade 6), Bank et al. (1993) found that mothers' use of ineffective parenting practices increased as their level of socioeconomic disadvantage increased.

Compared to their married counterparts, single mothers report greater psychological distress in the form of depression and decreased life satisfaction (Davies, Avison & McAlpine, 1997; McLanahan & Booth, 1989). In part, this is related to the greater economic hardships that they face in the form of unemployment and financial strain (McLoyd et al., 1994). Such psychological distress is likely to have negative implications for the types of parenting practices that have been associated with positive outcomes for children (McLoyd et al., 1994; Simons et al., 1993). For example, in an investigation by McLoyd et al. (1994) involving a sample of single, African American mothers, women who were more depressed expressed a more negative view of their maternal role, which increased their probability of using more abusive disciplinary strategies (e.g. yelling, hitting, and threatening the child).

Family structure is a proxy variable for other statuses and, more importantly, for family process variables that explain variations in outcomes (McLanahan & Garfinkel, 1995). By examining only single-mother families in the nationally representative sample of adolescents, the current study seeks to uncover the potential variability of family processes in response to neighborhood characteristics in a group that is typically considered homogeneous. Other researchers have sought to identify explanatory variables that are more endemic to the parent or the family system itself, such as the psychological well-being of the parent, that mediate the relationship between personal characteristics and parenting behavior in single mothers families. Consistent with ecological perspectives on parenting behavior (Belsky, 1984; Bogenschneider, Small & Tsay, 1997; Luster & Okagaki, 1993), the present investigation extends previous analysis by focusing on group effects in the form of neighborhood characteristics – social processes that exist outside the personal and family system. These neighborhood level social processes are expected to explain variation in the use of supportive parenting strategies by single mothers.

In the context of past research, three home status risk factors are entered as control variables in the analysis: poverty status (i.e. adolescent receives free or reduced price lunch at school), race/ethnicity (i.e. cultural minority), and number of adults in the household (i.e. one adult in household). These risk status variables are likely to be confounded with neighborhood characteristics, influencing the self-selection of single mothers into particular neighborhoods.

CONCEPTUAL MODEL

Figure 1 illustrates the model tested in the current study. Both neighborhood threats and collective efficacy are depicted in the model as having a direct influence on adolescents' perceptions about their mothers' supportive parenting practices. In addition, collective efficacy is shown as mediating the relationship between neighborhood threats and parenting practices. Neighborhood, a concept that has proved difficult to define and operationalize for research purposes (Furstenberg & Hughes 1997), is defined from a phenomenological perspective as "the part of town you live in" on the self-report questionnaire used in the study.

Collective efficacy is expected to promote supportive parenting practices by

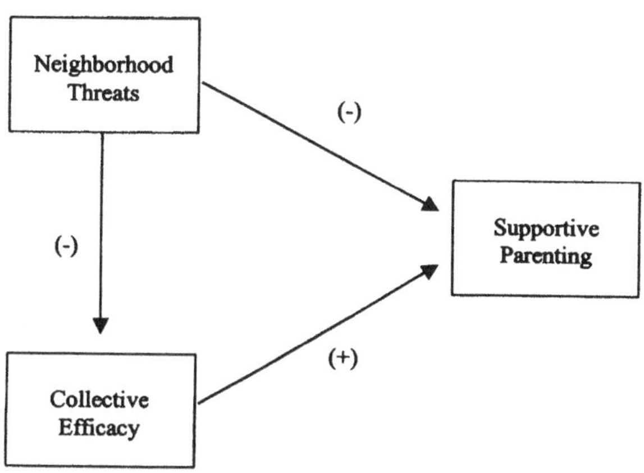

Figure 1. Conceptual Model: Predicted Relationships Between Neighborhood Threats, Collective Efficacy, and Supportive Parenting.

single mothers with adolescents by representing a supportive and protective neighborhood feature. Neighborhood threats, defined as adolescents' perceptions of the prevalence of crime and violence and negative peer behaviors, are manifestations of ecological disadvantage in the neighborhood. They represent stressors to parents, or risk factors that decrease the probability of supportive parenting. In addition, these neighborhood threats are expected to decrease levels of collective efficacy in the neighborhood and, as a consequence, have negative implications for supportive parenting practices.

LITERATURE REVIEW

The contexts in which families function are potentially sources of support or sources of stress (Belsky, 1984; Luster & Okagaki, 1993). Neighborhoods offer support to parents in the form of supportive and educational interactions with other adults, shared parenting roles (such as monitoring and regulating children's behavior), institutional sources of support, safe recreational facilities and opportunities for social participation, and positive values and behavior norms among neighborhood youth.

Neighborhoods represent stress to parents when crime and violence and negative youth behaviors and values encourage isolation and withdrawal from potentially supportive adults and institutions, or when support sources simply do not exist within the neighborhood. Unfortunately, families with the risk factors of poverty and single parenthood are both the most likely to live in neighborhoods that lack supportive structures and interactions, and the most highly in need of support. As concluded by Garbarino & Sherman (1980), these settings are least likely to have adults, including parents, who are 'free from drain' (p. 198).

Negative Neighborhood Effects

Much of the research examining neighborhood effects on youth outcomes has focused on negative characteristics of the neighborhood environment and aggregations of family risk factors among residents (e.g. Bowen & Chapman, 1996; Furstenberg, 1993; Kupersmidt et al., 1995; Simcha-Fagan & Schwartz, 1986). Aggregate measures of family risk factors, such as poverty, single parenthood, and unemployment, which are highly confounded with one another, have been consistently associated with negative neighborhood effects on youth behavior and functioning (e.g. Brooks-Gunn et al., 1993; Crane, 1991; Kupersmidt et al., 1995; Simcha-Fagan & Schwartz, 1986).

The majority of studies about neighborhood effects focus on outcomes related to youth behaviors (e.g. delinquency, school drop-out, pregnancy, aggression) and psychological functioning (e.g. adjustment, sense of hopelessness, and post-traumatic stress disorder). A few studies have also found relationships between neighborhood violence or danger and school-related outcomes (e.g. Bowen & Bowen, 1999; Coley & Hoffman, 1996). The effects of negative neighborhood characteristics on family interactions, such as parenting styles, have also been documented (Brodsky, 1996; Furstenberg, 1993; Garbarino & Sherman, 1980; Patterson, 1983). The research evidence supports the hypothesized direct and negative relationship between neighborhood threats and supportive parenting. Crime and violence and negative peer behavior in the neighborhood represent stressors that may impede supportive parenting actions by creating emotional strain or making parents irritable (McLoyd, 1990; Patterson, 1983).

Empirical support also exists for the indirect relationship between neighborhood threats and supportive parenting practices, as mediated by collective efficacy. Shaw & McKay (1969) were among the first to describe how situations of ecological disadvantage contribute to the breakdown of community organization and culture – reducing social integration among residents and weakening social control mechanisms. Threatening neighborhood conditions, such as violence and negative youth behaviors, may lead parents to limit the interactions of family members with neighbors and potentially supportive institutions in the neighborhood (Brodsky, 1996; Furstenberg, 1993; Garbarino & Sherman, 1980). Choosing isolation over physical danger, the parents in these studies, who were predominantly single mothers, eschewed the possibility of developing potentially supportive interactions with neighbors and community institutions. By enforcing restrictions on their children's behavior at a time when youth have a developmental need to seek extra-familial social interactions and independence, the mothers also increased the likelihood of parent-adolescent conflict.

Garbarino & Sherman (1980) examined the relationship between neighborhood characteristics and the extreme case of parenting breakdown – child abuse – and found that in neighborhoods with unexpectedly high rates of child abuse, parents were less likely to interact with each other, use available resources, and allow their children to play with neighborhood children. They also discovered that high-risk neighborhoods were especially threatening to the functioning of high-risk families: "ecological niches can 'make or break' risky families" (p. 196).

Positive Neighborhood Effects

Increasingly, there is evidence that positive characteristics of neighborhoods and communities have the opposite effect on youth and parent functioning as negative characteristics – they promote positive functioning. Research by Sampson (1997) has revealed that a positive neighborhood feature (collective regulation of youth behavior) mediated a substantial proportion of the effect of a negative feature (residential instability) on delinquency. The Search Institute (Benson, 1993; Blyth, 1993) and others (Bogenschneider, 1996; Hawkins & Catalano, 1992) have sought to identify community characteristics that are associated with positive youth outcomes and can be developed to reduce the prevalence of negative outcomes. Important findings of the research conducted by the Search Institute (Blyth, 1993) suggest that: (a) family, school, and peer-related strengths and opportunities for involvement in community institutions are associated with lower numbers of at-risk behaviors among youth, and (b) healthy communities (defined in terms of the proportion of youth engaged in low levels of at-risk behavior) have especially beneficial effects on their most vulnerable youthful residents, i.e. those who have low levels of support or few internal assets. This individual-level finding is analogous with Garbarino & Sherman's (1980) conclusion about the special vulnerability of at-risk families in at-risk neighborhoods. Children living in single parent homes, experiencing the risks commonly associated with this status, are likely to be among the more vulnerable youth in a given neighborhood and the most sensitive to negative or positive neighborhood conditions.

Intervention research provides evidence for the importance of neighborhood supportive mechanisms for effective parenting practices. Interventions that build social support networks among parents and increase parents' involvement with community institutions (e.g. school, workplace, adult education, volunteer opportunities) have resulted in improved family functioning and youth outcomes (McDonald et al., 1997; Webster-Stratton, 1997).

In summary, research focusing on neighborhood strengths supports the inclusion of a positive neighborhood construct in the model and predicts a direct and positive path between it and supportive parenting. The positive neighborhood construct used in the current study is discussed in more detail below.

Collective Efficacy — Positive Neighborhood Characteristic

Sampson et al. (1997) propose a definition of collective efficacy that encompasses social cohesion as a context for collective action and the willingness to act to promote the common good. Their more specific definition of neighborhood efficacy emphasizes a task-specific, informal phenomenon of concerted effort to regulate youth behavior, maintain order, and inhibit violence. According to Sampson et al. (1997), the willingness and ability of neighbors to work together for the common goals "depends in large part on conditions of mutual trust and solidarity among neighbors" (p. 919), or the social cohesion component of their definition of collective efficacy. The measure of collective efficacy used in the current study attempts to capture this neighborhood contextual feature that is conducive to collective efficacy.

Collective Efficacy as a Resource for Single Parents

Based on the literature discussed above, it was hypothesized that in neighborhoods where adults help each other out, express interest in youth, and are available to help youth, single parents will be more capable of employing supportive parenting practices. In addition to the general benefits of social support for psychological health, one explanation for this effect focuses on the provision of parenting support from neighboring adults to parents. It is likely that in the context of mutual interest and concern, neighboring parents exchange parenting information, values, strategies, and feedback, all of which have been linked to more effective parenting practices (Furstenberg, 1993; Garbarino & Benn, 1992; Luster & Okagaki, 1993; McLoyd, 1990; Taylor, Casten & Flickinger, 1993).

Another mechanism by which a socially cohesive neighborhood may represent a resource for single parents is through the availability of concerned adults who are willing to offer social support directly to youth. Werner's (e.g. Werner & Smith, 1982) longitudinal study of the entire 1955 birth cohort on the island of Kauai found consistent evidence of the protective role of extra-familial support systems (i.e. adults involved with children) on the outcomes of children exposed to multiple risk factors. Adults outside the family who are interested in and involved with youth contribute to positive developmental outcomes, according to a review by Scales & Gibbons (1996) and others (Bowen & Chapman, 1996) and they do so by fulfilling functions similar to the supportive functions of parents.

Supportive Parenting

The supportive parenting measures used in the current study contained indicators of adolescents' perceptions that are consistent with the acceptance-involvement component of Baumrind's authoritative parenting construct (Baumrind, 1978; Steinberg, Elmen & Mounts, 1989). Acceptance-involvement indicators assessed warmth, attention, and emotional support. This component of the construct has been studied independently and in conjunction with the other two components of Baumrind's authoritative parenting construct: psychological autonomy and behavior regulation. These three core components of authoritative parenting have been found to be moderately intercorrelated and individually predictive of outcomes (Barber & Olsen, 1997; Dornbusch et al., 1987; Eccles & Harold, 1996; Herman et al., 1997; Pettit, Bates & Dodge, 1997; Steinberg et al., 1989). In these studies, constructs of supportive parenting practices have been associated with positive school performance and other desirable developmental outcomes, including psychosocial development and avoidance of behavior problems. In a recent study involving nearly 9,000 14- to 18-year olds, Gray & Steinberg (1999) noted the consistent and positive contributions of acceptance-involvement on adolescent outcomes independent of the other two components of the authoritative parenting construct.

METHODS

Source of Data

Louis Harris and Associates, Inc. collected data from a nationally representative sample of 2,099 middle and high school students between October 31, 1996 and February 15, 1997 using a two-stage stratified sampling design. The sampling procedure was designed to ensure adequate representation of students by gender, race/ethnicity, urbanicity of location, and region. Data were collected using the School Success Profile (SSP) (Richman & Bowen, 1997), a self-administered survey instrument. Based on an ecological perspective, the SSP assesses students' perceptions of their family, school, peer, and neighborhood environments, and dimensions of individual adaptation, such as physical health, psychological well being, and school functioning. SSP data are self-report data. Researchers have argued convincingly that school-age children, particularly as they get older, are often the best informants about their own situations, behavior, and feelings (cf. Richman, Rosenfeld & Bowen, 1998).

Sample Profile

The current study focused on the subset of middle school-age youth (6th–8th grade) living in homes with single mothers (n = 150). Listwise deletion of cases with missing data on any of the analyzed indicators resulted in the elimination of 21 cases (14.0%), leaving 129 cases for analysis. The results from contingency table analysis reveal that the demographic characteristics of the analyzed cases did not differ significantly from all youth in single-mother homes in the sample. The target sample had a similar proportion of females (50.1%) and males

Table 1. Descriptive Statistics and Bivariate Correlations for Measures in the Analysis (N = 129)

Variable	Range	Mean	SD	Number of Items	Bivariate Correlations 1	2	3	4
NEIGHBORHOOD CHARACTERISTICS								
1. Crime and Violence	0–7	1.00	1.47	7				
2. Negative Peer Behavior	0–4	1.34	1.52	4	0.45*			
3. Collective Efficacy	0–4	2.35	1.22	4	–0.21*	–0.36*		
SUPPORTIVE PARENTING								
4. Family Integration	7–21	17.12	3.79	7	–0.04	–0.05	0.39*	
5. Parent Support	8–24	20.11	4.26	8	–0.03	–0.05	0.26*	0.71*

*p < 0.05

(49.9%). Almost one third of the youth with single mothers were African-American (32.0%); the rest were White (48.7%) or in other race/ethnic categories (19.3%). Almost three-quarters of the mothers were working full-time (72.1%). Another one in seven (14.2%) worked part-time. In spite of this high rate of maternal employment outside the home, more than one half of the youth (53.5%) participated in the free or reduced lunch program at school, a proxy indicator of poverty and socioeconomic disadvantage (see Bowen & Chapman, 1996).

Measures

Five measures were defined to test the conceptual model in Figure 1: neighborhood threats (two measures), collective efficacy (one measure), and supportive parenting (two measures). Measures of neighborhood threats were coded from low to high threat. Collective efficacy and the two measures of supportive parenting were coded such that the higher the value, the more positive the interpretation. Table 1 provides descriptive statistics for all measures in the analysis.

Table 2 reports mean variations in the two measures of neighborhood threats and the single measure of collective efficacy by five demographic and situational characteristics of respondents: gender, race/ethnicity, school lunch participation, number of adults in the home, and community type. With the exception of community type, respondents provided information on these variables. Louis Harris and Associates, Inc. coded community type for each respondent as either urban, suburban, or rural. Between group differences were examined using the means procedure in SPSS. A 0.05 level of statistical significance was used to evaluate results from this analysis. Although no hypotheses informed this

Table 2. Means And Standard Deviations for Measures of Neighborhood Threats and Collective Efficacy by Demographic Variables

	Crime and Violence		Negative Peer Behavior		Collective Efficacy	
	M	SD	M	SD	M	SD
Gender						
Female	0.88	1.32	1.36	1.55	2.50	1.22
Male	1.12	1.61	1.32	1.50	2.20	1.21
Race/Ethnicity						
White	0 48[a]	0.89	1.01[a]	1.33	2.47	1.21
Other	1.48[b]	1.73	1.65[b]	1.63	2.23	1.23
School Lunch						
Non-recipient	1.06	1 61	1.33	1.50	2.37	1.29
Recipient	0.94	1.35	1.35	1.54	2.33	1.16
Number of Adults in Home						
One Adult	0.97	1.48	1 35	1.53	2.38	1.21
More Than One Adult	1.07	1.48	1.32	1.51	2.26	1.25
Community Type						
Urban	1.78[a]	1.50	1.96[a]	1.55	1.50[a]	1.08
Suburban	1.13[a,b]	1.69	1.54[a,b]	1.61	2.39[b]	1.19
Rural	0.65[a]	1 12	1.97[c]	1.35	2.54[b]	1 20

Note: Unique superscripts indicate pairwise contrasts that are significantly different from each other (alpha < 0 05) An absence of superscripts or the same superscripts indicate that the mean contrasts were not significantly different

analysis, information about variation in neighborhood threats and collective efficacy is important in specifying potential target groups for intervention and prevention activities in the context of hypothesized links between neighborhood characteristics and supportive parenting.

Neighborhood Threats

Two subscales from the SSP were used as indicators of neighborhood threat: (a) crime and violence, and (b) negative peer behavior. Subscales were hypothesized based on preliminary exploratory factor analyses of theoretically related indicators. The crime factor was based on seven dichotomous (yes/no) questionnaire items assessing whether or not the following crimes had occurred in the respondent's neighborhood in the last 30 days: (a) "Someone in your neighborhood was robbed or mugged;" (b) "Someone broke into your home or a neighbor's home;" (c) "You heard gunshots;" (d) "You saw someone selling illegal drugs;" (e) "A person was murdered;" (f) "A fight broke out between two gangs;" and (g) "You saw someone threatened with a weapon such as a gun, knife, or club." The scale for neighborhood crime ranged from 0 to 7, with higher numbers representing more crime. The alpha reliability coefficient for the scale in the single mother sample was good (0.76).

Perceived crime and violence was low (M = 1.00) but there was significant variation among respondents (SD = 1.47). As indicated in Table 2, no significant variation was found in the level of perceived crime and violence by gender, school lunch receipt, or number of adults in the home. However, subgroup differences did appear by race/ethnicity and community type. Respondents from cultural minority groups reported greater crime and violence in their neighborhood than their White counterparts. In addition, respondents in urban areas perceived higher levels of crime and violence than those in rural areas. No statistically significant mean differences were found in levels of crime and violence reported by respondents in urban and suburban locations and by respondents in suburban and rural locations.

The second neighborhood threat component, negative peer behavior, was based on four dichotomous (likely/unlikely) questionnaire items assessing respondents' perceptions that youth in their neighborhoods were likely to: (a) "get in trouble with the police," (b) "use drugs," (c) "join a gang," and (d) "drink alcohol beverages." The scale ranged from 0 to 4, with higher numbers representing a greater number of 'likely' responses. For the analyzed sample, this scale had an alpha of 0.83.

The average number of negative behaviors for respondents was below the scale's midpoint (M = 1.34) – a fairly high average score given the seriousness of some of the items. Reports of negative peer behavior did not vary by respondents' gender, school lunch receipt, or number of adults in the home. However, race/ethnicity and community type were important factors in influencing variation in the number of negative peer behaviors reported by respondents. Respondents from cultural minority groups reported greater likelihood of negative peer behavior than students who reported their race/ethnicity as White. In addition, respondents in urban and suburban areas reported more negative peer behavior than their counterparts in rural areas. No significant mean differences appeared between respondents in urban and suburban areas.

Collective Efficacy

Collective efficacy was assessed with four dichotomous indicators. Informed in part by the work of Small (1991), respondents were asked whether they agreed or disagreed with the following statements about their neighborhood: (a) "Adults in my neighborhood are interested in what young people in the neighborhood are doing;" (b) "If I did something wrong, adults in my neighborhood who knew about it would probably tell the adults I live with;" (c) "If I had a problem, there are neighbors who could help me;" and (d) "People in my neighborhood really help one another out." Respondents received a count score from 0 to 4 that reflected the number of items with which they agreed. The alpha reliability for these four items in the analyzed sample was 0.52. Although the alpha coefficient for this measure was relatively low, the items loaded adequately (0.49 to 0.73) on a single factor and have demonstrated good concurrent validity in prior investigations (Bowen & Chapman, 1996; Nash & Bowen, 1999).

Students reported moderate levels of perceived collective efficacy (M = 2.35). As reported in Table 2, levels of collective efficacy did not vary significantly by gender, race/ethnicity, school lunch program participation, or number of adults in the home. However, levels of collective efficacy did vary by community type: students in suburban and rural locations agreed with a higher number of the collective efficacy indicators than those in urban areas.

Supportive Parenting

Supportive parenting was assessed by two subscales from the SSP, which were treated as separate dependent variables in the analysis: (a) family integration, and (b) parent support. Although moderately correlated with one another ($r = 0.71$), these subscales are considered to capture distinct aspects of the acceptance-involvement component of Baumrind's (1978) authoritative parenting construct. Adapted from Bowen's (1991) Family Value Profile, family integration was based on seven items that captured the extent to which family members support one another during difficulty times, communicate openly with one another, play together, provide one another with emotional support and attention, and work together to solve problems. Items were evaluated by respondents on a three-point scale that was reverse coded for purposes of scale construction: (1) "a lot like us," (2) "a little like us," and (3) "not like us." The alpha coefficient for this measure was high (0.91). Ranging from negative to positive, the mean score for respondents was 17.12 on a scale that ranged from 7 to 21.

The second supportive parenting subscale, parent support, was based on eight questionnaire items assessing respondents' perceptions about the various ways that parents provide support: (a) "Gave you encouragement," (b) "Let you know you were loved," (c) "Made you feel important," (d) "Made you feel appreciated," (e) "Told you that you did a good job," (f) "Made you feel special," (g) "Spent free time with you," and (h) "Provided you with loving support." Respondents evaluated each item on a three-point response scale: (1) "never,"

(2) "sometimes," and (3) "often." The scale had an alpha of 0.95. The summary measure ranged from 8 to 24, with high scores being more positive. On average, respondents felt generally positive about the support they received from their parents: the mean score was 20.11.

DATA ANALYSIS PROCEDURES

Two stages of data analysis were conducted to test the conceptual model in Fig. 1. In the first stage of analysis, separate simultaneous multiple regressions were run for each of the two dependent outcomes: family integration and parent support. These analyses involved two steps. The dependent outcome was first regressed on the three home risk status control variables, the two measures of neighborhood threats (crime and violence, negative peer behavior), and the measure of collective efficacy. Models with at least one statistically significant variable were subsequently trimmed based on these results, using a $p < 0.10$ cutoff for variables, and the multiple regression was re-computed to yield a more parsimonious explanation.

Procedures in the second stage of analysis paralleled those in the first stage. In this analysis, collective efficacy was regressed on the three home risk status variables and the two measures of neighborhood threats. As in the first stage of analysis, the resulting model was trimmed ($p < 0.10$) and the path estimates were re-computed. A 0.05 level of analysis was used to evaluate results in the final step of each regression procedure. Collinearity diagnostics revealed little dependency between independent and control variables in the analysis. All analyses were conducted using SPSS, version 9.0.

RESULTS

Table 3 shows the regression results across both stages of data analysis. Included in these tables are the standardized coefficients of the initial and final set of predictors. In the analysis regressing parent support, the initial model failed to explain a statistically significant proportion of variance in the dependent variable. Yet, because this initial model included a statistically significant regression coefficient for collective efficacy, the model was trimmed and model parameters were re-computed. In the regression analyses for family integration and collective efficacy, the initial models were statistically significant. In each of the three regressions, the values of all statistically significant path coefficients in the full models remained strong and statistically significant when the models were trimmed and re-run. As indicated in Table 3, the trimmed models explained between 7% and 15% of the variance in the respective endogenous variables.

Table 3. Multiple Regression Predicting Endogenous Variable (N = 129)

Variable	Family Integration		Parent Support		Collective Efficacy	
	Initial	Final	Initial	Final	Initial	Final
Poverty Status (1 = school lunch)	–0.01		–0.05		–0.02	
Race/Ethnicity (1 = cultural minority)	0.11		0.05		0.01	
Adults in Home (1 = one adult)	0.11		0.03		–0.05	
Collective Efficacy	0.43*	0.39*	0.28*	0.26*		
Crime and Violence	–.03		–0.02		–0.06	
Negative Peer Behavior	0.10		0.05		–0.34*	–0.36*
Constant	13.04	14.28	17.58	17.96	2.82	2.74
Multiple R	0.43	0.39	0.27	0.26	0.37	0.36
R squared	0.18*	0.15*	0.07	0.07*	0.14*	0.13*
F	4.52	22.50	1.63	9.36	3.90	19.21

* p < 0.05

The three home risk status variables were not statistically significant in any of the analyses. In addition, the crime and violence indicator of neighborhood threats failed to exert a statistically significant effect in any of the analyses in which it was included. This latter finding was counter to our expectations that crime and violence would decrease adolescents' perceptions of both supportive parenting practices and collective efficacy among neighbors.

Although the second indicator of neighborhood threats, negative peer behavior, was hypothesized to exert a direct effect on adolescents' perceptions of family integration and parent support, this hypothesis was also not supported. Yet, as hypothesized, negative peer behavior was found to have a significant negative and indirect effect via collective efficacy on middle school students' perceptions of family integration and parent support. Supportive neighbors appear to play a particularly important direct and positive role in influencing middle school students' perceptions of supportive parenting practices.

DISCUSSION

A growing body of literature suggests that neighborhoods characterized by crime and violence and negative youth behaviors have harmful effects on young residents and their families. Single mothers and their children are often overrepresented in threatening neighborhoods and may be most vulnerable to their harmful effects. There is evidence that the coping strategies many single mothers may be forced to adopt in dangerous neighborhoods may be detrimental to adolescent developmental needs (Brodsky, 1996; Furstenberg, 1993), in spite of their short-term relative desirability. Recent work by Sampson and others (Bowen & Chapman, 1996; Nash & Bowen, 1999; Sampson, 1997; Sampson et al., 1997) suggests that positive features of neighborhoods may also contribute to youth and family functioning. Understanding the potential role of collective efficacy as a direct influence on supportive parenting strategies, as well as a screen that filters the harmful effects of neighborhood risk factors, offers the promise of identifying targets for effective community-level interventions.

The current study analyzed the effects of positive and negative features of neighborhoods on the ability of single mothers to employ the kinds of parenting practices that have been associated with social and academic competence. One of the major findings indicates that youth perceptions of collective efficacy were associated with their perceptions of supportive parenting. The relationship between collective efficacy and supportive parenting is consistent with previous research, and may operate through the provision of emotional and/or instrumental support to parents, through the availability of adults to support and regulate youth, or through other mechanisms.

Collective efficacy also had the hypothesized role as a mediator between negative peer behavior and supporting parenting. Increases in negative peer behaviors decreased youth perceptions of collective efficacy among neighbors, which in turn had a positive and direct effect on youth perceptions of supportive parenting strategies. This finding suggests that adolescents are likely to evaluate the supportiveness of neighbors, in part, based on the activities of their peers. The role of peers as a comparison point for adolescents in evaluating their presenting situations is well established in the literature. Crime and violence, a second indicator of neighborhood threats, did not exert a statistically significant effect on adolescents' perceptions of neighborhood collective efficacy. Inferred from recent findings by Nash & Bowen (1999), crime and violence in the neighborhood may influence youth perceptions of collective efficacy indirectly rather than directly by leading adolescents to more negatively evaluate the behavior of their peers. Investigators are encouraged to explore this hypothesis in subsequent studies.

In many studies framed by social disorganization theory, indicators of neighborhood threats are studied as dependent outcomes rather than as predictor variables (Elliott et al., 1996; Nash & Bowen, 1999; Sampson et al., 1997). In these studies, indicators of neighborhood disadvantage (e.g. residential mobility, cultural heterogeneity, unemployment rates, and poverty) are often used as proxy variables for group effects. Consistent with the work of Chow (1998), this study treated the presence of neighborhood threats as a neighborhood social condition, and investigated the direct and indirect effects of crime and violence and negative peer behavior on youth perceptions of family integration and parent support. Although empirical justification can be developed from the literature for the role of neighborhood threats as both a predictor and a consequence of collective efficacy, the relationship between these variables is most likely reciprocal over time. This reciprocal expectation in the relationship between neighborhood social conditions and neighborhood social organization is consistent with Wilson's (1987) and Pandey & Coulton's (1994) analyses of community breakdown in inner city ghettos. Longitudinal data are required to understand better the causal paths in the relationship between neighborhood threats and patterns of social organization in neighborhoods.

It is possible that neighborhood threats affect parenting strategies indirectly through mediational processes other than collective efficacy, such as through mothers' psychological states (e.g. depression, feelings of 'drain' or stress), or other neighborhood features not examined in the present analysis. More research is needed to better understand the potential mediating effects of such influences for purposes of designing community interventions and mobilizing community resources to assist parents and youth in coping with neighborhood threats.

Contrary to previous research findings (e.g. Brodsky, 1996; Furstenberg, 1993; Garbarino & Sherman, 1980; McLoyd, 1990), the path from neighborhood threats to supportive parenting was not significant. Although the second indicator of neighborhood threats, crime and violence, neither had a direct nor an indirect effect on the dependent outcome indicators, the findings, as reviewed above, suggest that the relationship between neighborhood threats and youth perceptions of supportive parenting is indirect rather than direct. Although the path from neighborhood threats to supportive parenting was not significant, there is ample research evidence that numerous positive parent and child outcomes are associated with neighborhoods that are not characterized by the prevalence of crime and violence and negative youth behaviors. From this perspective, interventionists are encouraged to focus their work on both promoting neighborhood social organization, as reflected in processes assessed by the measure of collective efficacy, and reducing negative social conditions, such as neighborhood threats, that challenge the safety and security of families and their children. As stated by Garbarino (1995), we must "ultimately detoxify the social environment" (p. 88).

Three home status risk factors were entered into the regression analysis to control for the influence of family characteristics that are likely confounded with neighborhood conditions and social processes (i.e. factors that influence how single mothers select themselves into particular neighborhoods). None of these risk factors produced statistically significant regression coefficients in any of the models tested. Although these variables were entered in the analysis primarily to control for the potential effects of families' self selection into different types of neighborhoods, it was surprising in the context of the literature that they did not emerge as significant predictors of variation in the outcomes of interest. One explanation for these findings is that once relevant social processes are identified, proxy variables for these processes that are related to the dependent outcomes fail to remain significant. While a plausible explanation, zero order correlations between the three home status risk factors and the three neighborhood measures were modest. In the context of ecological and social capital theory (e.g. Coleman, 1988; Garbarino, 1992), the lack of a statistically significant relationship between the number of adults in the home and the two indicators of supportive parenting was particularly surprising. For example, Garbarino states that the quality of a microsystem depends on the number of adults, the density of interactions, and the quality of interactions between adults and children. It is possible, however, that the density and quality of interactions are the key beneficial aspects of microsystem quality. Recent research by Hunter et al., (1998) suggests that the role of multiple caretakers in single parent families must be understood in the context of racial and cultural differences.

According to the results of this study, interventions and policies targeting community institutions, resources, and norms that promote supportive interactions among neighborhood residents are likely to improve the quality of adolescents' family microsystems. Results from the means analysis suggest that neighborhoods in urban areas may be most in need of this type of social revitalization. Youth in urban areas reported the lowest levels of collective efficacy and the highest levels of crime and violence and negative peer behavior, as compared to their counterparts in suburban and rural areas. Unlike their older high school counterparts, middle school students typically have less freedom of movement to find ecological niches that are more supportive of their development and well being. Although these findings should be viewed with caution in the context of sample size constraints, middle school age adolescents in urban areas may need greater assistance from adults than students in suburban and rural areas in navigating neighborhood situations that potentially threaten their security and the safety and parenting effectiveness of the adults with whom they live.

The results from the means analysis also suggest that adolescents from cultural minority groups may be more vulnerable to crime and violence and negative peer behavior than Whites. Additional research is needed that examines the effects of race/ethnicity on neighborhood characteristics in the context of community types.

In summary, this study has sought to shed light on the relationship between neighborhood features and the parenting practices of single mothers with middle school age adolescents. Although sample respondents were identified from a nationally representative sample of youth, caution should be exercised in the interpretation of findings in the context of the relatively small number of middle school students in the sample from single mother families. Additional research is needed that attempts to replicate these findings, especially in the context of specific groups of adolescents. In addition, longitudinal research is needed to understand better the operation and sequencing of risk and protective features in neighborhoods on family processes and youth outcomes.

REFERENCES

Bank, L., Forgatch, M. S., Patterson, G. R., & Fetrow, R. A. (1993). Parenting Practices of Single Mothers: Mediators of Negative Contextual Factors. *Journal of Marriage and the Family, 55*, 371–384.

Barber, B. K. (1997). Introduction: Adolescent Socialization in Context – Connection, Regulation, and Autonomy in Multiple Contexts. *Journal of Adolescent Research, 12*, 173–177.

Barber, B. K., & Olsen, J. A. (1997). Socialization in Context: Connection, Regulation, and Autonomy in the Family, School, and Neighborhood, and With Peers. *Journal of Adolescent Research, 12*, 287–315.

Baumrind, D. (1978). Parental Disciplinary Patterns and Social Competence in Children. *Youth and Society, 9*, 239–276.

Belsky, J. (1984). The Determinants of Parenting: A Process Model. *Child Development, 55*, 83–96.

Benson, P. L. (1993). *The Troubled Journey*. Minneapolis, MN: Search Institute.

Blyth, D. A. (1993). *Healthy Communities, Healthy Youth*. Minneapolis, MN: Search Institute.

Bogenschneider, K. (1996). An Ecological Risk/Protective Theory For Building Prevention Programs, Policies, and Community Capacity to Support Youth. *Family Relations, 45*, 127–138.

Bogenschneider, K., Small, S. A., & Tsay, J. C. (1997). Child, Parent, and Contextual Influences on Perceived Parenting Competence Among Parents of Adolescents. *Journal of Marriage and the Family, 59*, 345–362.

Bowen, G. L. (1991). Navigating the Marital Journey. New York: Praeger.

Bowen, G. L., & Chapman, M. V. (1996). Poverty, Neighborhood Danger, Social Support, and Individual Adaptation Among At-Risk Youth in Urban Areas. *Journal of Family Issues, 17*, 641–666.

Bowen, G. L., Desimone, L. M., & McKay, J. K. (1995). Poverty and the Single Mother Family: A Macroeconomic Perspective. *Marriage and Family Review, 20*, 115–142.

Bowen, G. L., & Pittman, J. F. (Eds). (1995). *The Work and Family Interface: Toward a Contextual Effects Perspective*. Minneapolis, MN: National Council on Family Relations.

Bowen, G. L., Richman, J. M., & Bowen, N. K. (2000). Families in the Context of Communities Across Time. In: S. J. Price, P. C. McKenry, & M. J. Murphy (Eds), *Families Across Time:*

A Life Course Perspective, (117–128). Los Angeles, CA: Roxbury Publishers.
Bowen, N. K., & Bowen, G. L. (1999). Effects of Crime and Violence in Neighborhoods and Schools on the School Behavior and Performance of Adolescents. *Journal of Adolescent Research, 14*, 319–342.
Brodsky, A. E. (1996). Resilient Single Mothers in Risky Neighborhoods: Negative Psychological Sense of Community. *Journal of Community Psychology, 24*, 347–363.
Brodsky, A. E. (1999). Making It: The Components and Process of Resilience Among Urban, African-American, Single Mothers. *American Journal of Orthopsychiatry, 69*, 148–160.
Bronfenbrenner, U. (1979). *The Ecology of Human Development.* Cambridge: Harvard University Press.
Bronfenbrenner, U. (1986). Ecology of the Family as a Context for Human Development: Research Perspectives. *Developmental Psychology, 22*, 723–742.
Brooks-Gunn, J. (1995). Children in Families in Communities: Risk and Intervention in the Bronfenbrenner Tradition. In: P. Moen, G. H. Elder, Jr., and K. Luscher (Eds), *Examining Lives in Context: Perspectives on the Ecology of Human Development*, (467–519). Washington, DC: American Psychological Association.
Brooks-Gunn, J., Duncan, G. J., & Aber, J. L. (Eds). (1997a). *Neighborhood Poverty.* Vol 1, *Context and Consequences for Children.* New York: Russell Sage Foundation.
Brooks-Gunn, J., Duncan, G. J., & Aber, J. L. (Ed.) (1997b). *Neighborhood Poverty.* Vol 2, *Policy Implications in Studying Neighborhoods.* New York: Russell Sage Foundation.
Brooks-Gunn, J., Greg J. D., Klebanov, P. K., & Sealand, N. (1993). Do Neighborhoods Influence Adolescent Development? *American Journal of Sociology, 99*, 353–395.
Chow, J. (1998). Differentiating Urban Neighborhoods: A Multivariate Structural Model Analysis. *Social Work Research, 22*, 131–142.
Coleman, J. S. (1988). Social Capital in the Creation of Human Capital. American *Journal of Sociology 94 (Supplement)*, S95–S120.
Coley, R. L., & Hoffman, L. W. (1996). Relations of Parental Supervision and Monitoring to Childrens Functioning in Various Contexts: Moderating Effects of Families and Neighborhoods. *Journal of Applied Developmental Psychology, 17*, 51–68.
Coulton, C. J. (1997). Potential and Problems in Developing Community-Level Indicators of Childrens Well-Being. In: R. M. Hauser, B. V. Brown, & W. R. Prosser (Eds), *Indicators of Childrens Well Being*, (372–391). New York: Russell Sage Foundation.
Crane, J. (1991). The Epidemic Theory of Ghettos and Neighborhood Effects on Dropping Out and Teenage Childbearing. *American Journal of Sociology, 96*, 1226–1259.
Davies, L., Avison, W. R., & McAlpine, D. D. (1997). Significant Life Experiences and Depression Among Single and Married Mothers. *Journal of Marriage and the Family, 59*, 294–308.
Dornbusch, S. M., Ritter, P. L., Leiderman, P. H., Roberts, D. F., & Fraleigh, M. J. (1987). The Relation of Parenting Style to Adolescent School Performance. *Child Development, 58*, 1244–1257.
Duck, Steve, (Ed.) (1993). *Social Context and Relationships.* Newbury Park, CA: Sage.
Eccles, J. S., & Harold, R. D. (1996). Family Involvement in Childrens and Adolescents Schooling. In: A. Booth & J. F. Dunn (Eds), *Mahwah Family-School Links: How Do They Affect Educational Outcomes?* (3–34). NJ: Lawrence Erlbaum Associates.
Elliott, D. S., Wilson, W. J., Huizinga, D., Sampson, R. J., Elliott, A., & Rankin, B. (1996). The Effects of Neighborhood Disadvantage on Adolescent Development. *Journal of Research in Crime and Delinquency, 33*, 389–426.
Forehand, R., & Wierson, M. (1993). The Role of Developmental Factors in Planning Behavioral Interventions for Children: Disruptive Behavior as an Example. *Behavior Therapy, 24*, 117–141.
Furstenberg, F. F. Jr. (1993). How Families Manage Risk and Opportunity in Dangerous

Neighborhoods. In: W. J. Wilson (Ed.), *Sociology and the Public Agenda*, (231–258). Newbury Park, CA: Sage.

Furstenberg, F. F., Jr., & Hughes, M. E. (1997). The Influence of Neighborhoods on Childrens Development: A Theoretical Perspective and a Research Agenda. In: J. Brooks-Gunn, G. J. Duncan, & J. L. Aber (Eds), *Neighborhood Poverty*. Vol II, (23–47). New York: Russell Sage Foundation.

Garbarino, J. (1992). *Children and Families in the Social Environment*. (2d. ed.). New York: Aldine de Gruyter.

Garbarino, J. (1995). Raising Children in a Socially Toxic Environment. San Francisco, CA: Jossey-Bass Inc.

Garbarino, J., & Benn, J. L. (1992). The Ecology of Childbearing and Child rearing. In: J. Garbarino (Eds), *Children and Families in the Social Environment*, (2d ed.), (133–177). New York: Aldine de Gruyter.

Garbarino, J., & Kostelny, K. (1993). Neighborhood and Community Influences on Parenting. In: T. Luster and L. Okagaki (Ed.), *Parenting: An Ecological Perspective*, (203–226). Hillsdale, NJ: Lawrence Erlbaum Associates.

Garbarino, J., & Sherman, D. (1980). High-Risk Neighborhoods and High-Risk Families: The Ecology of Child maltreatment. *Child Development*, *51*, 188–198.

Gray, M. R., & Steinberg, L. (1999). Unpacking Authoritative Parenting: Reassessing a Multidimensional Construct. *Journal of Marriage and the Family*, *61*, 574–587.

Hawkins, J. D., & Catalano, R. F. (1992). *Communities That Care: Action for Drug Abuse Prevention*. San Francisco: Jossey-Bass Publishers.

Herman, M. R., Dornbusch, S. M., Herron, M. C., & Herting, J. R. (1997). The Influence of Family Regulation, Connection, and Psychological Autonomy on Six Measures of Adolescent Functioning. *Journal of Adolescent Research*, *12*, 34–67.

Hunter, A. G., Pearson, J. L., Ialongo, N. S., & Kellam, S. G. (1998). Parenting Alone to Multiple Caregivers: Child Care and Parenting Arrangements in Black and White Urban Families. *Family Relations*, *47*, 343–353.

Kupersmidt, J. B., Griesler, P. C., DeRosier, M. E., Patterson, C. J., & Davis, P. W. (1995). Childhood Aggression and Peer Relations in the Context of Family and Neighborhood Factors. *Child Development*, *66*, 360–375.

Luster, T., & Okagaki, L (1993). Multiple Influences on Parenting: Ecological and Life-Course Perspectives. In: T. Luster & L. Okagaki (Eds), *Parenting: An Ecological Perspective*, (227–250). Hillsdale, NJ: Lawrence Erlbaum.

McDonald, L., Billingham, S., Conrad, T., Morgan, A., Payton, N. O, & E. (1997). Families and Schools Together (FAST): Integrating Community Development with Clinical Strategies. *Families in Society*, 140–155.

McLanahan, S., & Booth, K. (1989). Mother-Only Families: Problem, Prospects, and Politics. *Journal of Marriage and the Family*, *51*, 557–580.

McLanahan, S., & Garfinkel, I. (1989). Single Mothers, the Underclass and Social Policy. *Annals of the American Academy of Political and Social Science*, *501*, 92–104.

McLanahan, S., & Garfinkel, I. (1995). Single-Mother Families and Social Policy: Lessons for the United States from Canada, France, and Sweden. In: K. McFate, R. Lawson & W. J. Wilson (Eds), *Poverty, Inequality, and the Future of Social Policy*, (367–383). New York: Russell Sage Foundation.

McLoyd, V. C. (1990. The Impact of Economic Hardship on Black Families and Children: Psychological Distress, Parenting, and Socioemotional Development. *Child Development*, *61*, 311–246.

McLoyd, V C., Jayaratne, T. E., Ceballo, R., & Borquez, J. (1994). Unemployment and Work Interruption Among African American Single Mothers: Effects on Parenting and Adolescent Socioemotional Functioning. *Child Development, 65*, 562–589.

Moen, P., Elder, G. H. Jr., & Luscher, K. (Eds). (1995). *Examining Lives in Context: Perspectives on the Ecology of Human Development.* Washington, DC: American Psychological Association.

Nash, J. K., & Bowen, G. L. (1999). Perceived Crime and Informal Social Control in the Neighborhood as a Context for Adolescent Behavior: A Risk and Resilience Perspective. *Social Work Research, 23*, 171–186.

Pandey, S., & Coulton, C. (1994). Unraveling Neighborhood Change Using Two-Wave Panel Analysis: A Case Study of Cleveland in the 1980s. *Social Work Research, 18*, 83–96.

Patterson, G. R. (1983). Stress: A Change Agent for Family Process. In: N. Garmezy & M Rutter (Eds), *Stress and Coping, and Development in Children*, (235–264). New York McGraw-Hill Book Co.

Pettit, G., S., Bates, J. E., & Dodge, K. A. (1997). Supportive Parenting, Ecological Context, and Childrens Adjustment: A Seven-Year Longitudinal Study. *Child Development, 68*, 908–923.

Richman, J. M., & Bowen, G. L. (1997). School Failure: An Ecological-Interactional-Developmental Perspective. In: M. W. Fraser (Ed.), *Risk and Resilience in Childhood: An Ecological Perspective*, (95–116). Washington, DC: NASW Press.

Richman, J. M., Rosenfeld, L. B., & Bowen, G L. (1998). Social Support for Adolescents at-Risk of School Failure. *Social Work, 43*, 309–323.

Sampson, R. J. (1997). Collective Regulation of Adolescent Misbehavior: Validation Results From Eighty Chicago Neighborhoods. *Journal of Adolescent Research, 12*, 227–244.

Sampson, R. J., Raudenbush, S. W., & Earls, F. (1997). Neighborhoods and Violent Crime: A Multilevel Study of Collective Efficacy. *Science, 277*, 918–924.

Scales, P. C , & Gibbons, J. L. (1996). Extended Family Members and Unrelated Adults in the Lives of Young Adolescents: A Research Agenda. *Journal of Early Adolescence, 16*, 365–389

Shaw, C. R., & McKay, H. D. [1942] (1969). *Juvenile Delinquency and Urban Areas*, (rev. ed.) Reprint, Chicago: The University of Chicago Press.

Simcha-Fagan, O., & Schwartz, J. E. (1986). Neighborhood and Delinquency: An Assessment of Contextual Effects. *Criminology, 24*, 667–703.

Simmons, R. G., Burgeson, R., Carlton-Ford, S., & Blyth, D. A. (1987). The Impact of Cumulative Change in Early Adolescence. *Child Development, 58*, 1220–1234.

Simons, R. L., Beaman, J., Conger, R. D., & Chao, W. (1993). Stress, Support, and Antisocial Behavior Trait as Determinants of Emotional Well-Being and Parenting Practices Among Single Mothers. *Journal of Marriage and the Family, 55*, 385–398.

Small, S. (1991). *Teenage Assessment Project Question Bank.* Madison, WI: University of Wisconsin-Madison, School of Family Resources and Consumer Sciences.

Steinberg, L., Darling, N. E., Fletcher, A. C., Brown, B. B., & Dornbusch, S. M. (1995). Authoritative Parenting and Adolescent Adjustment: An Ecological Journey. in Examining Lives In: P. Moen, G. H. Elder, Jr., & K. Luscher (Eds), *Context: Perspectives on the Ecology of Human Development*, (423–466). Washington, DC: American Psychological Association.

Steinberg, L., Elmen, J. D., & Mounts, N. S. (1989). Authoritative Parenting, Psychosocial Maturity, and Academic Success Among Adolescents. *Child Development, 60*, 1424–1436.

Taylor, R. D., Casten, R., & Flickinger, S. M. (1993). Influence of Kinship Social Support on the Parenting Experiences and Psychosocial Adjustment of African-American Adolescents. *Developmental Psychology, 29*, 382–388.

Webster-Stratton, C. (1997) From Parent Training to Community Building. *Families in Society*, 156–171.

Werner, E. E., & Smith, R. S (1982). *Vulnerable But Invincible: A Longitudinal Study of Resilient Children and Youth.* New York: McGraw-Hill Book Company.

Wilson, W.. J. (1987). *The Truly Disadvantaged: The Inner City, the Underclass, and Public Policy.* Chicago: University of Chicago Press.

GENDER DIFFERENCES IN THE EFFECT OF CHILD MALTREATMENT ON CRIMINAL ACTIVITY OVER THE LIFE COURSE

Ryan E. Spohn

Past research examining the link between child maltreatment and subsequent juvenile delinquency has provided evidence that juvenile delinquency, though not inevitable, is a common consequence of maltreatment. However, data inadequacies have precluded a careful analysis of the long-term detrimental effects of child maltreatment. In addition, most studies have failed to examine the possibility that the effects of maltreatment may differ for males and females. The present study is an attempt to increase our knowledge of the relationships between gender, maltreatment, and criminality.

REVIEW OF THE LITERATURE

Child Maltreatment and Juvenile Delinquency

A connection between child maltreatment and subsequent juvenile delinquency has been observed in a number of studies. The current quantitative research can be broken down into two broad categories: retrospective longitudinal studies and prospective longitudinal studies. *Retrospective longitudinal studies* of the effects of child maltreatment on delinquency attempt to test the hypothesis

Families, Crime and Criminal Justice, Volume 2, pages 207–231.
2000 by Elsevier Science Inc.
ISBN: 0-7623-0737-4

that juvenile delinquents experience abuse or neglect at a higher rate than their non-delinquent counterparts (Gray, 1981). Many retrospective studies have found high rates of child maltreatment among juvenile delinquents (Alfaro, 1981; Mouzakitis, 1981; Wick, 1981; Kratcoski, 1982). Three additional studies found that delinquents experienced higher rates of abuse than nondelinquent control groups (Glueck & Glueck, 1950; Lewis & Shanok, 1977; Scudder et al., 1993).

Prospective longitudinal studies investigating the link between maltreatment and delinquency identify samples of maltreated juveniles and attempt to determine the effects of maltreatment on subsequent delinquent activity. Prospective studies that have found notable amounts of delinquency in maltreated samples include Bolton, Reich & Gutierres (1977) and Alfaro (1981). A number of additional prospective studies found support for a relationship between maltreatment and delinquency through comparisons of maltreated youth and control groups that did not experience maltreatment (McCord, 1983; Widom, 1989c; Widom, 1991; Scudder et al., 1993; Smith & Thornberry, 1995; Kakar, 1996). In contrast, a study by Zingraff et al., (1993) found that, controlling for the number of maltreatment reports, age, gender, race, and family type, the experience of maltreatment was not related to juvenile delinquency.

Child Maltreatment and Adult Criminality

A final study has examined the effects of early childhood maltreatment on adult criminality (Widom, 1989b). Widom found that individuals who were abused and/or neglected in their youth were approximately 1.7 times more likely to have an adult record for criminal violence than individuals in the control group (1989b, 263). This study suggests that additional research is needed to determine the long-term effects of maltreatment on criminal activity.

Widom's research provides an important component of our current knowledge regarding the adverse effect of child maltreatment on criminal outcomes, but the existing research leaves a number of questions unanswered. First, most studies do not distinguish cases of abuse from cases of neglect. Such a distinction is important, as many authors suggest that the effects of abuse will differ from the effects of neglect in relation to juvenile delinquency because the nature of each type of maltreatment is quite different. Whereas neglect is an act of omission (i.e. a failure to adequately care for a child), abuse is an act of commission, in which an inappropriate physical or sexual act is directed toward a child (Conaway & Hansen, 1989). As Browne and Finkelhor suggest, it is important that we "disentangle sources of trauma" in dealing with child maltreatment (1986, 177). The unique effects of child neglect and child abuse should be examined separately for two essential reasons. First, being a victim of neglect may lead to increased vulnerability to the adverse effects of child abuse. In order to determine the true, non-spurious effects of abuse, then, it is important to control for the independent effects of neglect. Second, neglect is an important aspect of maltreatment in its own right. Official reports suggest that 52% of all victims of maltreatment suffer from neglect (U.S. Department of Health and Human Services, 1998). Moreover, a number of empirical studies provide evidence that neglect is more likely than abuse to result in antisocial behavior and delinquency (Martin & Beezley, 1980; Egeland & Sroufe, 1981; McCord, 1983; Zingraff et al., 1993; Zingraff et al., 1994; Spohn, 1998).

A second question left unanswered by existing research is whether juvenile delinquency *mediates* the relationship between child maltreatment and adult criminality. This suggests two possible pathways through which child maltreatment may influence criminality. First, maltreatment in early childhood may influence adult criminality indirectly, through a youth's involvement in juvenile delinquency. Through this 'life course' perspective, child maltreatment would increase involvement in juvenile delinquency, and delinquency would have a positive effect on adult criminality. In this fashion, involvement in criminal behavior is perpetuated through the process of 'cumulative continuity' (Caspi, Elder & Bem, 1987; Sampson & Laub, 1993; Browning & Laumann, 1997). *Cumulative continuity* refers to the process through which the cumulative consequences of maladaptive behaviors (such as delinquency) over time channel individuals into situations that provoke further maladaptive behaviors (Caspi et al., 1987). Thus, according to the life course perspective, maltreatment may increase adult criminality indirectly by increasing involvement in juvenile delinquency. If the delinquent is labeled by official agents of social control, initial deviance may be replicated in subsequent criminal behavior. As a result, a maltreated youth's involvement in juvenile delinquency may increase the likelihood of involvement in adult criminality. The direct effect between child maltreatment and criminality, however, would be mediated (see Fig. 1).

Another possible scenario is that maltreatment affects adult criminality directly, in addition to its indirect effect through involvement in juvenile delinquency. In this scenario, cumulative continuity may occur, but the initial impetus for deviant behavior (child maltreatment) continues to impact adult behavior through *interactional continuity*, a process whereby maladaptive behaviors are the result of "an interactional style carried through life that evokes reciprocal, maintaining responses from others in social interaction" (Caspi et al., 1987). Through this 'psychogenic' perspective, the psychological state resulting from experiences of victimization may persist into adulthood and increase one's propensity for criminal involvement independent of earlier involvement in

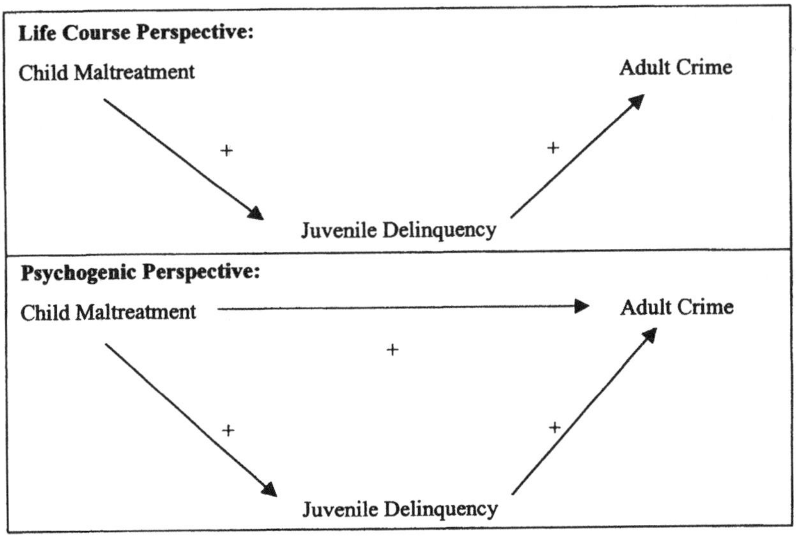

Figure 1. The Life Course and Psychogenic Perspectives.

juvenile delinquency (Browning & Laumann, 1997). Including juvenile delinquency as a control variable and comparing the life course versus psychogenic perspectives will provide a more adequate understanding of the maltreatment criminality relationship.

A final concern that has received inadequate attention is whether there are gender differences in the effects of child maltreatment on criminality. This question is suggested by many studies of female offenders indicating that victimization is a more common and more substantial impetus for female criminal involvement. Evidence of the dramatic impact of child maltreatment on female victims has become too compelling to ignore.

Maltreatment and the Criminalization of Women

Many researchers have documented gender differences in the relationship between maltreatment and various forms of delinquency. For example, status offenses are an outcome often associated with physical and sexual abuse, especially for females. Runaways often report high level of victimization (Widom & Ames, 1994) and this is especially true for female runaways (Rush, 1980; Chesney Lind & Rodriguez, 1983; McCormack, Janus & Burgess, 1986; Janus et al., 1995). Studies of prostitutes also disclose high rates of childhood victimization (James, 1976; James & Myerding, 1976; Silbert & Pines, 1981; Burgess, Hartman & McCormack, 1987; Widom & Ames, 1994). Similarly, in a study of women street hustlers, Miller (1986) found that sexual abuse was a precipitating factor for both running away and drug use, and had a considerable influence on driving women into street life. Hagan & McCarthy (1997) reported similar findings for both male and female youths in Canada.

Studies of Institutionalized Females

The most compelling evidence of the particularly detrimental effect of child maltreatment in the lives of girls and women has originated from studies of women in juvenile and adult correctional settings. For example, Chesney Lind & Rodriguez (1983) interviewed 16 women representing three quarters of the sentenced female felons in the Oahu Community Correctional Facility. Over 60% of these women reported histories of severe child abuse, often resulting from disciplinary spankings that escalated into violent beatings. Although this percentage is alarming, generalizations are limited by the fact that the researchers do not provide control groups of non-institutionalized women, institutionalized men, or non-institutionalized men. Without these important comparison groups, conclusions relating to gender differences in the relationship between maltreatment and criminality must be made with caution.

Richie (1996) reported similar findings in a study of 37 women incarcerated at Rikers Island Correctional Facility. This study focused on black inmates, but included a control group of white female inmates. Through the use of open ended interview questions, Richie (1996) found that about 20% of the women experienced physical abuse as children and about 35% experienced sexual abuse. For the white women in her control group, Richie suggests that the physical and sexual abuse was "compounded by serious childhood neglect" (1996, 65). Again, control groups of males and non-institutionalized individuals are absent from the study.

Kathleen Daly (1994) makes an important contribution to this literature by comparing incarcerated women in New Haven, Connecticut, to a matched control group of a random sample of incarcerated men. She found that one-third of the women were physically abused by fathers or stepfathers, or witnessed the abuse of their siblings or their mothers.[1] Of the male inmates, about 10% grew up in households in which they or their siblings were physically abused or neglected.

Daly (1994) presents paths of female initiation into criminal activity that are remarkably compatible with the life course and psychogenic perspectives presented above. Her first pathway, a route to crime that Daly labels 'street women', includes women who were pushed out of or ran away from an abusive home (1994, 47). This pathway is comparable to the life course perspective in which child maltreatment has an indirect effect on adult criminality through involvement in juvenile delinquency. Daly's second pathway, labeled 'harmed and harming', includes cases in which the trauma of maltreatment leads to violent aggressive attitudes, drug and alcohol addiction, and psychological problems throughout the life course. For these women, child maltreatment has a direct effect on adult criminality in addition to an indirect effect through juvenile delinquency.

These studies of institutionalized women provide initial evidence that maltreatment may influence subsequent criminal behavior differently for females and males, but their findings are not conclusive for several reasons. First, none of the studies of females in correctional facilities provide comparative rates of victimization for non-institutionalized/non-convicted individuals, and only one study (Daly, 1994) provides a male control group. Second, the samples for these studies are quite small and non-representative of the general population. Third, reports of victimization in these studies most often stem from personal interviews, in which social desirability factors and inaccuracy in reporting may influence their findings (Widom, 1989d). Just as subjects tend to overestimate 'socially desirable' responses such as church attendance (Hout & Greeley, 1998; Presser & Stinson, 1998), they tend to underreport undesirable responses such as delinquent behavior or the existence of family conflict in childhood (Henry et al., 1994). As an example, Williams (1995) located over 120 women with documented histories of sexual abuse and interviewed them as adults. About one-third of these women could not remember the abuse, and of those who could, about 20% of them reported that at one time or another in their lives they could not remember the abuse. The occurrence of such high levels of repression calls into question the validity of selfreports of maltreatment.

Levels vs. the Effects of Maltreatment

In addition to studies suggesting that female delinquents experience higher *levels* of abuse than males, one study suggests that abuse may have more detrimental *effects* on females as compared to males. McCormack et al., (1986) found that female runaways who had been sexually abused were more likely than their non-abused counterparts to commit subsequent delinquency and criminality. Among the males in their sample, however, no relationship was found between abuse and criminal activity. Thus, females may not only suffer from higher rates of victimization, but the effects of this victimization on subsequent criminality may be more substantial for females as compared to males.

Chesney-Lind & Shelden (1998) suggest explanations for why female offenders experience higher levels of child maltreatment than their male counterparts. They assert that the victimization of young females and their response to victimization is highly influenced by their status as young women. Girls are much more likely to suffer from sexual abuse due to existing gender and sexual scripts found in patriarchal families. Men in these traditional families are likely to consider daughters and stepdaughters as their sexual property. The vulnerability of girls to both physical and sexual abuse is intensified by traditional norms that keep wives and daughters in the home, where victimizers have greater access to them (Chesney-Lind & Shelden, 1998).

Criminologists examining the relationship between gender and crime have also suggested that maltreatment may have stronger *effects* on female criminality. In describing the white women in her study, Richie (1996) states that their gender and class position left them feeling alienated and outside the mainstream of social life. These institutional forces limited their opportunities. As suggested by Messerschmidt (1986), these individuals are marginalized as women in a patriarchal system in which men control both the labor power and sexuality of women. This marginalization interacts with the stigma of abuse and neglect. Moreover, the black women in Richie's (1996) study describe similar constraints that were often exacerbated by their minority status.

These insights are compatible with the assertions of Hagan and his colleagues regarding the control of girls in American society. Whereas boys have had greater control over their actions and relative freedom from adult supervision, girls have faced greater parental control (Hagan, Gillis & Simpson, 1985). As a result, when abuse is present in the household, males are more capable of escaping the abuse, whereas girls are less able to escape the abusive home environment.

Finally, Chesney-Lind & Sheldon (1998) assert that attempts by girls to escape from abusive families (such as running away) have been criminalized. When girls attempt to escape from their abusive homes, parents may encourage agencies of the state to enforce their return to the home. To the detriment of girls, they are more likely than boys to be referred to court by their parents instead of law enforcement agencies (Chesney-Lind & Sheldon, 1998). Persistence on the part of the victim leads to arrest and possible incarceration. This argument provides another path through which maltreatment may be more likely to result in an arrest record for girls, as compared to boys.

THE CURRENT STUDY

This examination of the child maltreatment/delinquency literature, combined with life course and feminist theoretical perspectives linking child maltreatment to criminality, provides a framework for the development of a number of hypotheses. First, based on previous findings (Spohn, 1998), Hypothesis 1 proposes that the lack of parental attachment and supervision suggested by neglect should increase juvenile delinquency. Second, Hypothesis 2 predicts that the strain of experiencing abuse should also increase juvenile delinquency. Third, based on the psychogenic perspective, Hypothesis 3 suggests that the detrimental effects of abuse and neglect should extend into adulthood, increasing rates of adult criminality. Fourth, from a life course perspective, controlling for involvement in juvenile delinquency should mediate the effects of child maltreatment on adult criminality. Findings inconsistent with hypothesis four will provide support for the psychogenic perspective. In addition to testing these hypotheses, the final segment of the analysis will explore the possibility of gender differences in the effects of child maltreatment on delinquency over the life course through an examination of the male and female subgroups.

METHOD

The data examined in this analysis are drawn from a prospective cohorts research design matched with a control group cohort (Widom, 1994). Victims of physical and sexual abuse and neglect were sampled from a metropolitan area in the Midwest. This maltreated sample was matched with a control group cohort on the basis of sex, race, age, and approximate family socioeconomic status (Widom, 1989a). The two samples are assumed to differ only in the attribute to be examined (child maltreatment).

The Maltreated Sample

The sample of maltreated youth consists of 908 physically abused, sexually abused, and/or neglected individuals. The subjects consist of any person ages 0 through 11 in 1967–1971 who appeared in the records of the juvenile court and juvenile probation department as victims of abuse and/or neglect. Only children ages 11 or less were included in the study to avoid the possibility that delinquency preceded, or might have caused, the abuse or neglect. Cases were excluded if the maltreatment occurred after age 11, if the child was adopted as an infant, if the incident involved 'involuntary' neglect only, if the incident involved 'placement' only, or if the incident was simply a failure to pay child support. Cases were included if they were validated and substantiated by the county juvenile court, juvenile probation department, or adult criminal court (Widom, 1989a).

The courts utilized the following definitions of abuse and neglect: (1) *Physical abuse* indicates cases in which an individual had "knowingly and willfully inflicted unnecessarily severe corporal punishment" or "unnecessary physical suffering" upon a child (Widom, 1989c, 162). (2) *Sexual abuse* refers to a variety of charges, including fondling or touching in an obscene manner, vaginal penetration, sodomy, and evidence of parental incest. (3) *Neglect* refers to instances of physical neglect, medical neglect, educational neglect, inadequate supervision, and abandonment by the mother or father.

The Control Sample

Controls were matched with the maltreated sample on the basis of age, sex, race, and approximate socio-economic background of the family. Any child with an official record of abuse or neglect was eliminated from this sample. Children under school age in the maltreated sample were matched with children of the same sex, race, date of birth (plus or minus one week), and hospitals of birth (Widom, 1989a). Children of school age were matched with children of the same sex, race, date of birth (plus or minus six months), and class in elementary school. Because busing did not exist during the period of the study, the elementary schools were found to represent very homogenous neighborhoods, and thus a proxy for socio-economic status. A total of 667 matches are included in the control sample.

Dependent Variables

The first dependent variable for this study consists of official juvenile arrest records. These records include status offenses, property offenses, and violent offenses, but exclude traffic offenses. The final dependent variable reflects an adult arrest record for a criminal offense, excluding traffic offenses. The information on delinquency was retrieved from juvenile probation department records for both the maltreatment and control groups and the adult criminal arrests were retrieved from a criminal record search at the local, state, and federal levels (Widom, 1989a). Because subjects could have arrest records in localities not searched by the investigators, the data may not be a complete record of arrests. These official measures of criminal activity are compatible with qualitative studies that focus on the histories of abuse reported by convicted/institutionalized women and men.

Independent Variables

Control variables are included for age, race, and gender. Approximately 99% of the sample was 18 years of age or older at the time of the data collection for juvenile and adult arrest records. As a result, age is not included in the models predicting juvenile delinquency. For the models predicting involvement in adult criminality, age is included as a control for 'time-at-risk' for criminality. The variable for abuse represents individuals suffering from officially reported abuse, but not neglect. Similarly, the variable for neglect represents individuals suffering from officially reported neglect, but not abuse. An additional variable is included in the models to represent subjects who suffered from both abuse and neglect.

To a certain extent, these official reports are more reliable than self-reported maltreatment and delinquency, because they avoid the self-report bias associated with sensitive or stigmatizing survey or interview questions. However, these official reports may be influenced by biases inherent in the child protection and criminal justice systems. To the extent that this is true, the current findings should be most similar to those of Daly (1994) whose measures of maltreatment for institutionalized men and women stem from reports of presentence investigators.

RESULTS

Table 1 provides the coding and descriptive statistics for the variables included in the statistical analyses. Differences in the percentage of individuals involved in both juvenile delinquency and adult criminality vary across the maltreated and control groups. Whereas 26.0% of the maltreated individuals have a juvenile arrest record, only 16.8% of the individuals in the control group have a juvenile arrest record. Similarly, 28.3% of those in the maltreated group have an adult arrest record, whereas only 20.2% of those in the control group have an adult arrest record. The multivariate analysis will examine these relationships controlling for the effects of age, gender, and race.

Table 1 indicates that the sample is approximately equally divided between males and females. Roughly two-thirds of the sample is white and about one-third of the subjects are African-American or Hispanic. The mean age for the full sample, maltreated group, and control group is approximately 26 years. Concerning the types of maltreatment, the majority of the maltreated youth suffered from neglect (66.5%). An additional 22.1% of the maltreated youth suffered only from abuse, and the remaining 10.7% suffered from both abuse and neglect.

Table 1. Descriptive Statistics

Variables	Coding	Pooled Group N = 1575 N	%	Maltreated Group N = 908 N	%	Control Group N = 667 N	%
Categorical Variables							
Race[a]	1 = Minority	519	33%	286	31.5%	233	34.9%
	0 = White	1041	66.1%	607	66.9%	434	65.1%
Sex	1 = Male	777	49.3%	443	48.8%	334	49.9%
	0 = Female	798	50.7%	465	51.2%	333	50.1%
Abuse Only	1 = Yes	201	12.8%	201	22.1%		
	0 = No	1374	87.2%	707	77.9%		
Neglect Only	1 = Yes	604	38.3%	604	66.5%		
	0 = No	971	61.7%	304	33.5%		
Abuse and Neglect	1 = Yes	97	6.2%	97	10.7%		
	0 = No	1478	93.8%	811	89.3%		
Juvenile Delinquency	1 = Yes	348	22.1%	236	26.0%	112	16.8%
	0 = No	1227	77.9%	672	74.0%	555	83.2%
Adult Criminality	1 = Yes	392	24.9%	257	28.3%	135	20.2%
	0 = No	1183	75.1%	651	71.7%	532	79.8%
		Mean	S.D.	Mean	S.D	Mean	S.D.
Continuous Variable							
Age		26.04	3.52	26.01	3.52	26.09	3.53

[a] The ethnicity of four of the subjects in the minority groups is described as Hispanic The ethnicity of the remaining minority cases is Black. Information on ethnicity is missing for 15 cases in the maltreated sample.

Table 2 includes the bivariate correlation matrix. Of particular interest is the relationship between juvenile delinquency, adult criminality, and abuse. Whereas neglect is positively related to delinquency and adult criminality, abuse and the combination of abuse and neglect are essentially unrelated to delinquency and adult criminality. In addition, being male, older, and of minority racial status are all positively related to delinquency and criminality.

Main Effects

The logistic regression model in Table 3 allows a test of the first two hypotheses. This model examines the effects of abuse, neglect, and the combination of abuse and neglect on juvenile delinquency, controlling for race and gender. In support

Table 2. Correlation Matrix for the Maltreated Group

	(1)	(2)	(3)	(4)	(5)	(6)	(7)	(8)
(1) Juvenile Delinquency	1.000							
(2) Adult Criminality	0.333*	1.000						
(3) Age	0.062	0.163*	1.000					
(4) Minority	0.125*	0.128*	0.039	1.000				
(5) Male	0.151*	0.303*	–0.072*	0.002	1.000			
(6) Abuse	–0.006	–0.014	0.066*	–0.049	–0.130*	1.000		
(7) Neglect	0.122*	0.095*	–0.056*	0.066*	0.086*	–0.303*	1.000	
(8) Abuse and Neglect	–0.009	0.008	0.003	–0.130*	–0.020*	–0.098*	0.203*	1.000

* $p \leq 0.05$

of hypothesis 1, experiencing neglect has a significant impact on juvenile delinquency, almost doubling the odds of a juvenile arrest record. Child abuse also has a significant impact on juvenile delinquency, which supports hypothesis 2. Surprisingly, the combination of experiencing both abuse and neglect does not have a significant effect on juvenile delinquency.

The model in Table 4 allows a test of the third hypothesis. The dependent variable for this model is adult criminality. In support of hypothesis 3, abuse, neglect, and the combination of abuse and neglect all have a significant impact on adult criminality, controlling for age (representing time at risk for an adult criminal record), race, and gender. The combination of abuse and neglect has the largest effect of the maltreatment variables, more than doubling the odds of an adult arrest record. Another important aspect of this model is the large gender effect. Controlling for the effect of all other independent variables, males are 5 times more likely than females to have an adult arrest record.

Table 3. Logistic Regression Model of the Effects of Child Maltreatment on Juvenile Delinquency

	B	S.E.	Exp(B)
Minority	0.640*	0.130	1.897
Male	0.759*	0.129	2.135
Neglect	0.677*	0.140	1.969
Abuse	0.569*	0.208	1.767
Abuse and Neglect	0.495	0.279	1.641
Intercept	–2.286	0.146	
	Model $\chi^2 = 84.69$, p = 0.000		

* $p < 0.05$

Table 4. Logistic Regression Model of the Effects of Child Maltreatment on Adult Criminality

	B	S.E.	Exp(B)
Age	0.147*	0.020	1.159
Minority	0.688*	0.133	1.989
Male	1.610*	0.139	5.000
Neglect	0.561*	0.142	1.753
Abuse	0.578*	0.211	1.782
Abuse and Neglect	0.765*	0.271	2.148
Intercept	–6.527	0.568	

Model $\chi^2 = 239.03$, $p = 0.000$

* $p \leq 0.05$

Table 5 presents a model that allows a test of the psychogenic perspective versus the life course perspective. This model includes a control variable representing involvement in juvenile delinquency. The inclusion of this variable does not mediate the effects of any of the maltreatment variables. This finding contradicts hypothesis 4, and provides support for the psychogenic perspective. That is, child maltreatment exerts a significant direct effect on adult criminal behavior, over and beyond the effects of cumulative continuity and the labeling processes that promote a criminal career.

Table 5. Logistic Regression Model of the Effects of Child Maltreatment on Adult Criminality, Controlling for Involvement in Juvenile Delinquency

	B	S.E.	Exp(B)
Age	0.146*	0.021	1.157
Minority	0.540*	0.140	1.715
Male	1.513*	0.145	4.539
Neglect	0.394*	0.149	1.483
Abuse	0.465*	0.220	1.592
Abuse and Neglect	0.714*	0.274	2.043
Juvenile Delinquency	1.398*	0.143	4.047
Intercept	–6.676	0.594	

Model $\chi^2 = 334.33$, $p = 0.000$

* $p \leq 0.05$

Gender Differences

The final segment of the analysis examines the possibility of gender differences in the effect of maltreatment on criminal activity. In order to test for gender differences, separate models are presented for the female and male subgroups, allowing a comparison of the corresponding odds ratios for each maltreatment variable. In addition, Tables 6, 7, and 8 present the relevant *z*-values[2] and significant levels testing the null hypothesis of equality of the logistic regression coefficients across gender. Table 6 shows that all three maltreatment variables have a significant impact on the juvenile delinquency of females. In comparison, only neglect has a significant impact on male delinquency, and neglect has a larger effect on the delinquency of males as compared to females. Of greatest interest in this model, though, is the effect of being both abused and neglected. This type of victimization increases the odds of female delinquency by nearly three times that of the control group. In contrast, the parameter estimate for abuse and neglect in the male sub-group is not significant. The corresponding *z*-value (2.191, $p < 0.05$) is significant, indicating that we can reject the null hypothesis of the equality of parameter estimates across gender.

The logistic regression model in Table 7 compares the differential effects of victimization on the adult criminality of females and males. Similar to the previous model, all three forms of victimization exert a significant effect on the criminality of females, but only neglect exerts a significant effect on male criminality. The coefficients representing maltreatment are larger for females as compared to males for each type of maltreatment, but the relevant *z*-values indicate that none of the gender differences are significant.

Table 6. Logistic Regression Models and *z*-Values of Differences in the Effects of Child Maltreatment on Juvenile Delinquency, Estimated Separately by Gender

	Females			Males			
Variables	B	S.E.	Exp(B)	B	S.E.	Exp(B)	Z-Values
Minority	0.505*	0.204	1.656	0.734*	0.170	2.082	
Neglect	0.592*	0.232	1.807	0 732*	0.177	2 080	–0.482
Abuse	0.606*	0.286	1.834	0 536	0.310	1.709	0.167
Abuse and Neglect	1.085*	0.370	2.958	–0.172	0.438	0.842	2.191*
Intercept	–2.248	0.194		–1.555	0 153		
	Model $\chi^2 = 17.02$, p = 0.002			Model $\chi^2 = 40.14$, p = 0.000			

* $p \leq 0.05$

Table 7. Logistic Regression Models and z-Values of Differences in the Effects of Child Maltreatment on Adult Criminality, Estimated Separately by Gender

Variables	Females B	Females S.E.	Females Exp(B)	Males B	Males S.E.	Males Exp(B)	Z-Values
Age	0.132*	0.034	1.141	0.155*	0.024	1.168	
Minority	0.647*	0.225	1.909	0.707*	0.166	2.028	
Neglect	0.834*	0.265	2.302	0.449*	0.170	1.567	1.225
Abuse	0.707*	0.321	2.028	0.574	0.296	1.775	0.305
Abuse and Neglect	1.199*	0.419	3.317	0.506	0.349	1.658	1.272
Intercept	−6.260	0.967		−1.555	0.153		
	Model $\chi^2 = 37.83$, p = 0.000			Model $\chi^2 = 72.99$, p = 0.000			

* $p \leq 0.05$

The final logistic regression models, presented in Table 8, examine gender differences in the effects of maltreatment on adult criminality, controlling for involvement in juvenile delinquency. Additional gender differences become apparent in these models. The male sub-group is most consistent with the *life course* perspective. The inclusion of juvenile delinquency in this model mediates the effect of neglect on adult criminality. However, involvement in juvenile delinquency increases the odds of an adult criminal record by nearly five times. Thus, none of the maltreatment variables exert a direct effect on adult criminality, but neglect influences this adult outcome indirectly through involvement in juvenile delinquency. Each of the variables representing maltreatment has a stronger effect on female criminality, but the *z*-values show that these are not significant differences between the subgroups. The female model is more compatible with the psychogenic perspective of the effects of maltreatment on adult criminality. With the inclusion of juvenile delinquency in the model, neglect and the combination of abuse and neglect maintain significant direct effects on adult criminality.

The differences between the male and female subgroups in the effects of maltreatment over the life course become most obvious through a comparison of figures combining the models in Tables 6 and 8. These figures include the regression models in Table 6, in which juvenile delinquency is the dependent variable and the regression models in Table 8, in which adult criminality is the dependent variable and delinquency is a control variable. When only significant paths are included in the diagrams in Figs 2 and 3, it is apparent that child maltreatment is a much greater influence on criminal involvement for the female subgroup.

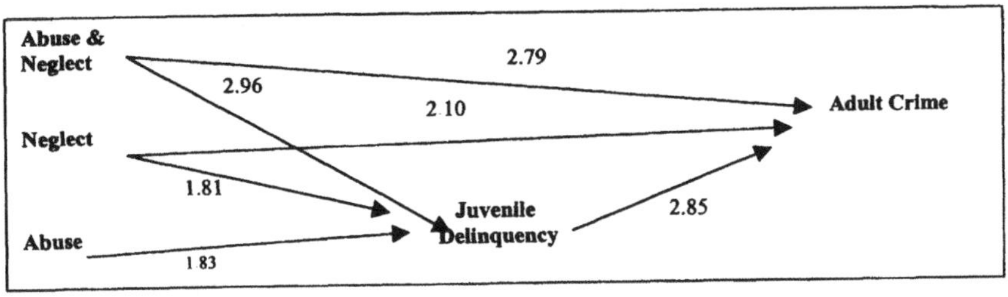

Figure 2 Significant Effects of Maltreatment on Subsequent Criminality of Females.

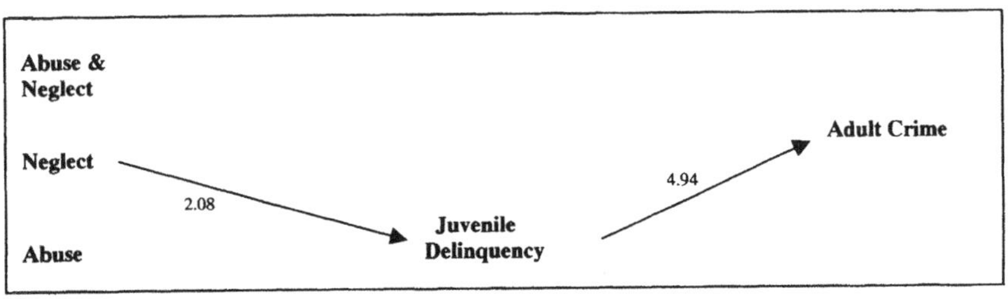

Figure 3 Significant Effects of Maltreatment on Subsequent Criminality of Males.

SUMMARY AND CONCLUSIONS

Summary

The purpose of this study was to use a prospective research design to model the effects of child maltreatment on criminal involvement over the life course. First, I examined the effects of substantiated child neglect and child abuse on having an official arrest record for juvenile delinquency and/or adult criminality. In addition, I tested for gender differences in this relationship that have been suggested by studies of institutionalized women. An examination of the full sample provided evidence for the first two hypotheses: both abuse and neglect significantly increased the odds of involvement in juvenile delinquency. In support of hypothesis 3 and consistent with Widom's (1989b) study that collapsed abuse and neglect into one measure of maltreatment, the child maltreatment variables also increased rates of adult criminality. In the full sample, support was found for a psychogenic perspective and the assertion that the psychological damage inflicted by child maltreatment continues to influence

Table 8 Logistic Regression Models and *z*-Values for Tests of Differences in the Effects of Child Maltreatment on Adult Criminality Controlling for Involvement in Juvenile Delinquency, Estimated Separately by Gender

Variables	Females			Males			
	B	S.E.	Exp(B)	B	S.E.	Exp(B)	Z-Values
Juvenile Delinquency	1.046*	0.246	0.2845	1.597*	0.182	4.936	–1.809
Age	0.127*	0.035	1.136	0.158*	0.026	1.171	
Minority	0.552*	0.230	1.737	0.530*	0.177	1.699	
Neglect	0.744*	0.269	2.104	0.229	0.181	1.257	1.588
Abuse	0.622	0.326	1.862	0.456	0.316	1.578	0.365
Abuse and Neglect	1.025*	0.427	2.787	0.600	0.361	1.822	0.761
Intercept	–6.260	0.967		–5.462	0.699		
	Model $\chi^2 = 54.63$, p = 0.000			Model $\chi^2 = 154.58$, p = 0.000			

* $p \leq 0.05$

adult criminality after controlling for involvement in juvenile delinquency and the associated cumulative continuity of negative consequences that results from one's involvement in the criminal justice system.

Through an analysis of the female and male subgroups in the sample, however, important gender differences emerge. When the female and male models are compared, it becomes apparent that maltreatment influences criminality through a greater number of significant paths for females as compared to males. The combination of abuse and neglect has a particularly detrimental effect on female delinquency, and this effect is significantly larger than the corresponding effect on male delinquency. Thus, this study finds moderate support for the assertion that child maltreatment exerts a stronger influence on the criminality of female victims. Finally, the relationship between the maltreatment of females on their subsequent criminality fits both a life course and a psychogenic perspective. Maltreatment exerts a direct effect on juvenile delinquency as well as a direct effect on adult criminality, controlling for involvement in juvenile delinquency. In addition, both abuse and neglect influence adult criminality indirectly through involvement in juvenile delinquency and association with the criminal justice system. These findings fit nicely with Daly's (1994) conception of 'harmed and harming' women for whom the trauma of maltreatment influences participation in criminalized activity throughout the life course. Adult criminality is not simply a result of the stigmatization of involvement in the criminal justice system as a youth. Rather, the psychological impact of maltreatment appears to maintain an influence on the victim's adult behavior through the process of interactional continuity.

A different pattern is observed for the male subgroup. For males, involvement in juvenile delinquency mediates the effects of neglect on adult criminality. Thus, child maltreatment does not exert a significant effect on adult criminality independent of previous involvement in juvenile delinquency. Maltreatment influences adult criminal behavior only through the indirect effect of juvenile delinquency and the effects of labeling resulting from associations with the criminal justice system.

Because maltreatment influences female criminality through multiple pathways (see Figs 2 and 3), this study provides moderate support for the assumption that maltreatment has a more substantial effect on female, as compared to male criminality. However, these findings suggest additional quantitative and qualitative research is needed to accurately portray the effects of child maltreatment over the life course for females and males. Qualitative studies of female offenders should be complemented by control groups of male offenders and non-institutionalized individuals. Only in this manner can we avoid the dangers of inappropriate generalizations and conclusions. Moreover, additional quantitative research is needed to examine the effects of maltreatment on self-reported delinquency and criminality. Smith & Thornberry (1995) have made an important contribution to this effort, but have not taken gender differences into account.

Conclusions

The empirical analyses provide a number of results that deserve additional theoretical attention. One surprising finding is that the combination of abuse and neglect was not found to have a significant effect on juvenile arrest records for the male subgroup or the full sample. In contrast, the combination of these forms of maltreatment has a strong, significant effect on arrests for the female sample. A possible explanation for this finding may be derived from Chesney-Lind & Shelden's (1998) examination of the criminalization of women. These authors assert that attempts by girls to escape from abusive families (in the form of drug and alcohol use, running away, etc.) have been criminalized. They suggest that the victimizers of these girls have the ability to invoke official agencies of social control in their efforts to keep the victims in the home. When girls attempted to escape from the abuse, parents historically encouraged agencies of the state to enforce their return to the home. To their detriment, girls were more likely than boys to be referred to court by their parents instead of law enforcement agencies (Chesney-Lind & Sheldon, 1998). Once labeled, these girls were at greater risk for subsequent offending. Based on this scenario, parents may be neglectful of their daughters' needs while simultaneously relying on abuse and official agents of the state to control their daughters. These circumstances would predict the significant effect of a combination of abuse and neglect on girls' juvenile arrest records.

In contrast to the female sample, the results suggest that neglect alone influences the likelihood of arrest records for males, but neglect combined with abuse does not influence arrest records of males. According to Chesney-Lind & Sheldon's (1998) argument, abusing parents are not as likely to rely on official agents to control their sons. As a result, whereas neglect may free boys to commit delinquency, abuse combined with neglect may serve as a control on boys' deviant behavior. Moreover, even if abuse and neglect does increase delinquency among boys, the criminalization argument is addressing *official* involvement with the criminal justice system, not delinquency per se. Additional research examining both official arrest rates and self-reports of criminal activity would be useful to further illuminate the processes resulting in the criminalization of women. If Chesney-Lind and Shelden are correct, then the gap between self-reported criminal behavior and arrests for criminal behavior should be narrower for females as compared to males. In other words, if parents assist official agents of social control in exerting greater control over abused girls, then female delinquency will be more likely than male delinquency to result in an arrest.

The finding of a non-significant effect of the combination of abuse and neglect on juvenile delinquency for the full sample and male sub-group is strikingly similar to an earlier research project examining the developmental sequelae of child maltreatment in infancy (Egeland & Sroufe, 1981; Egeland, Sroufe & Erickson, 1983). In a prospective study of the antecedents of child maltreatment for high-risk families, these researchers were confronted with a perplexing finding: multiple forms of maltreatment were less detrimental to children's socioemotional well-being than the experience of a single form of maltreatment. For instance, Egeland & Sroufe (1981) found that psychological unavailability (of parents) *alone* had more detrimental consequences on socioemotional well-being than the *combination* of psychological unavailability and physical abuse. Moreover, victims of neglect alone scored high on an anger rating, whereas the neglected and abused group were more likely to express less confrontational frustration and whining. In a follow-up study, Egeland et al., found that neglected children were the least flexible and least creative of all the maltreated groups, lacking the necessary self-esteem and agency "to cope effectively with the environment" (1983, 468). Finally, the neglected children that did not suffer from physical abuse received the lowest ratings of all maltreated groups in both agency and self-esteem, presented the least positive and most negative affect, were most dependent, and demonstrated the lowest ego control (Egeland et al.,

1983, 469). In summary, the research of Egeland et al., mirror the current findings suggesting that neglect alone has more substantial negative effects on social outcomes than neglect combined with physical abuse.

Although the authors do not distinguish between boys and girls in their analysis, they provide a possible explanation to this quandary that is applicable to the current findings. Egeland & Sroufe suggest that "the contact that occurred in the form of physical abuse was better than no contact at all" (1981, 89). In other words, physical abuse may not exacerbate the negative outcomes of neglect, because the victims prefer the 'negative' parent-child interaction of abuse to an alternative of neglectful parenting or an absence of interaction.

The empirical results and the arguments presented thus far provide substantial evidence that neglect should be recognized as a detrimental form of maltreatment that serves as a precursor for criminal activity. Neglect was found to predict juvenile and adult arrest records for both males and females. In stark contrast, abuse was not associated with juvenile or adult arrests for the male sample. A substantial impact of neglect on delinquency was also found in the research of Zingraff and his colleagues (Zingraff et al., 1994). These authors suggest two reasons for the consistently strong relationship between neglect and delinquency. First, the lack of parental attention and parental control associated with neglect is likely to have an especially adverse effect on a child's cognitive and normative development. Second, neglect is highly associated with racial and economic disadvantage (Zingraff et al., 1994), and neither their data, nor the Widom (1994) data examined in this study provide a satisfactory control for socioeconomic status. Thus, measures of neglect may be reflecting not only the impact of maltreatment but also the relationship between poverty and official arrest rates.

Although the strength of the relationship between neglect and delinquency is of great importance for a complete understanding of the etiology of delinquency and crime, the discovery of gender differences in the relationship between abuse and criminality is the most important contribution of this study. A number of difficulties confront an interpretation of this finding. First, in a previous analysis of this data focusing only on arrests for *violent* offenses, Widom (1989c) reports that more females are sexually abused than males, and that individuals experiencing only sexual abuse have the lowest probability of arrests for violent offenses (even lower than the control group). Similarly, Zingraff et al., (1994) found that sexual abuse did not have an important impact on delinquency in their sample of maltreated youth. If females experience more sexual abuse, and sexual abuse is essentially unrelated to criminality, then a strong relationship between abuse and female arrest rates (such as that found in this study) should not exist. A more adequate explanation is required to explain our findings of a persistent relationship between females' abuse and criminality.

A return to the findings of Daly's (1994) in-depth study of male and female inmates provides some clues to an explanation of the gender differences in the abuse-crime relationship. Within her category of 'harmed and harming women' that suffered from child maltreatment, three subgroups are distinguished based on mediating circumstances that linked the maltreatment to later criminality: women whose violence resulted from alcohol consumption, women whose criminality was necessitated through a need to support drug habits, and women who could not cope with their immediate circumstances as a result of psychological problems (Daly, 1994, 47–55). Thus, drug, alcohol, and psychological problems supported mechanisms of interactional continuity through which early abuse resulted in subsequent criminal activities. Consistent with a psychogenic perspective, adult criminality for these women is not simply a consequence of involvement with juvenile delinquency, but is a symptom of "a lingering psychic disorder whose etiology is traced directly to characteristics of the original (abuse) experience" (Browning & Laumann, 1997, 541).

In contrast to the female victims of child maltreatment, Daly states that the 'harmed and harming men' in her study "were not characterized by family members as 'out of control,' nor did they exemplify alcohol related violence as much as the women did" (1994, 71). Daly argues that victimization in the early years of these men's lives did not result in violent or acting out behavior characteristic of harmed and harming women, but resulted instead in a life without purpose and substance use ('retreatism' in Merton's (1938) sense of the word).

More complete explanations of the gender differences in the abuse-criminality relationship should be pursued because the results of Daly's (1994) research and the current analyses contrast with traditional images of female offending. For instance, Widom argues that, whereas male victims of abuse may express their aggression outwardly, females express subtler manifestations of abuse such as depression (1989b, 266). The results of this study and Daly's (1994) research deny the role of a 'cycle of violence' for males that links child abuse to their violent criminal behavior. Moreover, Daly's (1994) research concurs that abuse results in 'subtler' outcomes such as psychological problems and substance abuse for females, but these outcomes tend to link childhood victimization with serious criminal behaviors in her sample. Additional quantitative and qualitative research is necessary for a more thorough comprehension of the complex relationships between gender roles, childhood victimization, and crime.

Finally, as noted throughout this chapter, interpretations of this research must take into account the fact that the measures of maltreatment and the measures of criminal involvement stem from official sources. Because these data were collected before the passage of 'mandatory report' laws, the individuals in the maltreated group may represent victims of the most severe instances of abuse and neglect. In addition, subsequent interactions with the criminal justice system may in part reflect the stigmatization and labeling effects of official court processing associated with the incident of maltreatment. A final limitation of this study is the lack of information regarding the subjects' adult life situations. Measures of life transitions and trajectories in addition to juvenile delinquency would provide a stronger test of the psychogenic perspective versus the life course perspective.

NOTES

1. Evidence of child maltreatment in this study was determined through the examination of pre-sentence investigations, rather than through face-to-face interviews. As a result, the findings of this study may be influenced by the perceptions and biases of the presentence investigators. See Daly (1994) for details.

2. The formula used to test the null hypothesis of equality of regression coefficients across subgroups is given by Clogg, Petkova & Haritou (1995, 1276): ($z = (\beta_1 - \beta_2)/[s^2(\beta_1)+s^2(\beta_2)]^{1/2}$) Absolute values greater than 1.96 represent a significant difference in regression coefficients across subgroups.

3. Values are changes in odds. Only significant paths are included in the diagram. The models also include background variables (not shown). Because the values in Fig. 2 and Fig. 3 are odds-ratios derived from logistic regression models, it is not possible to compute indirect effects.

ACKNOWLEDGEMENTS

The data used in this publication were obtained through the Inter-University Consortium for Political and Social Research (ICPSR 9480). The data were made available by the Data Resources Program of the National Institute of Justice, Sociometrics Corporation, 170 State Street, Suite 260, Los Altos, CA 94022–2812. The study entitled Child Abuse, Neglect, and Violent Criminal Behavior was conducted by Cathy Spatz Widom, Departments of Criminal Justice and Psychology, Indiana University, 302 Sycamore Hall, Bloomington, Indiana 47405. Data collection was funded by the National Institute of Justice (Award No. 86–IJ-CX-0033). Funding support for preparing the revised documentation for public distribution was provided by a contract (OJP-89–C-008) between the U.S. Office of Justice Programs and Sociometrics Corporation. Neither the original investigators, nor the funding agency, nor the Data Resources Program bear any responsibility for the analyses or interpretation presented here.

REFERENCES

Alfaro, J. D. (1981). Report on the Relationship between Child Abuse and Neglect and Later Socially Deviant Behavior. In: R. J. Hunner & Y. E. Walker (Eds), *Exploring the Relationship between Child Abuse and Delinquency,* (175–219). Montclair, NJ: Allanhel, Osmun.

Bolton, F. G., Reich, J. W., & Gutierres, S. E. (1977). Delinquency Patterns in Maltreated Children and Siblings. *Victimology, 2,* 349–357.

Browne, A., & Finkelhor, D. (1986). Initial and Long-Term Effects: A Review of the Research. In: D. Finkelhor (Ed.), *A Sourcebook on Child Sexual Abuse,* (143–179). Beverly Hills: Sage.

Browning, C. R., & Laumann, E. O. (1997). Sexual Contact Between Children and Adults: A Life Course Perspective. *American Sociological Review, 62,* 540–560.

Burgess, A. W., Hartman, C. R., & McCormack, A. (1987). Abused to Abuser: Antecedents of Socially Deviant Behaviors. *American Journal of Psychiatry, 144,* 1431–1436.

Caspi, A., Elder, G. H. Jr., & Bem, D. J. (1987). Moving against the World: Life-Course Patterns of Explosive Children. *Developmental Psychology, 23,* 308–13.

Chesney-Lind, M., & Rodriguez, N. (1983). Women under Lock and Key. *The Prison Journal, 63,* 47–65.

Chesney-Lind, M., & Shelden, R. G. (1998). *Girls, Delinquency, & Juvenile Justice.* (2nd ed.). Belmont, CA: Wadsworth Publishing.

Clogg, C. C., Potkova, E., & Haritou, A. (1995). Statistical Methods for Comparing Regression Coefficients between Models. *American Journal of Sociology, 100,* 1261–1293.

Conaway, L. P., & Hansen, D. J. (1989). Social Behavior of Physically Abused and Neglected Children: A Critical Review. *Clinical Psychology Review, 9,* 627–652.

Daly, K.. (1994). *Gender, Crime, & Punishment.* New Haven, CT: Yale University Press.

Egeland, B., & Sroufe, A. (1981). Developmental Sequelae of Maltreatment in Infancy. *New Directions for Child Development, 11,* 77–92.

Egeland, B., Sroufe, L. A., & Erickson, M. (1983. The Developmental Consequence of Different Patterns of Maltreatment. *Child Abuse & Neglect, 7,* 459–469.

Glueck, S., & Glueck, E. (1950). *Unraveling Juvenile Delinquency.* Cambridge, MA: Harvard University Press.

Gray, E. (1981). The Link Between Child Abuse and Delinquency: What We Know and Recommendations for Policy and Research. In: G. T. Hotaling, D. Finkelhor, J. T. Kirkpatrick & M. A. Straus (Eds), *Family Abuse and Its Consequences: New Directions in Research,* (109–123). Newbury Park, CA: Sage.

Hagan, J., Gillis, A. R., & Simpson, J. (1985). The Class Structure of Gender and Delinquency: Toward a Power-Control Theory of Common Delinquent Behavior. *American Journal of Sociology, 90,* 1151–1178.

Hagan, J., & McCarthy, B. (1997). *Mean Streets: Youth Crime and Homelessness.* New York: Cambridge University Press.

Henry, B., Moffitt, T. E., Caspi, A., Langley, J., & Silva, P. A. (1994). On the 'Remembrance of Things Past: A Longitudinal Evaluation of the Retrospective Method. *Psychological Assessment, 6,* 92–101.

Hout, M., & Greeley, A. (1998). What Church Officials Reports Don't Show: Another Look at Church Attendance Data. *American Sociological Review, 63,* 113–119.

James, J. (1976). Motivations for Entrance into Prostitution. In: L. Crites. (Ed.), *The Female Offender,* (177–205). MA: Lexington Books.

James, J., & Myerding, J. (1977). Early Sexual Experience and Prostitution. *American Journal of Psychiatry, 134*, 1382–1385.

Janus, M.-D., Archambault, F. X., Brown, S. W., & Welsch, L. A. (1995. Physical Abuse in Canadian Runaway Adolescents. *Child Abuse & Neglect, 19*, 433–447.

Kakar, S. (1996). *Child Abuse and Delinquency*. Lanham, MD: University Press of America

Kratcoski, P. C. (1982). Child Abuse and Violence Against the Family. *Child Welfare, 61*, 435–444.

Lewis, D. O., & Shanok, S. S. (1977). Medical Histories of Delinquent and Nondelinquent Children: An Epidemiological Study. *American Journal of Psychiatry, 134*, 1020–1025.

Martin, H. P., & Beezley, P (1980). Behavioral Observations of Abused Children. In: J. V. Cook & R. T. Bowles (Eds), *Child Abuse: Commission and Omission*, (435–450). Butterworths.

McCord, J. (1983). A Forty Year Perspective on Effects of Child Abuse and Neglect. *Child Abuse and Neglect, 7*, 265–270.

McCormack, A., Janus, M-D., & Burgess, W. A. (1986). Runaway Youths and Sexual Victimization: Gender Differences in an Adolescent Runaway Population. *Child Abuse & Neglect, 10*, 387–395.

Merton, R. K. (1938). Social Structure and Anomie. *American Sociological Review, 3*, 672– 682.

Messerschmidt, J. W. (1986). *Capitalism, Patriarchy, & Crime: Towards a Socialist Feminist Criminology*. Totowa, NJ: Rowman and Littlefield.

Miller, E. M. (1986). *Street Woman*. Philadelphia: Temple University Press

Mouzakitis, C. M. (1981). An Inquiry into the Problem of Child Abuse and Juvenile Delinquency. In: R. J. Hunner & Y. E. Walkers (Eds), *Exploring the Relationship Between Child Abuse and Delinquency*. Montclair, NJ: Allenheld, Osmun.

Presser, S., & L. Stinson. (1998). Data Collection Mode and Social Desirability Bias in Self-Reported Religious Attendance. *American Sociological Review, 63*, 137–145.

Richie, B. E. (1996). *Compelled to Crime: The Gender Entrapment of Battered Black Women* New York: Routledge.

Rush, F.. (1980). *The Best Kept Secret: Sexual Abuse of Children*. New York: McGraw–Hill.

Sampson, R. J., & Laub, J. H. (1993). *Crime in the Making: Pathways and Turning Points through Life*. Cambridge, MA: Harvard University Press.

Scudder, R. G., Blount, W. R., Heide, K. M., & Silverman, I. J. (1993). Important Links Between Child Abuse, Neglect, & Delinquency. *International Journal of Offender Therapy and Comparative Criminology, 37*, 315–323.

Silbert, M. H., & Pines, A. M. (1981). Sexual Child Abuse as an Antecedent to Prostitution. *Child Abuse & Neglect, 5*, 407–411.

Smith, C., & Thornberry, T. P. (1995). The Relationship Between Childhood Maltreatment and Adolescent Involvement in Delinquency. *Criminology, 33*, 451–482.

Spohn, R. E. (1998). *A Theoretical Development and Empirical Assessment of the Effects of Child Abuse and Neglect on Juvenile Delinquency*. Unpublished Masters Thesis, Texas A&M University.

U.S. Department of Health and Human Services. (1998). *Child Maltreatment 1996. Reports from the States to the National Child Abuse and Neglect Data System*. Washington, DC: U.S. Government Printing Office.

Wick, S C. (1981). Child Abuse as Causation of Juvenile Delinquency in Central Texas. In: R. J. Hunner & Y E. Walker (Eds), *Exploring the Relationship between Child Abuse and Delinquency*, (233–239). Montclair, NJ: Allenheld, Osmun & Co. Press.

Widom, C. S. (1989a). Child Abuse, Neglect, and Adult Behavior: Research Design and Findings on Criminality, Violence, and Child Abuse *American Journal of Orthopsychiatry, 59*, 355–367.

Widom, C. S. (1989b). Child Abuse, Neglect, and Violent Criminal Behavior. *Criminology, 27*, 251–271.

Widom, C. S (1989c). The Cycle of Violence. *Science, 244*, 160–166.

Widom, C. S. (1989d). Does Violence Beget Violence? A Critical Examination of the Literature. *Psychological Bulletin, 106*, 3–28.

Widom, C. S. (1991) Avoidance of Criminality in Abused and Neglected Children. *Psychiatry, 54*, 162–174.

Widom, C. S. (1994). *Child Abuse, Neglect, and Violent Criminal Behavior in a Midwest Metropolitan Area of the United States, 1967–1988* [MRDF]. Compiled by the Departments of Criminal Justice and Psychology, Indiana University. ICPSR ed. Ann Arbor, MI: Inter-university Consortium for Political and Social Research [producer and distributor].

Widom, C. S., & Ames, M. A. (1994). Criminal Consequences of Childhood Sexual Victimization. *Child Abuse & Neglect, 18*, 303–318.

Williams, L. M. (1995). Recovered Memories of Abuse in Women with Documented Child Sexual Victimization Histories. *Journal of Traumatic Stress, 8*, 649–673.

Zingraff, M. T., Leiter, J., Myers, K. A., & Johnson, M. C. (1993). Child Maltreatment and Youthful Problem Behavior. *Criminology, 31*, 173–202

Zingraff, M. T., Leiter, J., Johnson, M. C., & Myers, K. A. (1994). The Mediating Effect of Good School Performance on the Maltreatment Delinquency Relationship. *Journal of Research in Crime and Delinquency, 31*, 62–91.

THE EFFECT OF EARLY DELINQUENCY AND SUBSTANCE USE ON PRECOCIOUS TRANSITIONS TO ADULTHOOD AMONG ADOLESCENT MALES

Carolyn A. Smith, Marvin D. Krohn, Alan J. Lizotte, Cynthia Perez McCluskey, Magda Stouthamer-Loeber and Anne Weiher

INTRODUCTION

The life course perspective on human development expressly focuses on development across the entire life span, including transitions into, out of, and through life trajectories (Baltes, 1987). Trajectories are long-term, age-graded patterns of development with respect to major social institutions such as family, education, and occupation (Thornberry, 1996). Transitions are events or short-term changes with respect to social roles that can mark turning points in a life

***This work was supported by grants to each of the three projects from the Office of Juvenile Justice and Delinquency Prevention, Office of Justice Programs, U.S. Department of Justice. Additional funding was provided by the National Institute on Drug Abuse (Denver and Rochester projects), the National Institute of Mental Health (Pittsburgh project), and the National Science Foundation (Rochester project). Points of view in this document are those of the authors and do not necessarily represent the official position or policies of the funding agencies.**

Families, Crime and Criminal Justice, Volume 2, pages 233–253.
2000 by Elsevier Science Inc.
ISBN: 0-7623-0737-4

trajectory (Cowan, 1991; Elder, 1991). There are cultural and developmental expectations about the expected sequence of life transitions that occurs in the course of becoming an adult. For example, completing one's education should precede developing a career, marrying, and starting a family; similarly, marriage should precede parenthood (Kamerman, 1981; Rindfuss, Swicegood & Rosenfeld, 1987). However, the timing and sequence of these transitions are often disrupted, and at least half of the population does not follow this normative sequence of life cycle transitions (Rindfuss et al., 1987).

Life span theory suggests that out-of-order or off-time transitions can be associated with disruptions in normal developmental sequences, leading to short- and long-term consequences in later development (Elder, 1985). Transitions can be either out of order, such as having children before marriage, or 'off-time', such as having children at young ages, or both. Precocious transitions, those that occur off-time, include the assumption of adult roles like curtailing education, leaving home, entering the job market, or becoming a parent, before the normatively expected age. Adolescents who enter adult roles prematurely are typically not prepared for the obligations that accompany those roles (Newcomb & Bentler, 1988). In addition, precocious transitions can increase the probability of economic burdens and of unsuccessful or failed transitions such as divorce or job instability. The behavioral disruption, stress, and structural disadvantage caused by precocious transitions can, in turn, increase the probability of deviance, and thus continued maladjustment during the adult years (Krohn, Lizotte & Perez, 1997). Because of these long-term, negative consequences it is important to understand the precursors of precocious transitions so that they may be prevented.

In this study, we examine the effect of early adolescent substance use and delinquency among urban male adolescents on three precocious transitions: dropping out of school, causing a pregnancy, and leaving home prematurely. A unique feature of the current analysis is that we use panel data from high-risk samples located in three cities, Denver, Pittsburgh, and Rochester, that were expressly collected to facilitate such comparative analyses. Three companion projects, the Denver Youth Survey, the Rochester Youth Development Study, and the Pittsburgh Youth Study, have gathered data over a span of at least six years that allow us to investigate the development and consequences of problem behaviors among urban youth (Huizinga, Loeber & Thornberry, 1995). The longitudinal design of such studies enables us to understand the consequences of deviance in the life course as it unfolds in a number of developmental arenas.

SUBSTANCE USE, DELINQUENCY, AND PRECOCIOUS TRANSITIONS

Substance use and delinquency in early adolescence can have a number of problematic consequences in the adolescent life course. Involvement in deviant activities may increase tensions within the family, affect schoolwork, and increase the probability that adolescents will associate with others who are involved in a broad spectrum of deviant activities. Both the use of drugs and alcohol, and involvement in delinquent behavior during the adolescent years can increase the precarious tension between the desire to exit adolescent roles and the need to properly prepare for entry into adult roles.

One of the most significant 'off-time' transitions in the adolescent life course is teen pregnancy, since this has consequences for the next generation, as well as immediate and long-term consequences for the adolescent. Teen pregnancy is therefore among the transitional events that can have a powerful influence on the life course, for adolescent fathers as well as for young mothers (Furstenberg, Brooks-Gunn & Morgan, 1987; Lerman & Ooms, 1993).

Early delinquent behavior and substance use are related to sexual behavior among adolescents (Elliott & Morse, 1989; Jessor & Jessor, 1977; Newcomb & Bentler, 1988; Pugh et al., 1990; Smith, 1997). Moreover, Pugh et al. (1990) and Elliott & Morse (1989) found that delinquency and drug use are causally prior to sexual intercourse. Such behaviors in early adolescence, in turn, increase the probability of precocious transitions such as teenage pregnancy and teenage parenthood (Newcomb & Bentler, 1988; Pirog-Good, 1988; Pugh et al., 1990; Thornberry, Smith, & Howard, 1997). Hence, research indicates that early participation in delinquent behavior and the use of drugs or alcohol increases the probability of either becoming pregnant or impregnating someone and also of parenthood while still in the teenage years (Newcomb & Bentler, 1988; Pugh et al., 1990; Thornberry et al., 1997).

Education is considered a necessary prerequisite to entering adult roles and statuses. At a minimum, it is expected that adolescents will complete high school before entering career and family trajectories. A youth's decision not to stay in school can be characterized as a key turning point that can determine life chances long into the future (Entwhistle, 1990).

There is some evidence that both adolescent substance use and involvement in delinquency are related to dropping out of school (Elliott & Voss, 1974; Fagan & Jones, 1984; Friedman, Glickman & Utada, 1985; Hawkins & Lam, 1987; Kelly & Pink, 1971; Krohn et al., 1995; Mensch & Kandel, 1988; Thornberry, Moore & Christenson, 1985). However, this relationship may only be indirect (Kaplan & Liu, 1994) or even spurious. Fagan & Pabon (1990) have suggested that drug use, delinquency, and dropping out of high school all share similar correlates such as school and family disruption and, therefore, once these factors are introduced, the impact of delinquency and drug use on dropping out may be significantly reduced or disappear. They find support for their argument when they control for social development factors such as family and school variables. On the other hand, Friedman et al. (1985) and Mensch & Kandel (1988) control for similar factors and still find a direct relationship between drug use and dropping out of school.

Moving out of one's parental home before graduating from high school can also be considered a precocious transition to adult status. Residential independence is an important sign of autonomy and independence from parents. At later ages it is usually negotiated between child and parents in an orderly sequence. If it takes place early, however, it can involve strong conflict of expectations, which can add to its later costs. The consequences of non-family living, particularly at young ages, include mistimed marriage and parenthood, and poor educational and vocational planning (Goldscheider & Goldschieder, 1993).

Leaving one's family home prematurely is likely to be affected by delinquent behavior or substance use. Participation in delinquency or substance use can cause a rift in the relationship between parents and their adolescent children. Indeed, numerous studies have found that the quality of the relationship between parents and children is related to delinquency and substance use (see Loeber & Stouthamer-Loeber, 1986, for a review of some of this research). Moreover, several studies have shown that deviant behavior may have a stronger effect on attachment to parents than attachment to parents has on deviant behavior (Jang & Smith, 1997; Liska & Reed, 1985; Thornberry et al., 1991). A rift between adolescents and their parents may, in turn, lead to adolescents moving out of the house to live alone, with a friend or partner, or in some other situation where they are independent of parental supervision. Newcomb & Bentler (1988) have suggested that certain behaviors such as substance use and delinquency accelerate entry into adult roles, and they review a number of theories and mechanisms through which this might occur. One of the main mechanisms is the 'truncation of vital developmental sequences and the premature immersion into adult roles' (1988, 37). This further detaches these adolescents from adult guidance and support, and leaves them freer to indulge in deviant activities, and associate with similarly inclined peers.

Making the transition from adolescence to adulthood is a difficult and often disruptive process in the life course. Adolescents are expected to prepare themselves for the roles that they will play as adults while, at the same time, conforming to normative expectations and parental expectations in regard to both their current behavior and the timing and order of the transitions they make. When adolescents deviate from behavioral expectations, a process can be set in motion whereby they become more deeply entrenched in a deviant lifestyle. As part of this process, adolescent deviant behavior may result in adolescents trying to make the transition to adult roles and statuses prematurely. These precocious transitions can in turn have significant effects on later life events and life chances.

The current analysis examines the effect of early delinquent behavior and substance use on precocious transitions among young adolescent males. While some prior research has found relationships between adolescent deviance and precocious transitions, the current study is unique in that it (1) examines the impact of both delinquency and substance use on precocious transitions; (2) focuses on several precocious transitions rather than only one; (3) uses a sample that is at high risk for both delinquency and precocious transitions; (4) controls for variables that may be associated with both early delinquency, drug use and precocious transitions; and (5) examines these relationships with data from three cities that were collected with the express purpose of replicating results across research sites.

METHODS

Sample and Design

This study reports on data from the Program of Research on the Causes and Correlates of Delinquency, which consists of three research projects: the Denver Youth Survey, the Pittsburgh Youth Study, and the Rochester Youth Development Study. All three projects use longitudinal designs to study the developmental pathways that lead to youth offending and problem behaviors. Taken together, the three projects selected a total of 4,000 inner-city youth for study (Huizinga, Loeber & Thornberry, 1995). Because of the low base rate for serious delinquency, youth at high risk were overrepresented in all three samples; however, all three samples are probability samples which can be weighted to represent the populations from which they were drawn. All results reported in this study are based on weighted data.

In the first phase of the studies, subjects were interviewed over a period of at least four years. Face-to-face interviews were completed with subjects and parents or caretakers at regular intervals. All projects also collected data from agencies such as police, courts, schools, and social service agencies. In all projects, interviews lasted between one and two hours and were conducted in private settings. Respondents who moved were followed and interviewed in their new location. Sample retention has been excellent, with at least 84% of the subjects retained at each of the three sites. All sites examined the potential for bias due to differential attrition among different population groups, and concluded there was no evidence that the remaining samples were biased.

A common measurement strategy was devised, which has enabled us to replicate results across sites. Common measures exist on key dependent variables such as drug use and delinquency, as well as on key predictive variables such as school performance, school attitudes, parent-child relationships, demographic characteristics of the family, and so forth. Each project also contains measurement of critical adolescent developmental transitions, although these vary somewhat in specifics as discussed below.

While there is great commonality across the projects of the Program of Research on the Causes and Correlates of Delinquency, each project has some unique design features. To make the subjects of this analysis comparable across cities, only males of approximately the same age are included. A brief description of each design follows in order to indicate the key characteristics of each study.

The *Denver Youth Survey* (DYS) is based on a probability sample of households in 'high-risk' neighborhoods of Denver, Colorado. The survey respondents include approximately equal numbers of boys and girls, aged 7, 9, 11, 13, and 15 years old in 1988, and one of their parents, who lived in one of the randomly selected households. Each child or youth and parent were interviewed annually five times from 1988 to 1992. Only the oldest two cohorts are included in this analysis. Data from the first through fourth annual interviews are utilized, covering the ages 14 through 17. The Denver sample includes 296 males, including 39% African American, 47% Latino, 5% white, and 9% other males. Latino subjects from Denver are usually of Mexican descent.

The *Pittsburgh Youth Study* (PYS) started in the Spring of 1986. Boys attending the first, fourth, and seventh grades in the Pittsburgh Public School system were randomly selected for participation in an initial screening which consisted of a retrospective assessment of problem behaviors, as reported by the boy, his parent, and his teacher. The top 30%, approximately 250 boys, with the highest rates of disruptive behavior were selected from each of the three samples, and an equal number of the remaining 70% were randomly selected. Thus, about 500 boys from each of the three grade samples qualified for the follow-up assessments. Data utilized spanned eight half-yearly assessments covering four years of the study. In this study, only data from the oldest cohort of 506 boys is included, and their ages ranged from 13.2 at the first assessment to 17 at the eighth. The sample includes 57% African American and 43% white males.

The *Rochester Youth Development Study* (RYDS) started with a sample of 1,000 Rochester public school students in 1988. To maximize the number of serious, chronic offenders available for the study, the sample includes more youth from high-crime neighborhoods and fewer from low-crime neighborhoods. In addition, the sample was selected to be 75% male and 25% female. Interviews were conducted at six-month intervals (waves) between 1988 and 1992, providing nine data collection points over this 4½–year period. At the first assessment students were in the Spring semester of their seventh or eighth grade year, and the average age was 13.5. Data span the period from the first to the ninth assessment, when subjects were 17.5 years of age on average. The sample for this analysis includes all 729 males, of whom 56% are African American, 27% are white and 17% are Latino. Latino subjects in the Rochester sample are most often of Puerto Rican descent.

Measurement

Due to the design and purposes of the overall Program of Research, many measures are similar across sites, and a general description will be used in these cases. Where particular projects use a different measurement approach, this will be noted. The variables are grouped and will be discussed in three categories: delinquency and substance use, precocious transitions, and control variables. The coding of variables is summarized in Table 1.

Delinquency and Substance Use. Common measures from student interviews are used for these variables across sites. A measure of general delinquency and a measure of substance use are used as indicators of early adolescent deviance. These measures cumulate the frequency of responses to delinquency items, and instances of use of alcohol, marijuana, and other drugs across the first two years of data collection in the three projects, a period when subjects were generally 13–15 years old. Measurement is therefore temporally prior to the onset of the precocious transitions in all but a handful of cases. In the Pittsburgh and Rochester projects, data are cumulated across the first four biannual interviews; in Denver, they are cumulated across the first two annual interviews. The natural logarithm of drug and alcohol use and delinquency frequencies is used owing to skewness in the data. Means for general delinquency and substance use are comparable across sites.

The general delinquency measure is a 31-item omnibus index, including status, property, violent, and public disorder offenses. Early substance use is measured by summing responses to items asking whether the respondent had used alcohol, marijuana, crack, cocaine, hallucinogens, tranquilizers, amphetamines, or barbiturates.

Precocious Transitions. The most likely precocious transitions experienced during adolescence are causing a pregnancy, dropping out of school, and leaving the parental home early.

Table 1. Variable List and Descriptive Statistics.*

Variable	Categories	Denver	Pittsburgh	Rochester
Early Delinquency and Substance Use				
General Delinquency – unlogged	Continuous	49.11	39.43	41.67
		(122.46)	(103.77)	(90.64)
General Delinquency – logged	Continuous	2.44	2.29	2.11
		(1.73)	(1.63)	(1.83)
Alcohol and Drug Use – unlogged	Continuous	48.20	13.84	21.62
		(202.86)	(51.97)	(79.65)
Alcohol and Drug Use – logged	Continuous	1.55	1.24	1.16
		(1.83)	(1.39)	(1.64)
Precocious Transitions				
Impregnate	0,1	0.15	0 18	0.24
Dropout	0,1	0.18	0.19	0.27
Independent Living	0,1	0.10	0.04	0.13
Total Number of Transitions	0–3	0.43	0.40	0.62
		(0.73)	(0.70)	(0 88)
Control Variables				
Family Structure	0,1	0.67	0.64	0.60
African American	0,1	0.39	0.57	0.56
Public Assistance	0,1	0.56	0.36	0.41
Attachment to Parents	Continuous	3.85[c]	2.56[a]	2.68[a]
		(0.57)	(0.35)	(0.33)
School Commitment	Continuous	3.68[c]	2.38[a]	3.05[b]
		(0.50)	(0.33)	(0.38)

**Note:* Standard deviation of continuous variables presented in parentheses
[a] Scale range 1–3; [b] scale range 1–4; [c] scale range 1–5

Causing a pregnancy is assessed by a positive response to questions asking if they had ever gotten a girl pregnant. In the DYS, males are asked at each interview if they had gotten a girl pregnant, and 'yes' responses at any wave are coded as 1. In the PYS, males are asked about impregnation at the seventh and eighth interview, when they are aged 16–17. A positive response at either assessment is coded as 1. In RYDS, impregnation questions are included in each interview after and including the fifth, which occurred at about age 15. Again, any positive response is coded as 1. In most cases, subjects report pregnancies occurring before the end of high school. Percentages of male respondents reporting that they caused a pregnancy are 15% in Denver, 18% in Pittsburgh, and 24% in Rochester.

Dropping out is measured by subjects' self-report of leaving school before high school graduation. If subjects indicated that they were not currently in school and had not graduated, they were given a score of 1 on the variable. In all three sites, dropout is assessed at a comparable age, when the subjects are about 17, at the fourth interview in Denver, the eighth in Pittsburgh, and the ninth in Rochester. In Denver, 18% of the boys dropped out of school, in Pittsburgh the proportion of dropouts is 19%, and in Rochester it is 27%.

The third precocious transition is *living independently*, that is, without adult supervision, before graduating from high school. Adolescents who leave the home of their parents or guardian prior to graduating from high school are considered to have left home prematurely. This measure is taken from questions in the student interview asking where the student lived, and whether or not this was with parents or caretakers. In comparing the proportion of youth across sites that have moved out of the parental home, we find that students in the PYS are different from youth in the other sites. Specifically, fewer males in the PYS live on their own, compared to their DYS and RYDS counterparts (4% compared to 10% and 13%, respectively). This may be because responses to the comparable question in the Pittsburgh interview do not identify whether the 'adult caretaker' of the youth who reports living at home is in fact a specific parental figure or some other adult. This may have resulted in a relatively higher count of those reporting living at home, versus away, in the Pittsburgh sample, and fewer reporting living elsewhere.

Finally, a global *total number of transitions* variable was created which sums the number of precocious transitions. The total number of transitions ranges from 0 to 3, depending on how many of the above precocious transitions were experienced by the subject.

Control Variables. A number of measures are included in the analysis as control variables either because of their suspected relationship with precocious transitions, or because they represent basic demographic dimensions. With the exception of the low social class measure, all variables are from student interviews, and are measured when subjects were about 14.

Family structure is assessed at the initial interview and comes from questions asking the respondent to describe whether they lived with both biological parents or in another type of family structure. Subjects are coded as 1 if they did *not* live with both biological parents at Wave 1. Across all sites, between 60% and 67% of subjects live in homes without both biological parents present.

African American is measured using a dummy variable coded 1 for African American subjects: the omitted category refers to adolescent males who are Latino, white, or some other race. In Denver, 39% of the sample males are African American, compared to 57% in Pittsburgh, and 56% in Rochester.

To measure low social class, we asked parents to indicate whether they were receiving public assistance. This is used rather than a more standard measure because of the overall disadvantaged status of our sample. The percentage of subjects receiving public assistance is 56% in Denver, 36% in Pittsburgh, and 41% for Rochester subjects.

Attachment to parents is assessed at each site with slightly different sets of items measuring the child's liking, lack of hostility, and warmth towards parents. In Rochester and Pittsburgh, the specific parent asked about is the mother or female caretaker; in Denver, all items except one refer to 'parents.' The Denver scale consists of 6 items and has an alpha reliability coefficient of 0.68. The Pittsburgh attachment scale consists of 8 items, and has a reliability coefficient of 0.77. The Rochester scale contains 10 items, and has an alpha reliability of 0.83. Attachment means cannot be compared because of the somewhat different items in each site.

School commitment is measured through a set of 7 items at each site assessing the degree to which the subject likes school, teachers, and is involved in schoolwork. Items are highly comparable, but the range of responses for each item in the scale varies from 3 to 5. The respective reliabilities of this scale in Denver, Pittsburgh, and Rochester are 0.65, 0.56, and 0.70. School commitment means cannot be directly compared because the response categories vary across sites.

RESULTS

In this analysis we evaluate the impact of early delinquency and substance use on precocious transitions among adolescent males. We expect early delinquency and substance use to increase both the probabilities of precocious transitions occurring and the overall number of precocious transitions experienced. The impact of early delinquency and substance use on precocious transitions is expected to be independent of family structure, attachment to parents, commitment to school, low social class, and race/ethnicity.

Table 2 shows correlations between early delinquency and substance use and the measures of precocious transitions for boys at each of the three sites. The table shows that both early delinquency and substance use are consistently and significantly associated with all measures of precocious transitions for the Denver and Rochester research sites. However, not all of these relationships are replicated in Pittsburgh. Five of the eight correlations are statistically significant for Pittsburgh. Early delinquency is associated with dropout and the total number of transitions, as in the other sites. Similarly, alcohol and drug use is associated with causing a pregnancy, dropout, and the total number of transitions in Pittsburgh as they are in the other two locations. However, general delinquency is not related to causing a pregnancy, and neither delinquency nor early alcohol and drug use are associated with independent living in Pittsburgh.

Of course, the correlations in Table 2 do not hold other variables constant. Tables 3, 4, and 5 show logistic regression equations predicting each of the three precocious transitions and Table 6 contains an ordinary least squares equation predicting the total number of precocious transitions for boys in each of the three sites. These equations control for family structure, attachment to parents, school commitment, low social class, and race/ethnicity. Model improvement chi-square tests show that the final equations are significantly better than equations with only the intercept included.

Table 3 shows that in Denver and Rochester early delinquency and substance use are strong and statistically significant predictors of causing a pregnancy. However in Pittsburgh only substance use is associated with causing a pregnancy. The values in brackets under the coefficients and their standard errors in the table are the predicted probabilities of each transition given a one unit change in the amount of delinquency or substance use.[1] For example, for Denver the difference in committing zero and 3 delinquencies results in about an 11% increase in the probability of causing a pregnancy when evaluated at an initial 0.5 probability of pregnancy. In other words, the probability of pregnancy moves from 0.50 to 0.61, a substantial change.

The table also shows the effects of control variables on precocious transitions. Receiving public assistance significantly increases the likelihood of causing a pregnancy for the alcohol equation in Denver and for both equations in

Table 2. Pearson Correlation Coefficients Between Early Delinquency and Substance Use, Precocious Transitions, and Total Number of Transitions.

Transition	Denver	Pittsburgh	Rochester
		General Delinquency	
Impregnate	0.26*	0.04	0.35*
Dropout	0.22*	0.21*	0.33*
Independent Living	0.25*	0.07	0.26*
Total Number of Transitions	0.35*	0.17*	0.43*
		Alcohol and Drug Use	
Impregnate	0.33*	0.09*	0.27*
Dropout	0.21*	0.23*	0.29*
Independent Living	0.29*	0.04	0.31*
Total Number of Transitions	0.39*	0.18*	0.40*

$p < 0.05$, one-tailed test

Table 3. Logistic Regression Coefficients Predicting Impregnation.[1]

Variable	Denver (n = 263)		Pittsburgh (n = 397)		Rochester (n = 610)	
Early Delinquency and Substance Use						
General Delinquency	0.44*		0.08		0.48*	
	(0.12)		(0.10)		(0.06)	
	[0.11]				[0.12]	
Alcohol and Drug Use		0.46*		0.19*		0.36*
		(0.10)		(0.10)		(0.06)
		[0.12]		[0.05]		[0.09]
Control Variables						
Family Structure	–0.05	–0.01	0.09	0.12	0.08	0.13
	(0.44)	(0.45)	(0.35)	(0.35)	(0.24)	(0.23)
Attachment to Parents	0.13	0.14	0.18	0.08	–0.10	–0.23
	(0.32)	(0.33)	(0.43)	(0.41)	(0.32)	(0.31)
School Commitment	–0.18	–0.11	0.09	0.18	0.05	–0.36
	(0.38)	(0.39)	(0.47)	(0.47)	(0.31)	(0.30)
Public Assistance	0.63	0.71*	1.25*	1.25*	0.35	0.32
	(0.40)	(0.42)	(0 30)	(0.30)	(0 22)	(0.22)
		[0.18]	[0.34]	[0.34]		
African American	0.42	0.54	0.69*	0.64*	0.77*	1 00*
	(0.39)	(0.40)	(0.34)	(0.33)	(0.23)	(0.23)
			[0.16]	[0.14]	[0.19]	[0.25]
Improvement in X^2	21 5*	30.3*	35.2*	38.8*	95.9*	68.5*

Note: Standard errors presented in parentheses
*p ≤ 0.05, one-tailed test

[1]Probabilities (in square brackets) are only estimated for statistically significant predictors Probabilities are estimated at the 50% likelihood of the precocious transition occurring.

Pittsburgh. Being African American increases the likelihood of causing a pregnancy for both equations in Pittsburgh and Rochester.

Table 4 shows the same analysis for school dropout in the three cities. Early delinquency and substance use significantly predict school dropout in all three locations. Early delinquency and alcohol and drug use increase the probability of dropout by between 7 and 10% depending upon the location. Commitment to school significantly and substantially decreases the likelihood of dropout in both Pittsburgh and Rochester. For example, in Pittsburgh a one unit increase in commitment to school decreases school dropout between 21 and 23%. Being on public assistance significantly increases the probability of dropping out in

Table 4. Logistic Regression Coefficients Predicting School Dropout.[1]

Variable	Denver (n = 244)		Pittsburgh (n = 380)		Rochester (n = 576)	
Early Delinquency and Substance Use						
General Delinquency	0.32*		0.28*		0.40*	
	(0.11)		(0.10)		(0.06)	
	[0.08]		[0.07]		[0.10]	
Alcohol and Drug Use		0.26*		0.33*		0.31*
		(0.10)		(0.10)		(0.06)
		[0.07]		[0.08]		[0.08]
Control Variables						
Family Structure	0.14	0.20	0.39	0.48	0.12	0.16
	(0.41)	(0.41)	(0.37)	(0.36)	(0.23)	(0.23)
Attachment to Parents	0.35	0.31	0.36	0.03	0.47	0.45
	(0.32)	(0.32)	(0.46)	(0.42)	(0.33)	(0.33)
School Commitment	–0.28	–0.37	–0.85*	–0.91*	–0.59*	–0.92*
	(0.37)	(0.37)	(0.47)	(0.47)	(0.32)	(0.31)
			[–0.21]	[–0.23]	[–0.15]	[–0.23]
Public Assistance	0.39	0.43	1.12*	1.16*	0.29	0.29
	(0.37)	(0.37)	(0.31)	(0.31)	(0.23)	(0.22)
			[0.28]	[0.29]		
African American	–0.19	–0.14	–0.02	–0.10	–0.63*	–0.40*
	(0.38)	(0.37)	(0.33)	(0.33)	(0.22)	(0.21)
					[–0.16]	[–0.10]
Improvement in X^2	14.1*	13.0*	39.0*	43.6*	74.3*	57.0*

Note Standard errors presented in parentheses
* $p \leq 0.05$, one-tailed test

[1]Probabilities (in square brackets) are only estimated for statistically significant predictors. Probabilities are estimated at the 50% likelihood of the precocious transition occurring.

Pittsburgh. Being African American significantly decreases the probability of dropout in Rochester by between 10 and 16%.

Table 5 shows that early delinquency and substance use have a significant impact on independent living for males in Denver and Rochester but not in Pittsburgh. In Pittsburgh family structure has a significant and substantial impact on independent living. It changes the probability of independent living by about 45% at the 50% likelihood. In other words, if a boy in Pittsburgh has a 50% chance of independent living and his family is not intact his likelihood of independent living becomes 95%. The more attached one is to one's parents,

Table 5. Logistic Regression Coefficients Predicting Independent Living.[1]

Variable	Denver (n = 247)		Pittsburgh (n = 380)		Rochester (n = 555)	
Early Delinquency and Substance Use						
General Delinquency	0.37* (0.14) [0.09]		0.09* (0.17)		0.45* (0.09) [0.11]	
Alcohol and Drug Use		0.36* (0.12) [0.09]		0.01* (0.19)		0.50* (0.08) [0.13]
Control Variables						
Family Structure	0.52 (0.57)	0.58 (0.58)	1.80* (0.97) [0.45]	1.83* (0.97) [0.46]	–0.11 (0.32)	–0.10 (0.33)
Attachment to Parents	0.78* (0.38) [–0.20]	0.82* (0.39) [–0.21]	–1.11* (0.67) [–0.28]	–0.99 (0.64)	–0.67* (0.39) [–0.17]	–0.63 (0.41)
School Commitment	–0.17 (0.44)	–0.17 (0.44)	–1.3 (0.84)	–0.25 (0.83)	0.34 (0.42)	0.22 (0.43)
Public Assistance	0.08 (0.47)	0.16 (0.48)	0.76 (0.56)	0.74 (0 56)	0.71* (0.32) [0.18]	0.75* (0.32) [0.19]
African American	–0.18 (0.48)	–0.31 (0.49)	–0.40 (0.58)	–0.42 (0.58)	–0.88* (0.31) [–0.22]	–0.65* (0.31) [–0.16]
Improvement in X^2	21.7*	24.7*	13.8*	12.8*	51.0*	61.7*

Note. Standard errors presented in parentheses
* $p < 0.05$, one-tailed test

[1]Probabilities (in square brackets) are only estimated for statistically significant predictors. Probabilities are estimated at the 50% likelihood of the precocious transition occurring

the lower the likelihood of independent living in all three cities. However, this effect is not significant for the alcohol and drug use equations in Pittsburgh and Rochester. In Rochester, receiving public assistance significantly increases the likelihood of independent living, while being African American significantly decreases the likelihood.

Table 6 shows ordinary least squares equations predicting the number of precocious transitions from early delinquency, early alcohol and drug use, and the control variables. Here we find that early delinquency and alcohol and drug

Table 6. Ordinary Least Squares Regression Coefficients Predicting Total Number of Precocious Transitions

Variable	Denver (n = 247)		Pittsburgh (n = 380)		Rochester (n = 555)	
Early Delinquency and Substance Use						
General Delinquency	0.32*		0.12*		0.42*	
	(0.14)		(0.02)		(0.20)	
Alcohol and Drug Use		0 38*		0.15*		0.37*
		(0.16)		(0.03)		(0.20)
Control Variables						
Family Structure	0.02	0.05	0.06	0.07	0.00	0 01
	(0.03)	(0 07)	(0.08)	(0.08)	(0.00)	(0.02)
Attachment to Parents	0.35	0.31	0.36	0.03	–0.01	–0.02
	(0.32)	(0.32)	(0.46)	(0.42)	(–0.03)	(–0.04)
School Commitment	–0.04	–0.04	–0.07	–0.07	–0.03	–0.08*
	(–0.5)	(–0.06)	(–0.11)	(–0.11)	(–0.07)	(–0.18)
Public Assistance	0.08	0.09	0.29*	0.29*	0.07*	0.07*
	(0.12)	(0.13)	(0.07)	(0.07)	(0.12)	(0.13)
African American	0 04	0.06	0.04	0.02	–0.04	0.02
	(0.05)	(0.09)	(0.07)	(0.07)	(–0.07)	(0.03)
R^2	13*	0.17*	0.13*	0.15*	0.20*	0.18*
Adjusted R^2	0.10	0.15	0.12	13	0.19	0.17

Note: Standard errors presented in parentheses
* $p < 0.05$, one-tailed test

use are statistically significant predictors of the number of precocious transitions for all three sites. In addition, school commitment significantly decreases the number of precocious transitions for the alcohol and drug equation in Rochester. Finally being on public assistance significantly increases the number of precocious transitions in Pittsburgh and Rochester, but not in Denver.

Further Exploration of Pittsburgh Findings

While general delinquency and alcohol and drug use significantly impact the probability of all three precocious transitions in Denver and Rochester, they do not significantly impact the probability of independent living in Pittsburgh. In addition, general delinquency does not predict causing a pregnancy in Pittsburgh, although substance use does. To identify the potential source(s) of variation, a number of factors were considered, including racial/ethnic differences across sites and measurement issues.

In the RYDS, we note that Latino males are most likely to move out of the parental home by Wave 9; at age 17, 22% of Latino males report that they live independently of parents, while only 10% of white and African American males have a similar living arrangement. We anticipated that the inclusion of Latinos in the RYDS sample might explain, in part, the difference between PYS and RYDS measures. To some extent, this notion is supported; the percentage of white and African American males in the RYDS who live independently of parents is lower than that of the larger sample (10% compared to 13%). However, the distinction across sites is not fully explained by ethnic composition. Latino subjects in the DYS are similar to other ethnic groups in their likelihood of living away from the parental home, and the difference between PYS and RYDS remains when Latino subjects are excluded from the analysis.

To further explore differences across sites, we reviewed the PYS independent living measure. As we noted in Table 1, only 4% of boys in Pittsburgh report independent living compared to 10% in Denver and 13% in Rochester. It is important to reiterate the difference in the way the independent living variable is measured in Pittsburgh compared to the other two sites. The measure is taken from Phase I of the PYS (1992) and covers a one year period, which is consistent with DYS and RYDS data. The PYS measure differs from the other sites: during the interview, PYS students are asked to describe their living arrangements, and indicate whether they currently live with adult caretakers or on their own. Adult caretakers are not specifically identified in the PYS, which may result in relatively more youth specifying that they live at home, and fewer reporting that they live away from home. In the other sites, students identify parental figures: in the DYS, students are asked about biological, adoptive, and step parents, and other adult caretakers present in the home. In the RYDS, students provide detailed information about all those living in the home and parent figures are identified by investigators. Unfortunately, we cannot explore further whether the differences in questions about living at home versus away are responsible for the different percentages of youth reporting living out of the parental home at different sites, and the differential impact of early deviance on this transition.

SUMMARY AND DISCUSSION

This chapter set out to examine the consequences of early delinquent behavior and substance use for making precocious transitions to adult roles and statuses. The precocious transitions examined include causing a pregnancy, dropping out of school, and living independently. The research involved a unique set of three samples of high-risk youth and controlled for variables that may be associated with both early delinquency, drug use and precocious transitions in all three sites.

To summarize briefly the results of this study, we do find support for our main hypothesis that deviance in early adolescence is associated with precocious transitions net of other key variables. The findings are consistent across sites, but lack statistical significance for some estimates in Pittsburgh. In Rochester and Denver, delinquency and substance use predict each of the transitions separately as well as the total number of transitions experienced by the youth. These findings are partly replicated for the Pittsburgh males: early delinquency and drug use predict dropping out of school and the total number of transitions, and drug use predicts impregnation. As we discussed above, the restricted definition and the resulting lower percentage of those living independently in Pittsburgh may contribute to the lack of statistical significance in the effect of early deviance on this particular transition. In addition to this measurement difference, there are also differences in the impact of the control variables, although all effects make good theoretical sense. These intriguing findings raise the possibility that the urban contexts and populations in these study sites may differ in ways not captured in our analysis, and we plan to investigate this further.

Although deviance predicts more precocious transitions in general, it is clear that for some adolescents early delinquency and substance use have little effect on transitions, indicating change as well as continuity in life course patterns (Rutter, 1992). Some youth are able to stay on track in major ways despite their early delinquency and substance use. This is an important and encouraging finding, and we need to continue to study the discontinuities in problematic behavioral trajectories for clues relating to turning points when interventions can be maximally effective.

In general, early deviance does seem to be interwoven with the tendency to embark prematurely on an adult-like life course, with poor skills preparation. This seems to suggest, as problem behavior theory indicates, that a general pattern of early adolescent deviance is continued in the violation of norms about age-appropriate transitions. Whatever the origins of deviance (and relevant contributing factors are controlled in these analyses), adolescents may be embarking at as early as 13 or 14 on a set of trajectories which increasingly deviates from expected developmental norms. These findings are consistent with the life course perspective, which argues that there is continuing interaction between the individual and the trajectories in which they are embedded, and that various life events can lead to short- and long-term consequences in the life course. There is some evidence that teenagers who are involved in early deviance have an increased probability of moving precociously to adult roles for which they may be poorly prepared. The tendency for earlier problems to persist in later problem behaviors is partly due to what Elder & Caspi (1990) have called the 'accentuation principle,' which suggests that stress and adversity in early life tend to lay the groundwork for vulnerability to later stress. For example, dropping out of school, moving away from home, and becoming a parent are likely to affect educational and occupational opportunities, making life more economically and emotionally stressful, and potentially further straining relationships with significant adults whose continued support is important (Maughan & Rutter, 1998).

As the significance of some of the control variables suggests, early deviance acts in concert with other early life disadvantages such as low attachment to parents, living in single-parent homes, and low social class. Adolescents who display patterns of early deviance must additionally cope with the potentially stressful after effects of the life disruptions associated with early parenthood and interrupted educational and residential trajectories. Where all these negative consequences collide, these late adolescents are likely to continue to need access to systems that foster skills development and positive parenting, lest these negative consequences become themselves proximate causes of further deviance in and out of the family. We have mounting evidence, for example, that adolescent delinquency is linked with later family violence (Magdol, Moffitt, Caspi & Silva, 1998; Capaldi & Clark, 1998).

The significant contribution of this study is to indicate the disruptive effect of early deviance on the young adult life course, including the impact on the formation of new families. Early life transitions can have developmental consequences long into the future as cumulative advantages and disadvantages in the life course are set in motion, and, in fact, these transitions can presage intergenerational continuities in unstable problem behaviors (Elder, 1998). The impact of multiple transitions is likely to be the most disruptive to the later life course, as youth lose the groundwork of family and school support at the same time they must deal with parenthood at young ages. Findings are replicated for three samples of youth in three very different urban contexts, suggesting that this is a general mechanism whereby deviance perpetuates itself, possibly into another generation.

NOTES

1. These probabilities are evaluated at the 50% likelihood of the particular precocious transition occurring. It is important to note that since the delinquency and substance use measures are logged, a change from zero to one on either scale is equivalent to about 3 (2.7) delinquencies. One unit change in higher values of the scale represent dramatically

increasing changes in the number of delinquencies that would be attributed to a one unit change in the logged variable (see Petersen, 1985).

REFERENCES

Baltes, P. B. (1987). Theoretical propositions of life-span developmental psychology: On the dynamics between growth and decline. *Developmental Psychology, 23,* 611–626.

Capaldi, D. M., & Clark, S. (1998). Prospective family predictors of aggression toward female partners for at-risk young men. *Developmental Psychology, 34,* 1175–1188.

Cowan, P. A. (1991). Individual and family life transitions: A proposal for a new definition. In: P. A. Cowan & M. Hetherington (Eds), *Family Transitions,* (3–30). Hillsdale, NJ: Lawrence Erlbaum.

Elder, G. H. Jr. (1985). Perspectives on the life course. In: Glen H. Elder, Jr. (Ed.), *Life Course Dynamics: Transitions and Trajectories,* (23–49). Ithaca: Cornell University Press.

Elder, G. H. Jr. (1991). Family transitions, cycles and social changes. In: P. A. Cowan & M. Hetherington (Eds), *Family Transitions,* (31–58). Hillsdale, NJ: Lawrence Erlbaum.

Elder, G. H., Jr. (1998) The life course and human development. In: R. M. Lerner (Ed.), *Handbook of Child Psychology, Volume 1: Theoretical Models of Human Development,* (939–991). New York: Wiley.

Elder, G. H. Jr. & Caspi, A. (1990). Studying lives in a changing society· Sociological and personological explorations. In: A. I. Rabin, R. A. Zucker, S. Frank & R. A. Emmons (Eds), *Studying Persons and Lives.* New York: Springer.

Elliott, D. S. & Morse, B. J. (1989). Delinquency and drug use as risk factors in teenage sexual activity. *Youth & Society, 21,* 32–60.

Elliott, D S., & Voss, H. L. (1974). *Delinquency and Dropout.* Lexington, MA: D.C. Heath.

Entwhistle, D. R. (1990). Schools and the adolescent. In: S. Feldman & G. R. Elliott (Eds), *At The Threshold: The Developing Adolescent,* (197–224). Cambridge, MA: Harvard University Press.

Fagan, J., & Jones, S. J. (1984). Towards a theoretical model for intervention with violent juvenile offenders. In: R. Mathias, P. Demuro, & R. Allinson (Eds), *Violent Juvenile Offenders: An Anthology.* San Francisco, CA· National Council on Crime and Delinquency.

Fagan, J., & Pabon, E. (1990). Contributions of delinquency and substance use to school dropout among inner city youths. *Youth and Society, 21,* 306–354.

Friedman, A. S., Glickman, N. & Utada, A. (1985). Does drug and alcohol use lead to failure to graduate from high school? *Journal of Drug Education, 15,* 353–364.

Furstenberg, F. J. Jr., Brooks-Gunn, J., & Morgan, S. P. (1987) *Adolescent Mothers in Later Life.* Cambridge: Cambridge University Press.

Goldschieder, F., & Goldschieder, C. (1993). *Leaving Home Before Marriage: Ethnicity, Familism and Generational Relationships.* Madison, WI: University of Wisconsin Press.

Hawkins, J. D., & Lam, T. (1987). Teacher practices, social development and delinquency. In: J. Burchard, & S. Burchard (Eds), *Prevention of Delinquent Behavior,* (242–274). Newbury Park, CA: Sage.

Huizinga, D., Loeber, R., & Thornberry, T. P. (Eds) *Recent Findings from the Program of Research on the Causes and Correlates of Delinquency.* Prepared for the Office of Juvenile Justice and Delinquency Prevention, U.S. Department of Justice, July 1995.

Jang, S. J., & Smith, C. A. (1997). A test of reciprocal causal relationships among parental supervision, affective ties, and delinquency. *Journal of Research in Crime and Delinquency 34,* 307–336.

Jessor, R., & Jessor, S. L. (1977). *Problem Behavior and Psychosocial Development: A Longitudinal Study of Youth.* New York: Academic Press.

Kamerman, S. (1981). Young people as individuals and as family members: The implications for public policy. In: A. C. Eurich (Ed.), *Major Transitions in the Human Life Cycle,* (101–114). Lexington, MA: Lexington Books.

Kaplan, H. B., & Liu, X. (1994). A longitudinal analysis of mediating variables in the drug use-dropping out relationship. *Criminology, 32,* 415–439.

Kelly, D., & Pink, W. (1971). School commitment and school careers. *Youth & Society, 3,* 224–235.

Krohn, M. D., Lizotte, A. J., & Perez, C. M. (1997). The interrelationship between substance use and precocious transitions to adult statuses. *Journal of Health and Social Behavior, 38,* 87–103.

Krohn, M., Thornberry, T. P., Collins-Halls. L., & Lizotte, A. J. (1995). School dropout, delinquent behavior, and drug use: An examination of the causes and consequences of dropping out of school. In: H. B. Kaplan (Ed.), *Drugs, Crime, and Other Deviant Adaptations: Longitudinal Studies,* (163–183). New York: Plenum Press.

Lerman, R. I., & Ooms, T. J. (1993). Introduction: Evolution of unwed fatherhood as a policy issue. In: R. I. Lerman & T. J. Ooms (Eds), *Young Unwed Fathers: Changing Roles and Emerging Policies,* (1–26). Philadelphia, PA: Temple University Press.

Liska, A. E., & Reed, M. D. (1985). Ties to conventional institutions and delinquency: Estimating reciprocal effects. *American Sociological Review, 50,* 547–560.

Loeber, R., & Stouthamer-Loeber, M. (1986). Family factors as correlates and predictors of juvenile conduct problems and delinquency. In: M. Tonry & N. Morris (Eds), *Crime and Justice: An Annual Review of Research, Vol. 7,* (20–149). Chicago: University of Chicago Press.

Magdol, L., Moffitt, T.E., Caspi, A., & Silva, P.A. (1998). Developmental antecedents of partner abuse: A prospective-longitudinal study. *Journal of Abnormal Psychology, 107,* 375–389.

Maughan, B., & Rutter, M. (1998). Continuities and discontinuities in antisocial behavior from childhoood to adult life. In: T. H. Ollendick & R. J. Prinz (Eds), *Advances in Clinical Child Psychology, Volume, 20* (1–47). New York: Plenum Press.

Mensch, B., & Kandel, D. B. (1988). Dropping out of high school and drug involvement. *Sociology of Education, 61,* 95–113.

Newcomb, M., & Bentler, P. M. (1988). *Consequences of Adolescent Drug Use: Impact on the Lives of Young Adults.* Newbury Park: Sage.

Petersen, T. (1985). A Comment on Presenting Results from Logit and Probit Models. *American Sociological Review, 50,* 130–31.

Pirog-Good, M. (1988). Teenage paternity, child support, and crime. *Journal of Substance Abuse,* 69, 527–546.

Pugh, M., DeMaris, A., Giordano, P., & Groat, T. (1990). Delinquency as a risk factor in teenage pregnancy. *Sociological Focus, 23,* 89–100.

Rindfuss, R. R., Swicegood, C. G , & Rosenfeld, R. A. (1987). Disorder in the life course: How common and does it matter? *American Sociological Review, 52,* 785–801.

Rutter, M. (1992). Adolescence as a transition period: Continuities and discontinuities in conduct disorder. *Journal of Adolescent Health, 13,* 451–560.

Smith, C. (1997) Factors associated with early sexual activity among urban adolescents. *Social Work, 42,* 334–346.

Thornberry, T. P. (1996). Introduction: Some advantages of developmental and life-course perspectives on crime and delinquency. In: T. P. Thornberry (Ed.), *Advances in Criminological Theory: Developmental Theories of Crime and Delinquency.* New Brunswick, NJ: Transaction Publishers.

Thornberry, T. P., Lizotte, A. J., Krohn, M. D., Farnworth, M., & Jang, S. J. (1991). Testing interactional theory: An examination of reciprocal causal relationships among family, school, and delinquency. *Journal of Criminal Law and Criminology, 82*, 3–35.

Thornberry, T. P., Moore, M., & Christenson, R.L. (1985). The effects of dropping out of high school on subsequent criminal behavior. *Criminology, 23*, 3–18.

Thornberry, T. P., Smith, C., & Howard, G. J. (1997). Risk factors for teenage fatherhood. *Journal of Marriage and the Family, 59*, 505–522.

PART III

FAMILIES IN THE CRIMINAL JUSTICE SYSTEM

THE EFFECTS OF DELINQUENCY AND LEGAL SANCTIONS ON PARENTING BEHAVIORS

Eric A. Stewart[1], Ronald L. Simons[2] and Rand D. Conger[2]

Research from many disciplines continues to demonstrate the importance of family context for children's development and behavior (Barber & Rollins, 1990; Conger & Conger, 1994; Conger & Simons, 1997; Patterson Reid & Dishion, 1992; Simons, Johnson, Conger & Elder, 1998). Criminologists have long recognized the importance of family influences in the etiology of delinquent behavior (Gottfredson & Hirschi, 1990; Loeber & Stouthamer-Loeber, 1986; Patterson et al., 1992; Sampson & Laub, 1993). Several theoretical statements identify parental mismanagement as a major cause of delinquent behavior (Conger and Simons, 1997; Gottfredson & Hirschi, 1990; Patterson et al., 1992; Sampson & Laub, 1993). In addition, a variety of studies have produced empirical support for the proposition that specific dimensions of parental behavior can either increase or decrease an adolescent's risk for antisocial

During the past several years, support for this research has come from multiple sources, including the National Institute of Mental Health (MH00567, MH19734, MH43270, MH48165, MH51361), the National Institute on Drug Abuse (DA05347), the Bureau of Maternal and Child Health (MCJ109572), the MacArthur Foundation research network on successful adolescent development among youth in high-risk settings, and the Iowa Agriculture and Home Economics Experiment Station (project number 3320).
[1]Georgia State University, Department of Sociology
[2]Iowa State University, Department of Sociology

Families, Crime and Criminal Justice, Volume 2, pages 257–279.

ISBN: 0-7623-0737-4

behavior (Elliott, Huizinga & Menard, 1989; Loeber & Stouthamer-Loeber, 1986; Patterson et al., 1992; Sampson & Laub, 1993; Simons et al.,1998). Most of this theorizing and research assumes that ineffective parenting leads to delinquent behavior. However, these perspectives typically fail to consider the importance of delinquency on parents' and family functioning.

A few researchers have argued that delinquency may not merely be an outcome variable, but rather delinquency and problem behavior may contribute to the disruption of parenting (Ambert, 1997; Jang & Smith, 1997; Lytton, 1990; Patterson et al., 1992; Thornberry, 1996, 1997). For example, interactional theorists would posit a reciprocal relationship between parenting and delinquency such that increased delinquency predicts decreases in effective parenting and vice versa. Empirical research supports this hypothesis (Jang & Smith, 1997; and Thornberry, Lizotte, Krohn, Farnworth & Jang, 1991).

In this chapter, we contribute to this line of research by examining the impact of delinquent behavior on parenting. The model extends previous investigations of this issue both theoretically and methodologically. Theoretically, we incorporate legal sanctions as a mediator of the impact of delinquency on parenting skills. Methodologically, we avoid the problem of shared-method variance by using multiple informants (i.e. parent, child, sibling, and observer reports) to measure theoretical constructs.

DELINQUENCY AND LEGAL SANCTIONS

The effects of official sanctioning may serve as a stigma or label for adolescents who are arrested for delinquent behavior. Labeling theory predicts that formal labels from social agents in the criminal justice system can typecast an individual as essentially deviant, even if the individual is otherwise nondeviant (Garfinkel, 1956; Lemert, 1951; Matza, 1969; Paternoster & Iovanni, 1989; Scheff, 1966). Sampson & Laub (1997) have provided a developmental view of labeling theory by making legal sanctions a central feature of their theoretical perspective. They note that delinquency often results in encounters with the legal system, and that such encounters tend to have a number of deleterious consequences. Their discussion of negative consequences focuses upon the manner in which arrest, official labeling, and incarceration undermine a person's educational and occupational opportunities, thereby increasing the probability of future criminal behavior.

Drawing on the thesis of cumulative disadvantage, Sampson & Laub (1997) suggest that the process generated by criminal offending and official sanctions significantly decreases the possibility for positive life chances. The theory specifically suggests a 'snowball' effect – that adolescent delinquency and its negative consequences increasingly 'mortgage' one's future, especially later life chances molded by education and employment. The logic of this theoretical perspective points to the possible indirect role of delinquency and legal sanctions in 'knifing off' positive prosocial avenues and generating future crime.

In the present study, we extend this perspective by examining the impact of legal sanctions on family life. Encounters with the legal system may disrupt the quality of family life, especially parenting. In particular, encounters with the legal system create pressures and embarrassment for parents as their child's problem behavior becomes public and they have to invest time and usually, economic resources to deal with such difficulties. Moreover, legal sanctions can produce adolescents who are more hostile and less willing to obey parental authority and punishment (Ambert, 1997). This suggests that some delinquent adolescents who come into contact with the criminal justice system may behave in a way that disrupts effective parenting practices. Rose & Clear (1998) note that overrepresentation of formal control weakens family structures and increases the negative behaviors they were intended to address. If this is the case, one can reasonably expect legal sanctions to damage relationship networks consisting of family members, such as parents and siblings.

DELINQUENCY, PROBLEM BEHAVIORS, AND PARENTING

Delinquency research has assumed until recently that parents exercise a predominantly one-way influence on their children. However, several studies by developmental psychologists suggest that parent-child interactions are more dynamic. This research indicates that in addition to 'parental effects' on children, there are 'child effects' on parents (Lytton, 1990). Not only does ineffective parenting increase the probability of child conduct problems, but hostile, obstinate child behavior is often followed by a reduction in parental efforts to monitor and discipline (Moffitt, 1993; Patterson et al., 1992; Simons et al., 1998).

Patterson's research on cycles of coercion suggests that antisocial children cause parents to be irritable, ineffective in discipline, and to withdraw their support and attention. This process accelerates the child's antisocial behavior, which, in turn, precipitates further deterioration in parenting. In a similar investigation, Anderson, Lytton & Romney (1986) compared the reactions of mothers of conduct-disordered children with those of mothers of normal children. Results suggested that it was characteristics of the child, rather than characteristics of the mother, which determined how mothers behaved toward their child. When paired with conduct-disordered children, both mothers of conventional and conducted disordered children showed more negative behaviors with conduct-disordered children. Such studies indicate that difficult children influence parenting styles likely to further exacerbate their antisocial tendencies.

It is important to note, however, that this research focuses on parent-child interaction during the childhood years, and the antisocial behavior displayed by the child consisted of oppositional/defiant behavior directed toward the parent. Hence, it is not clear that these findings can be generalized to the parent-child relationship during adolescence when the child's deviant behavior consists of delinquent behaviors exhibited outside of the home. Unfortunately, only a few studies (Ambert, 1999; Jang & Smith, 1997; Thornberry et al., 1991) have examined the manner in which adolescent delinquency influences parental behavior.

In an investigation of this issue, Ambert (1999) provided evidence that parents of delinquents suffered a wide spectrum of negative consequences that resulted from their adolescent's delinquent behaviors. Parents felt ashamed and had to seek help as a result of their child's inappropriate behavior. Both parents suffered reductions in overall health and happiness. Although both parents were affected by their adolescent's delinquency, mothers were uniformly more affected than fathers. Mothers were more likely to experience difficulties with other people (relatives, friends, and coworkers), more likely to seek professional counseling, and more likely to be blamed and feel shame. This point was extremely salient if the adolescent exhibited multiple behavioral problems and arrests.

Thornberry & colleagues (1991) found similar patterns. They found that delinquent behavior attenuates the strength of adolescent bonds to parents. As a result of this attenuation, the adolescent's delinquent transgressions increased over time, especially in early adolescence (Thornberry et al., 1991). In looking at mid adolescence, Jang & Smith (1997) found that delinquency often fosters a reduction in parental supervision. And, as adolescents engaged in a consistent pattern of delinquency, they reduced their emotional ties to parents.

These studies suggest that unidirectional models of the relationship between parental and family behaviors and adolescent delinquency are overly simplistic. Although it may be true that ineffective parenting and negative family interactions increase an adolescent's risk for delinquent behavior, it also appears that delinquency disrupts the parent-child relationship. The present study provides new information regarding this hypothesized process.

THE PROPOSED THEORETICAL MODEL

The theoretical model to be tested is shown in Figure 1. The model proposes that delinquency at Time 1 leads to legal sanctions at Time 2. Legal sanctions,

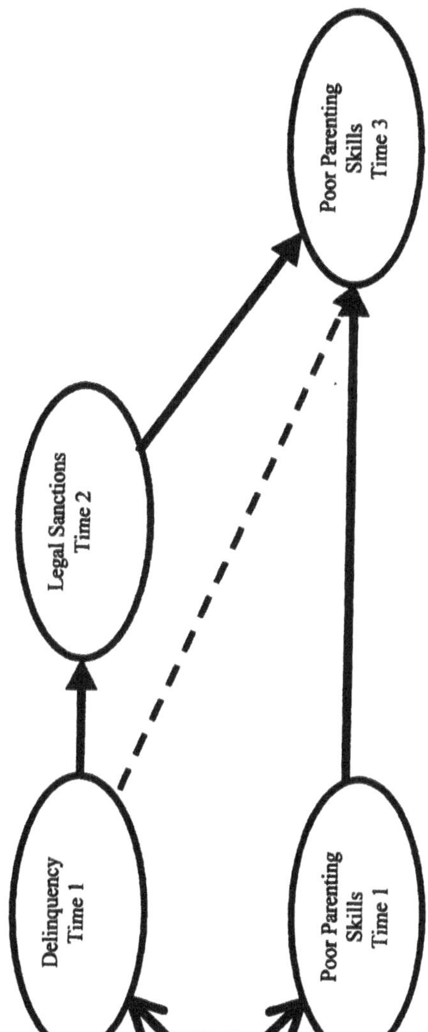

Figure 1. A SEM Predicting Poor-Parenting Skills as a Function of Delinquency and Legal Sanctions.

in turn, predict poor parenting skills at Time 3. The model also proposes that poor-parenting skills will be relatively stable over time.

An abundance of empirical work suggests that poor parenting skills predict antisocial and delinquent behavior (Gottfredson & Hirschi, 1990; Loeber & Stouthamer-Loeber, 1986; Patterson et al., 1992; Simons, Wu, Conger, Lorenz, 1994a). Thus, we expect poor parenting skills and delinquency to be correlated at Time 1. The dashed arrow between delinquency at Time 1 and poor parenting skills at Time 3 indicates that this path is expected to be statistically significant and positive prior to entering legal sanctions into the model. Such a finding would be consistent with prior studies that have found that delinquent behavior affects parenting (Ambert, 1997; Jang & Smith, 1997; Thornberry et al., 1991). We posit, however, that delinquency is most likely to produce these effects when it leads to legal sanctions. Thus, our model indicates that the effect of delinquency on poor parenting skills is indirect through legal sanctions.

There are several reasons for believing that legal sanctions might mediate the impact of delinquency on poor parenting skills. First, parents may not be aware of their child's behaviors outside of the home. That is, parents may not be cognizant of their adolescent's delinquent behavior because the adolescent may be successful at preventing parents from finding out about such inappropriate activities. Indeed, during this period parents typically begin to increase the amount of unsupervised time that their adolescent is allowed to spend away from home with peers (Simons, Whitbeck, Beaman & Conger, 1994b). These physical changes and new freedoms may contribute to experimentation with delinquent acts without their parents' knowledge. However, once an adolescent has been involved with the legal system, parents become knowledgeable about their adolescent's delinquent behavior. Second, legal sanctions most likely follow from more serious delinquent behaviors that are likely to be especially troubling for parents. Finally, response by legal authorities is embarrassing for parents as they are often viewed as having some responsibility for the misdeeds of their children. The following analyses test the empirical adequacy of the theoretical proposed model.

METHODS AND PROCEDURES

Sample

The data for these analyses come from the Iowa Youth and Families Project (IYFP), a 4-wave longitudinal investigation. A sample of 451 two-parent families was recruited through a cohort of male and female seventh-grade students living in eight north central Iowa counties. These students were enrolled in public or private schools during winter and spring, 1989. An additional criterion for inclusion in the study was the presence of a sibling within four years of age of the target seventh-grade student. Slightly fewer than half of the cohort of target youths had families whom met these criteria. Seventy-eight percent of the eligible families agreed to participate in the study; this rate is comparable to the response rates reported by other studies that attempt to recruit multiple family members (Capaldi & Patterson, 1987). Furthermore, substantial remuneration appears to be necessary to obtain participation by multiple family members in studies involving intensive assessment procedures (Capaldi & Patterson, 1987). Families in the present project received $250 for taking part, which translated into about $10 per hour for each family member's time.

The families in the sample lived on farms (about one-third) or in small towns. Because of the rural nature of the sample, in an area with almost no minority families, all of the participating families were White. The annual income ranged from $0 to $135,000, with a median of $33,700. The fathers' education ranged from eight to 20 years, with a mean of 13.5 years; for mothers the range was eight to 18 years, with a mean of 13.4 years. The fathers ranged in age from 31 to 68, with a median of 40 years; mothers' ages ranged from 29 to 53, with a median of 38 years. Because families of fewer than four were excluded from the sampling frame, the families were larger on average than one would expect from a general population survey. Families ranged from four to 13 members, with an average of 4.9.

In the present study, a total of 407 adolescents (189 boys and 218 girls) and their parents participated. Three waves of data were used at which time the target children's average ages were 13.5, 14.5, and 15.5. The boys and girls who were used in these analyses were in the 8th, 9th, and 10th grade. This time period was chosen because the literature identifies this age period as a time for escalating rates of delinquency (LeBlanc, 1990; Moffitt, 1993; Simons et al., 1994a; Thornberry, 1997).

Procedures

Data were collected annually from families when the target child was in grades 7, 8, 9, and 10. Essentially the same procedures and instruments were used at all four points. Each wave of data collection involved two visits to each of the study families. During the first visit, each of the four family members completed a set of questionnaires focusing on family processes, individual family members' characteristics, and socioeconomic circumstances. On average, approximately two hours were needed to complete the first visit. Between the first and the second visit, family members completed questionnaires left with them by the first interviewer. These questionnaires dealt with information on the extended family, the parents' developmental history, and plans for the future. Each family member was instructed to place his or her completed questionnaire in an envelope, seal it, and give it to the interviewer at the time of the second visit.

During the second visit, which normally occurred within two weeks of the first, the family was videotaped while engaging in several different structured interaction tasks. At the beginning of the visit, each individual completed a short questionnaire designed to identify issues of concern or disagreement within the family (e.g. chores, recreation, money). The family members were then gathered around a table and were given a set of cards to read and discuss. All four family members were asked to discuss among themselves each of the items listed on the cards and to continue talking until the interviewer returned. The interviewer remained out of earshot during the family discussion, in which family members essentially interviewed one another about the daily dynamics of their lives. The items on the cards concerned family issues such as discipline, chores, and the children's friends and school performance. The second task, 15 minutes in length, also involved all four family members. For this task, the family was asked to discuss and try to resolve the most significant issue or disagreement that they had cited in the questionnaires earlier in the visit. The third task, which involved only the two youths, took 15 minutes. The youths were given a set of cards listing questions about how they got along, how their parents treated them and their friends, and their future plans. The fourth task involved the married couple and lasted 30 minutes. Spouses were asked to discuss issues related to aspects of their relationship, areas of agreement and disagreement (e.g. parenting, finances), and their plans for the future.

As mentioned earlier, the family's interactions during these four tasks were videotaped. The interviewer explained each task and then left the room while the family members discussed issues raised by the task cards. The interviewer could not hear the discussion of other family members during this time. When a family member was not involved in a videotaped interaction task, he or she completed an additional questionnaire on significant life events, attitudes toward sexuality, and personal characteristics. The second visit lasted approximately two hours.

The videotapes were coded by project observers using the Iowa Family Interaction Rating Scales (Melby, Conger, Book, Rueter, Lucy, Repinski, Rogers, Rogers & Scaramella, 1998). These scales focus on the quality of behavioral exchanges between family members. Independent observers coded each interaction. Interobserver reliability estimates were computed on 25% of the videotapes using intraclass correlations for the observer reports that were coded by two independent observers. The independent observers were staff members who had received nearly three months of intense training on rating family interactions and who had specialized in coding one of the four interaction tasks. Before coding tapes, observers had to successfully pass four written exams in which they had to score at least 90% or higher. Once observers successfully passed the written portion of their training, they had to pass several observational tests. Observational coders had to independently rate practice interaction tasks and achieve at least 85% agreement with more experienced coders before they observed project videotapes.

Measures

An important issue in social science research is the manifestation of systematic measurement error or method variance bias (Bank, Dishion, Skinner & Patterson, 1990; Kandel, 1996; Lorenz, Conger, Simons, Whitbeck & Elder, 1991). Problems with method variance bias arise when two or more concepts are assessed using one source of information. That is, the correlation between two or more concepts can be a function of a single respondent reporting on all of the concepts of interest. For example, Kandel (1996) argues that much of the empirical work examining parents' and peers' influences on adolescent deviant behavior overestimates the influence of peers and underestimates the influence of parents. In part, this overestimation of peer effects is due to measurement problems. Adolescents often report on their perceptions of their friends' deviance, which is then used to predict their perceptions of their own deviance. Adolescents' perceptions of their peers' deviance are typically more similar to their self-reported deviance than with peers' reports of their own deviance (Kandel, 1996). Kandel reports that the correlation between an adolescent's report of their own and peer deviance is several times larger than the correlation between adolescent report of own deviance and peer report of own deviance. That is, when an adolescent perceives self as delinquent, they are likely to perceive friends as more delinquent than they actually are (see Scaramella, Conger, Spoth & Simons in press).

Thus, when a single person reports on all of the measures in a study, the relationship between a predictor and outcome variable may be biased, and usually it is inflated by reliance on a single method of measurement. In order to reduce method variance bias, we used different informants (mother, father, adolescent, sibling, observer) to develop each of the latent constructs in our model. Two empirical indicators were used to estimate the delinquency construct in Figure 1 (delinquent behavior and association with delinquent peers) one indicator to estimate legal sanctions, and three indicators to estimate poor parenting skills (poor supervision, harsh discipline, and harsh parenting).

Delinquent Behavior. A self-report, 20–item delinquency checklist adapted from the National Youth Survey (Elliott, Huizinga & Ageton, 1985; Elliott et al., 1989) was summed to form an indicator of delinquency. Targets were asked to indicate how often during the past 12 months they had engaged in each of 20 delinquent activities (1 = never, 5 = 6 or more times). The activities varied from relatively minor offenses, such as skipping school, to more serious offenses, such as attacking someone with a weapon or stealing something worth more than $25. Coefficient alpha for the scale was .78.

Delinquent Peers. Both a sibling report and a self-report of association with delinquent peers were combined to form a second indicator of delinquency. The correlation between sibling and target reports was .36 ($p < 0.01$). We included the sibling report instrument in an effort to reduce the degree of shared method variance between the self report delinquent behavior instrument and other constructs in the model. The sibling report scale consisted of three items. Siblings were asked to indicate the extent to which they agreed (1 = strongly disagree, 5 = strongly agree) that their brother/sister's close friends got into a lot of trouble, drank and smoked a lot, and sometimes broke the law. Coefficient alpha was approximately .72.

We adapted the self-report measure from the National Youth Survey (Elliott et al., 1985, 1989). Respondents were asked how many of their close friends had engaged in each of 15 delinquent acts (1 = none, 3 = half, 5 = all). These acts ranged from relatively minor offenses, such as skipping school, to more serious violations, such as fighting. We summed the responses to the items to obtain a total score concerning the extent to which the respondents' friends engaged in delinquent behavior. Coefficient alpha for the scale was .77.

Legal Sanctions. The variable legal sanctions was measured by eight items adapted from the National Youth Survey (Elliott et al., 1985, 1989) to form a composite measure of involvement with the criminal justice system. Targets were asked to indicate how often during the preceding year (1 = never, 5 = 6 or more times) they had been picked up by police for something they had done, had been picked up by police while on drugs, had been placed in juvenile detention or jail, had gone to court, or had been placed on probation. Coefficient alpha for the measure was .76.

Harsh Parenting. An 8-item, parent report scale was used to form an indicator of poor parenting skills at Time 1 and at Time 3. Fathers and mothers independently rated how often (1 = always, 5 = never) they had engaged in harsh and inconsistent parenting. Some of the questions were as follows: When the target child does something wrong, how often do you lose your temper and yell at him/her? How often do you spank or slap the target child when he/she does something wrong? When punishing the target child, how often do you hit him/her with a belt, paddle, or something else? When the target child does something wrong, how often do you tell him/her to get out or lock him/her out of the house? Fathers' and mothers' scores correlated at .46 at Time 1 and .49 at Time 3. Their scores were summed and combined for each wave to form an indicator of poor parenting skills. Coefficient alpha was .85 for Time 1 and .87 for Time 3.

Poor Supervision. Observational coders rated this scale. It assessed the extent to which parents displayed a limited range of knowledge regarding the child's behaviors and activities. It also focused on the degree to which the parents showed little interest in the child's endeavors and did not actively pursue information regarding the child's life and daily activities. This scale was rated using a 9-point scale ranging from 'no evidence of the behavior' to 'extreme evidence of the behavior.'

Harsh Discipline. We combined several observational scales (harsh punishment, hostility, physical attack, and inconsistent discipline) to form a measure of harsh discipline, which served as an indicator for poor parenting skills. This allowed us to examine various dimensions of parenting. *Harsh punishment* focused on the extent to which parents respond to the child's misbehavior or violation of parental standards by using punitive or severe disciplinary techniques (e.g. yelling, screaming, hitting). The *hostility scale* assessed the extent to which the parents are angry, critical, disapproving, and rejecting in their interactions with the child. *Physical attack* assessed any aversive physical contact by the parent toward the child, which included hitting, slapping, punching, shoving, and grabbing. *Inconsistent discipline* assessed the degree of parental inconsistency and lack of follow-through in maintaining and adhering to rules and standards of conduct for the child's behavior. Furthermore, past studies have shown that they predict adolescent internalizing and externalizing problems (Conger, Conger, Elder, Lorenz, Simons & Whitbeck, 1992; Simons et al., 1998). Each dimension was rated using a 9-point scale ranging from 'no evidence of the behavior' to 'extreme evidence of the behavior.' The correlations between these scales ranged from .68 to .79. Coefficient alphas were .84 (Time 1) and .85 (Time 3).

RESULTS

Table 1 reports the factor loadings of the indicators over time. The factor loadings ranged from .57 to .87 and show that all of the indicators chosen at Time 1 and 3 are acceptable in terms of the extent to which they reflect their underlying concepts. The indicators remain significant while demonstrating slight fluctuations in factor loadings across time.

Table 1 Standardized Factor Loadings for the Indicators of the Latent Constructs

Variables	Figure 2 Factor Loadings Time 1	Figure 2 Factor Loadings Time 3	Figure 3 Factor Loadings Time 1	Figure 3 Factor Loadings Time 3
Delinquent Behavior	0.83**		0.73**	
Delinquent Peers	0.57**		0.64**	
Poor Supervision	0.85**	0.86**	0.85**	0.87**
Harsh Discipline	0.66**	0.67**	0.66**	0.67**
Harsh Parenting	0.67**	0.60**	0.67**	0.59**

Significant at 0 01

Errors were correlated between like indicators at Time 1 and Time 3

Table 2 reports the zero-order correlations, means, and standard deviations for the indicator variables. Delinquent behavior has a mean of 29.6 and a standard deviation of 2.6. The delinquent peers' indicator has a mean of 21.5 with a standard deviation of 6.3. The mean score for legal sanctions is 10.7 with a standard deviation of 5.9. Of particular interest is the significant change in the mean levels of harsh discipline ($t = 89.4$; $p < 0.01$) and harsh parenting ($t = 67.5$; $p < 0.01$) from Time 1 to Time 3. Both parents and observational coders, on average, reported increases in harsh discipline and harsh parenting from Time 1 to Time 3. For example, Time 1 harsh discipline had a mean of 21.6. By Time 3, the average level of harsh discipline increased to 26.8. For harsh parenting, the mean increased from 18.0 at Time 1 to 25.4 at Time 3. Also of notable interest is the mean for poor supervision ($t = -25.9$; $P < 0.01$), which significantly decreased from 7.1 at Time 1 to 5.5 at Time 3.

Table 2 also presents the zero-order correlations for the indicators of the study variables. The relatively large correlations between the same indicators reflect the strong stability between indicators at Times 1 and 3. For example, the correlation between poor supervision at Time 1 and Time 3 is .79. The correlation between harsh parenting at Time 1 and Time 3 is .64. In addition, the legal sanctions measure shows relatively strong associations with the parenting and delinquency indicators. Legal sanctions correlated .44 with delinquent peers and .46 with delinquent behavior. Most of the indicators are statistically significant at the .05

Table 2 Zero-Order Correlations, Means, and Standard Deviations of the Indicator Variables

	1	2	3	4	5	6	7	8	9	Means	SD
1. Delinquent Behavior	1									29.6	2.6
2. Delinquent Peers	.47**	1								21.5	6.3
3. Legal Sanctions	.46**	.44**	1							10.7	5.9
4. Poor Supervision Time 1	.31**	.18**	.25**	1						7.1	2.1
5. Poor Supervision Time 3	.36**	.23**	.38**	.79**	1					5.5	1.7
6. Harsh Discipline Time 1	.21**	.11*	.19**	.55**	.46**	1				21.6	2.0
7. Harsh Discipline Time 3	.24**	.19**	.29**	.44**	.57**	.80**	1			26.8	1.6
8. Harsh Parenting Time 1	.12*	.07	.08	.57**	.45**	.46**	.35**	1		18.0	2.7
9. Harsh Parenting Time 3	.12*	.08	.13*	.42**	.51**	.36**	.43**	.64**	1	12	5.4

*p < 0 05 **p < 0 01 N = 407
Errors were correlated between like indicators at Time 1 and Time 3

and .01 levels in the expected direction. This suggests that all indicators chosen for this study are empirically as well as theoretically valid. The measures assumed to be indicators of a common construct were correlated with each other and were correlated in a similar fashion with other indicators of constructs.

In sum, the overall assessment of the hypothesized measurement model demonstrates that the variables selected to measure the latent constructs of interest in fact reflect those constructs in a theoretically as well as statistically valid and stable manner, despite some variations across indicators. As expected, the results also show that all indicators should be included in the final model. Given this positive evaluation of the measurement model, we now turn to an evaluation of the structural relationships.

Structural Equation Analyses

Structural equation modeling was used to evaluate the theoretical model. The analysis was performed using Amos 4.0 (Arbuckle & Wothke, 1999). All analyses included parental education and socioeconomic status as demographic controls. We began the structural equation modeling by estimating the models separately for boys and girls. However, subgroup comparisons showed that there were no significant differences in the models by gender. Therefore, we present the results obtained for the combined sample.

Figure 2 shows the effect of delinquency at Time 1 on poor parenting skills at Time 3. The effect of Time 1 poor parenting skills has been controlled so that the model estimates the effect of Time 1 delinquency on poor parenting skills from Time 1 to Time 3. The standardized regression coefficient from delinquency at Time 1 to poor parenting skills at Time 3 is .16 ($t = 3.5$). The association between poor-parenting skills and delinquency at Time 1 is .38. The stability of poor parenting skills between Time 1 and Time 3 is .75. The model's fit indices suggest that the model fits the data (GFI = .99; AGFI = .97). The structural path coefficients in Figure 2 are consistent with the thesis that delinquency affects parenting behaviors (Ambert, 1997; Jang & Smith, 1997; Patterson et al., 1992; Thornberry et al., 1991).

The next step in evaluating the theoretical model involved adding legal sanctions to the predictive equations. If the theoretical model is correct, legal sanctions will mediate the association between Time 1 delinquency and poor parenting skills at Time 3. If this pattern maintains, we can conclude that legal sanctions explain how it is that delinquency has an adverse effect on parenting behaviors. The results are presented in Figure 3.

Figure 3 shows that delinquency has a significant positive effect on legal sanctions at Time 2 ($\beta = .65$, $t = 9.1$). In turn, legal sanctions significantly predicted poor parenting skills ($\beta = .16$, $t = 3.1$) at Time 3. As expected, once the effects of legal sanctions were entered into the model, the path from delinquency to poor parenting skills at Time 3 was no longer significant. This finding is consonant with the hypothesis that legal sanctions mediate the relationships between delinquency and poor parenting skills.

DISCUSSION AND CONCLUSION

Delinquency research has assumed until recently that parents exercise a predominantly one-way influence on adolescents. However, in recent years, social scientists have adopted a more reciprocal view of the parent-child relationship. Research based on this perspective indicates that in addition to 'parental effects' on children, there are 'child effects' on parents (Lytton, 1990). While it seems true that ineffective parenting increases the chances that a child will develop conduct problems, it also appears to be the case that hostile, obstinate child behavior is often followed by a reduction in parental efforts to monitor and

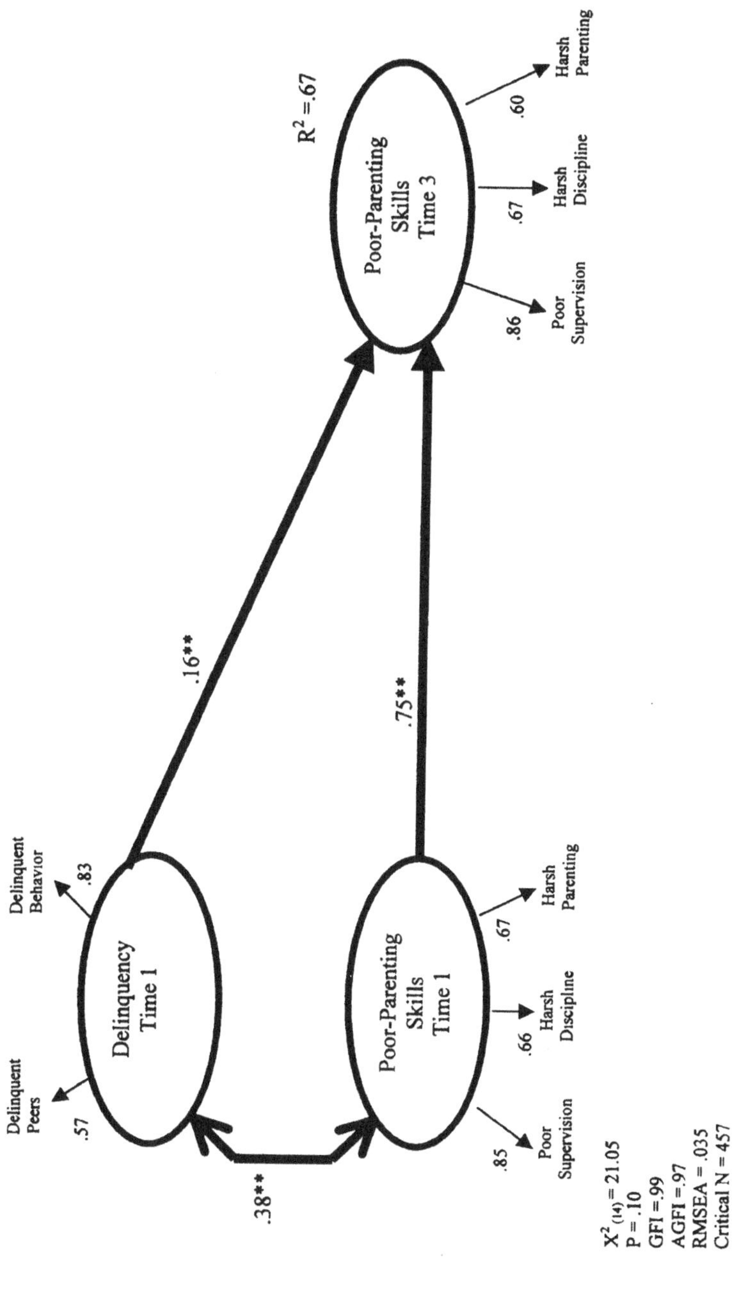

Figure 2. A SEM Predicting Poor-Parenting Skills as a Function of Delinquency.

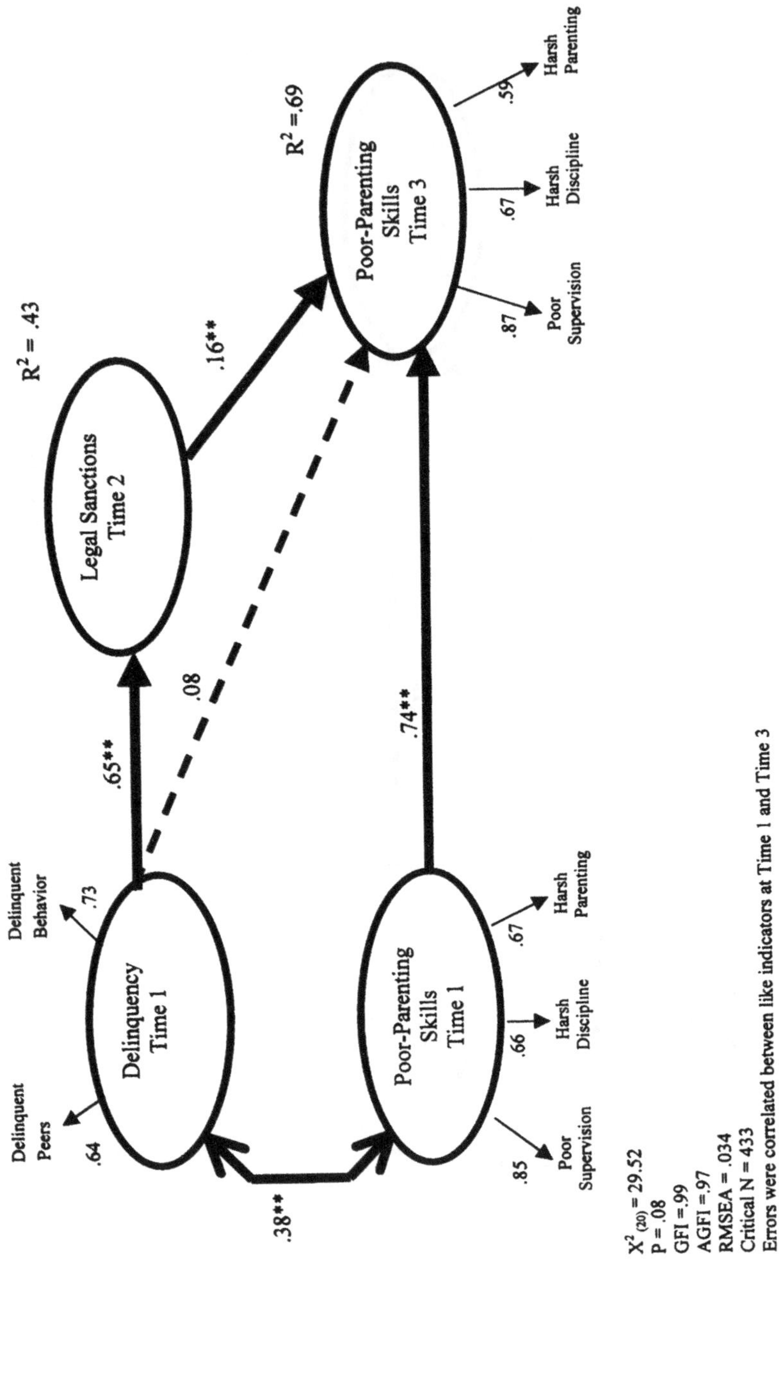

Figure 3. A SEM Predicting Poor-Parenting Skills as a Function of Delinquency and Legal Sanctions.

discipline (Moffitt, 1997; Patterson et al., 1992; Simons et al., 1998). Thus, difficult children encourage parenting practices that amplify their antisocial tendencies.

Research has shown that difficult children may be so uncontrollable that parenting skills break down as a result, which in turns lead to poor parenting practices, as well as poor parent-child relations (Ambert, 1997; Belsky, 1984; Forgatch, 1989; Patterson et al., 1992). In the present study, we extended this research by switching the focus from parental influences on delinquency to the impact of delinquency on parenting behaviors. The objective of this research has been to evaluate the relationships among delinquency, legal sanctions, and poor parenting skills. The analysis has considered the consequences of delinquency and legal sanctions for family processes.

Our results indicated that delinquent behavior during early adolescence influenced parenting skills during the ensuing year. The results support previous findings that implicate delinquency in disrupting family life (Ambert, 1999; Anderson et al., 1986; Lytton, 1990; Patterson et al., 1992). The significant path from delinquency to poor parenting skills supported this notion. Furthermore, delinquency was found to increase the probability of legal sanctions, and legal sanctions predicted poor parenting skills. Most importantly, adding legal sanctions to the model eliminated the association between delinquency and poor parenting skills. This pattern remains significant when controlling for poor parenting skills at an earlier time. Overall, the findings suggest that delinquency is most apt to have a negative impact upon family processes when it results in legal sanctions.

We believe there are several reasons why this is the case. First, parents are often unaware of their adolescent's delinquent behavior. Adolescence is a period when parents begin to increase the amount of unsupervised time that their child is allowed to spend away from home with peers (Simons et al., 1994b). Adolescents have greater access to their own transportation, which increases their ability to move throughout their neighborhood and city. This further expands the number and diversity of social relationships. Adolescents dramatically increase the extent to which they interact with others away from their parents and other adults. Adolescents, for example, begin to leave the house on their own in the evening and on weekends. This increased freedom allows the child to experiment with delinquent acts without parental knowledge. Parents may suspect that their child is engaging in deviant behaviors of various sorts, but it is often difficult to prove that this is the case. Once an adolescent becomes involved with the legal system, however, parents typically become fully aware of their adolescent's illegal behavior. Furthermore, an adolescent is most likely to become involved with the legal system as a result of engaging in serious illegal acts. Thus, the adolescent's encounter with the legal system corroborates parents' worst fears regarding their offspring. In addition, parents are likely to experience encounters with legal authorities as embarrassing. This is a reaction, in part, to the perception by the legal system as well as others in the community that parents have some responsibility for the misdeeds of their children. Finally, juvenile involvement in the justice system may be expensive for parents in terms of both time and money as they seek professional assistance to deal with such problems.

As a result of these factors, parents experience a high level of distress (Ambert, 1999). This distress is likely to have a disruptive effect on the parent-child relationship and the quality of parenting. This pattern was exemplified in Ambert's (1997) qualitative study of delinquents and their parents. She found that although the relationship between parents and delinquent adolescents was often strained, it usually became significantly worse after the parents became aware of their child's multiple behavior problems and encounters with the law. "Parents no longer trusted their youngster who had lied, disobeyed, betrayed their trust and behaved in a way that they had never expected from him or her" (Ambert, 1997, 210).

Our findings, and those of Ambert, are consonant with Sampson & Laub's (1997) cumulative disadvantage theory of delinquency and crime. They argue that the stability of antisocial behavior across the life course is a consequence of deviant behavior at early stages of development undermining role relationships that are important social controls during later stages. Adolescent antisocial behavior, for example, is seen as disrupting parenting, fostering affiliation with deviant peers, and reducing school commitment, with these changes, in turn, serving to amplify involvement in delinquent behavior. Several studies have provided support for this view (Patterson et al., 1992; Sampson & Laub, 1993, 1997; Simons et al., 1998; Thornberry et al., 1991).

Furthermore, Sampson & Laub (1997) note that delinquency often results in encounters with the legal system, and that such encounters tend to have a number of deleterious consequences. Their discussion of negative consequences focuses upon the manner in which arrest, official labeling, and incarceration undermine a person's educational and occupational chances, thereby increasing the probability of future criminal behavior. They cite a number of studies supporting their position (see Sampson & Laub, 1997). Findings from the present study extend this perspective by identifying another way in which legal sanctions operate inadvertently to amplify an adolescent's antisocial tendencies. Encounters with the legal system tend to disrupt parenting and the quality of the parent-child relationship. These disturbances in family processes are likely to amplify involvement in delinquent behavior.

Our research supports the growing recognition of the importance of the multidimensional effects of delinquency on family life (Ambert, 1999; Jang & Smith, 1997; Lytton, 1990; Patterson et al., 1992; Thornberry, 1997). It seems clear that researchers should consider the dynamic consequences of delinquency and the complex problems it presents. The effect of adolescent behavior on parents is understudied even though it has implications for engaging and maintaining parents in treatment directed at strengthening parental management of antisocial behavior. Acknowledging the disruptive effects of antisocial behavior on parents and other family members provides a more complete picture of the relationship between adolescent conduct problems and family life.

In this study, we addressed several methodological problems in the current research on family life and delinquency. We recognized that there is growing evidence of problems in measurement and interpretation posed by differing perceptions of family informants (Krohn, Stern, Thornberry & Jang, 1992), as well as by differences in the meaning and use of language in family measurement (Smith & Krohn, 1995). First, we used a longitudinal, multi-informant design. A longitudinal design provides greater confidence in the temporal ordering of constructs than does a cross-sectional research design. Thus, we were able to identify the mechanism through which delinquency affects family life. Second, we avoided the problem of shared method variance by using multiple sources of data to develop measures of constructs (Bank et al., 1990; Kandel, 1996; Lorenz et al., 1991). It is also important to note that the elimination of shared method variance in the present analyses means that the findings are extremely conservative compared to studies that use single informants to assess two or more theoretical constructs.

Although this study demonstrated several methodological strengths, it also suffered from certain limitations. First, only a small portion of the sample reported involvement in delinquent behavior or with the criminal justice system. Thus, the results obtained here may not be generalized to a more urban sample. Delinquent behavior is likely to be more prevalent in cities. It is possible, for example, that the greater availability of deviant peers and adults in urban areas exert more influence on adolescent delinquency than we found in the present study.

Second, the sample consisted entirely of White, two-parent families living in the rural Midwest. Presumably the same processes operate in single-parent families, but this possibility could not be examined in the present study. On the other hand, it could also be reasonably argued that single-parent families might be affected more by their adolescent's delinquency. Indeed, Ambert (1999) suggested that single mothers are generally more affected than mothers in two-parent structures regarding their adolescent's misbehavior. Compared with married mothers, and all fathers, single mothers experienced more problems with people (relatives, friends, neighbors, coworkers) and greater blame as a result of their adolescents' difficulties (Ambert, 1999). Importantly, research has demonstrated that single-parent families are associated with higher rates of welfare, financial hardship, greater family disorganization, and child behavior problems (Sampson, 1993; Wilson, 1996). With this in mind, delinquent behavior and criminal justice involvement can be added stressors for single-parent families because the parent will have to be absent from work to attend court and counseling sessions.

Third, although most of the families experienced financial hardship, they did not suffer extreme persistent poverty as found in ethnic and minority families (Wilson, 1996). It is possible that adolescents from families characterized by persistent poverty and low education levels might be more affected by official labeling. For example, Hagan's (1991) research suggests that the deleterious effect of adolescent deviance on negative future outcomes is greatest among lower class boys, especially as mediated by police contacts. On the other hand, middle-class boys who escaped the negative consequences of official labeling did not suffer any future impairment as a result of their adolescent delinquency.

In some important aspects, such as official labeling, the formal labeling process may not be uniform across ethnic groups. For instance, groups who have been discriminated against may view formal sanctions as part of an overall pattern of mistreatment. Jerome Miller (1996) suggests that many young, ethnic minority males are routinely socialized to expect unfair and discriminatory acts against them by police and the judicial system. This has led Elijah Anderson (1999) to identify a street culture that has developed in the inner city as a result of this treatment. He observed that the code of the streets is actually a cultural adaptation to a profound lack of faith in the police and the judicial system. The police are most often seen as representing the dominant and discriminatory society and as not caring enough to protect inner city residents.

Thus, it is possible that routine arrests, bookings, detention, and imprisonment are not seen as stigmatizing by some parents. Rather, they may view the criminal justice system as being unfair and discriminating against their child. Therefore, any generalizations about the formal labeling process across ethnic groups should be made with caution. Therefore, the findings from this study should be replicated with samples that include adolescents from inner-city urban areas.

Overall, the results support the notion that delinquency has consequences. Our research supports the growing recognition of the importance of the multi-dimensional effects of delinquency on family life (Ambert, 1999; Jang & Smith, 1997; Lytton, 1990; Patterson et al., 1992; Thornberry, 1996, 1997). It seems clear that researchers should consider the dynamic consequences of delinquency and the complex problems it presents. The effect of adolescent behavior on parents is understudied even though it has implications for engaging and maintaining parents in treatment directed at strengthening parental management of antisocial behavior. Acknowledging the disruptive effects of antisocial behavior on parents and other family members provides a more complete picture of the relationship between adolescent conduct problems and family life.

REFERENCES

Ambert, A.-M. (1997). *Parents, Children, and Adolescents: Interactive Relationships and Development in Context.* New York: Haworth Press.

Ambert, A.-M. (1999). The Effect of Male Delinquency on Mothers and Fathers: A Heuristic Study. *Sociological Inquiry, 69,* 621–640.

Anderson, E. (1999). *Code of the Street: Decency, Violence, and the Moral Life of the Inner City.* New York, N.Y.: W. W. Norton and Company.

Anderson, K., Lytton, H., & Romney, D. (1986). Mothers' Interactions with Normal and Conduct-Disordered Boys. Who Affects Whom? *Developmental Psychology, 22,* 604–609.

Arbuckle, J. L., & Wothke, W. (1999). Amos Users' Guide Version 4.0. Chicago, IL: Smallwaters Corporation.

Bank, L., Dishion, T., Skinner, M., & Patterson, G. R. (1990). Method Variance in Structural Equation Modeling. Living with GLOP. In: Gearld R. Patterson (Ed.), *Aggression and Depression in Family Interactions.* Hilldale, NJ.: Lawrence Erlbaum Associates.

Barber, B., & Rollins, B. C. (1990). *Parent Adolescent Relationships.* Lanham, MD: University Press of America.

Belsky, J. (1984). The Determinants of Parenting: A Process Model. *Child Development, 55,* 83–96.

Capaldi, D., & Patterson, G. R. (1987). An Approach to the Problem of Recruitment and Retention Rates for Longitudinal Research. *Behavioral Assessment, 9,* 169–177.

Conger, K. J., & Conger, R. D. (1994). Differential Parenting and Change in Sibling Differences in Delinquency. *Journal of Family Psychology, 8,* 287–302.

Conger, R. D., Conger, K. J., Elder Jr., G. H., Lorenz, F. O., Simons, R. L., & Whitbeck, L. B. (1992). A Family Process Model of Economic Hardship and Influences on Adjustment of Early Adolescent Boys. *Child Development, 63,* 526–541.

Conger, R. D., & Simons, R. L. (1997). Life Course Contingencies in the Development of Adolescent Antisocial Behavior: A Matching Law Approach. In: T P. Thornberry (Ed.), *Developmental Theories of Crime and Delinquency,* (55–99). New Brunswick, N.J.: Transaction.

Elliott, D. S., Huizinga, D., & Ageton, S. S. (1985). Explaining Delinquency and Drug Use. Beverly Hills, CA.: Sage.

Elliott, D. S., Huizinga, D., & Menard, S. (1989). *Multiple Problem Youth: Delinquency, Substance Use, and Mental Health Problems.* New York: Springer-Verlag.

Forgatch, M. S. (1989) Patterns and Outcomes in Family Problem Solving: the Disrupting Effect of Negative Emotion. *Journal of Marriage and the Family, 51,* 115–124.

Garfinkel, H. (1956). Conditions of Successful Degradation Ceremonies. *American Journal of Sociology, 61,* 420–426

Gottfredson, M. R., & Hirschi, T. (1990). *A General Theory of Crime.* Stanford, CA: Stanford University Press.

Hagan, J. (1991). Destiny and Drift: Subcultural Preferences, Status Attainments, and the Risks and Rewards of Youth. *American Sociological Review*, *56*, 567–582.

Jang, S. J., & Smith, C. A. (1997). A Test of Reciprocal Causal Relationships among Parental Supervision, Affective Ties, and Delinquency. *Journal of Research in Crime and Delinquency*, *34*, 307–336.

Kandel, D. B. (1996). The Parental and Peer Contexts of Adolescent Deviance: Algebra of Interpersonal Influences. *Journal of Drug Issues*, *26*, 289–315.

Krohn, M. D., Stern, S. B., Thornberry, T. P., & Jang, S. J. (1992). The Measurement of Family Process Variables: An Examination of Adolescent and Parent Perceptions of Family Life and Delinquent Behavior. *Journal of Quantitative Criminology*, *8*, 287–315.

LeBlanc, M. (1990). Two Processes of the Development of Persistent Offending: Activation and Escalation. In: L. Robins, & M. Rutter (Eds), *Straight and Devious Pathways from Childhood to Adulthood*, (82–100). New York: Cambridge University Press.

Lemert, E. (1951). *Social Pathology*. New York: McGraw-Hill.

Loeber, R., & Stouthamer-Loeber, M. (1986). Family Factors as Correlates and Predictors of Juvenile Conduct Problems and Delinquency. In: M. Tonry, & N. Morris (Eds), *Crime and Justice: An Annual Review of Research* vol. *7*, (29–149). Chicago, IL: University of Chicago Press.

Lorenz, F. O., Conger, R. D., Simons, R. L., Whitbeck, L. B., & Elder, G. H. Jr. (1991). Economic Pressure and Marital Quality: An Illustration of the Method Variance Problem in the Causal Modeling of Family Processes. *Journal of Marriage and the Family*, *53*, 375–388.

Lytton, H. (1990). Child and Parent Effects in Boys' Conduct Disorder: A Reinterpretation. *Developmental Psychology*, *26*, 683–697.

Matza, D. (1969). *Becoming Deviant*. Englewood Cliffs, NJ: Prentice Hall.

Melby, J., Conger, R. D., Book, R., Rueter, M., Lucy, L., Repinski, D., Rogers, S., Rogers, B., & Scaramella, L. (1998). *The Iowa Family Interaction Rating Scales*. Ames: Iowa Youth and Families Project

Miller, J. G. (1996). *Search and Destroy: African-American Males in the Criminal Justice System*. New York, NY: Cambridge University Press.

Moffitt, T E. (1993). Adolescence-Limited and Life-Course-Persistent Antisocial Behavior: A Developmental Taxonomy. *Psychological Review*, *100*, 674–701

Moffitt, T. E. (1997). Adolescent-Limited and Life-Course Persistent Offending: A Complementary Pair of Developmental Theories. In: T. P. Thornberry (Ed.), *Developmental Theories of Crime and Delinquency*, (11–54). New Brunswick, N.J.: Transaction.

Patterson, G. R., Reid, J. B., & Dishion, T. J. (1992). *Antisocial Boys*. Eugene, OR: Castalia.

Paternoster, R., & Iovanni, L. (1989). The Labeling Perspective and Delinquency: An Elaboration of the Theory and An Assessment of the Evidence. *Justice Quarterly*, *6*, 359–394.

Rose, D. R., & Clear, T. R. (1998). Incarceration, Social Capital, and Crime: Implications for Social Disorganization Theory. *Criminology*, *36*, 441–479

Sampson, R. J. (1993). The Community Context of Violent Crime. In: W. J. Wilson (Ed.), *Sociology and the Public Agenda*, (259–286). Newbury Park, CA: Sage.

Sampson, R. J., & Laub, J. (1993). *Crime in the Making. Pathways and Turning Points through Life*. Cambridge, MA: Harvard University.

Sampson, R. J., & Laub, J. (1997). A Life-Course Theory of Cumulative Disadvantage and the Stability of Delinquency. In: T. P. Thornberry (Ed.), *Developmental Theories of Crime and Delinquency*, (133–161). New Brunswick, N.J : Transaction.

Scaramella, L., Conger, R. D , Spoth, R., & Simons, R. L (Forthcoming). Evaluation of A Social Contextual model of Delinquency. A Cross-Study Replication. *Child Development*

Scheff, T. H. (1966). *Becoming Mentally Ill*. Chicago, IL: Aldine.

Simons, R. L., Wu, C.-I., Conger, R. D., & Lorenz, F. O. (1994a). Two Routes to Delinquency: Differences between Early and Late Starters in the Impact of Parenting and Deviant Peers. *Criminology, 32*, 247–275.

Simons, R. L., Whitbeck, L. B., Beaman, J., & Conger, R. D. (1994b). The impact of mothers' parenting, involvement by nonresidential fathers, and parental conflict on adjustment of adolescent children. *Journal of Marriage and the Family, 556*, 356–374.

Simons, R. L., Johnson, C., Conger, R. D., & Elder, G. Jr. (1998). A Test of Latent Trait Versus Life-Course Perspectives on the Stability of Adolescent Antisocial Behavior. *Criminology, 36*, 217–244.

Smith, C. A., & Krohn. M. D. (1995). Delinquency and Family Life among Male Adolescents: The Role of Ethnicity. *Journal of Youth and Adolescence, 24*, 69– 93.

Thornberry, T. P., Lizotte, A. J., Krohn, M. D., Farnworth, M., & Jang, S. J. (1991). Testing Interactional Theory: An Examination of Reciprocal Causal Relationships among Family, School, and Delinquency. *Journal of Criminal Law and Criminology, 82*, 3–35.

Thornberry, T. P. (1996). Empirical Support for Interactional Theory; A Review of the Literature. In J. D. Hawkins (Ed.), *Delinquency and Crime: Current Theories*, (198–235). New York: Cambridge University Press.

Thornberry, T. P. (1997). Introduction. Some Advantages of Developmental and Life-Course Perspectives for the Study of Crime and Delinquency. In: T. P. Thornberry (Ed.), *Developmental Theories of Crime and Delinquency*, (1–10). New Brunswick, N.J.: Transaction.

Wilson, W. J. (1996). When Work Disappears: The World of the New Urban Poor. New York, NY: Knopf.

COMING HOME: THE TRANSITION FROM INCARCERATED TO PAROLED YOUNG FATHER

Anne M. Nurse

> Miguel, a seventeen-year-old father of one child, was 15 years old when he was arrested. At the time of this arrest, his daughter Michelle was three months old. For the first half of his two-year sentence Miguel did not see his daughter because her maternal grandfather refused to allow her to come to prison. After Miguel pleaded with him, the grandfather relented and Michelle's mother brought her to visit once a month. Since his release, Miguel has generally seen his daughter every day, except when he is having trouble getting along with Michelle's mother.

> John is a twenty-year-old father of three. He has been in and out of prison three times since the birth of his oldest child. During his stays in prison, his son's mother brought the boy to visit about once a month. John was released from prison almost two years ago. At first he saw his son often, but soon the frequency of his visits began to decrease. He now has no contact with his child.

In recent years, young men like Miguel and John have received increased public attention. As juvenile delinquents, and as young fathers, they represent two separate categories of national concern. Public fears about adolescent crime

This research could not have been done without the generous support of three granting agencies: the National Science Foundation (Dissertation Improvement Award number SBR-9633153), the U.S. Department of Justice (OJP-97-017-M), and the Ford Foundation (#975-1299). The author also expresses appreciation to Mary Jackman, Carole Joffe, Lawrence Cohen, and James Cramer, who provided helpful comments. Many thanks also to the California Youth Authority parolees and staff who gave generously of their time and knowledge.

Families, Crime and Criminal Justice, Volume 2, pages 281–308.
2000 by Elsevier Science Inc.
ISBN: 0-7623-0737-4

have sparked debate as well as criminal law reform and changes to the juvenile prison system. At the same time, perceptions of irresponsible fathering have led to the modification of welfare and child-support policies. Miguel and John stand at the intersection of these two social problems – they are both criminals and young fathers. Interestingly, little attention has been paid to the fact that large and growing numbers of our nation's young fathers, like Miguel and John, are either in prison or are on parole.

In the last twenty years the incarceration rate of young men has risen dramatically. Between 1982 and 1992, violent-crime arrest rates for males under the age of eighteen more than doubled, from 2,000 per 100,000 youths to 4,200 per 100,000 youths. The rates also increased for men ages nineteen to twenty-four. Estimates suggest this rate may continue to rise for some time (Hunzecker, 1995). Currently, there are close to 100,000 juveniles in public and private correctional institutions, camps, and treatment centers nationwide. In 1990, the rate of admission to such facilities was 3,213 per 100,000 eligible youth. This means that over 3.2% of all American teenagers were incarcerated for some period during that year (Austin, Krisberg, DeComo, Rudenstine & Del Rosario, 1995).

There are several reasons to believe that the juvenile prison system houses a large number of fathers. First, rates of both incarceration and young out-of-wedlock childbearing are high in impoverished African-American and Latino communities. African Americans in the United States have six times the rate of incarceration, and four times the rate of out-of-wedlock childbearing when compared to whites (Hagan, 1995; Lerman, 1993). Latinos also have an out-of-wedlock birth rate that is higher than the white rate and, while Latinos constitute about 8% of the national population, they represent over 16% of the population in state prisons (Garwood, 1992; Lerman, 1993; Maguire & Pastore, 1994).

Research suggests that the overrepresentation of African Americans and Latinos in prison is a result of both discrimination and poverty. Race appears to have direct and indirect effects on decisions made by police, judges, and courts in juvenile justice cases (Bishop & Frazier, 1996; Conley, 1994; Wordes, Bynum & Corley, 1994). As a result of these decisions, nonwhites are more likely than whites to serve time in prison, even when the severity of the crime and the defendant's past record are taken into account. In the United States, race is also highly correlated with poverty. African Americans and Latinos tend to be poorer than whites and less able to afford lawyers and alternatives to prison, such as private treatment centers (Bishop & Frazier, 1996; Bortner & Williams, 1997). Discrimination and poverty also help to explain the racial/ethnic disparity in rates of young out-of-wedlock fatherhood. Researchers have found, at least in the case of African Americans, that out-of-wedlock fathering is related to a complex set of factors including low employment rates, wages, and education (for a review of the literature as well as for recent original research on this question see Tucker & Mitchell-Kernan, 1995).

The intersection of race/ethnicity, poverty, and young parenthood leads to an overrepresentation of fathers in the juvenile prison population. This overrepresentation is further pronounced by the fact that fathers, regardless of their backgrounds, appear to be more likely than their nonfather counterparts to engage in delinquent behaviors and to go to prison (Christmon & Lucky, 1994; Elster, Lamb & Tavare, 1987; Lerman, 1993). These delinquent behaviors range from alcohol and substance abuse to school problems. In a large-scale study of the relationship between delinquency and young fatherhood, Stouthamer-Loeber & Wei (1998) found that fathers are more than twice as likely to engage in delinquent behaviors as are nonfathers. The fathers are also more likely to abuse alcohol and drugs, have school problems, and exhibit aggressive behaviors. In their conclusion, Stouthamer-Loeber & Wei comment that, "Compared to males of the same neighborhood, race, and age, young fathers tended to be troubled young men who were significantly more likely than their matched controls to have engaged in varied serious acts of delinquency in the year of fatherhood and in the year after" (p. 64).

The overlap of risk factors for incarceration and young parenthood means that time in prison has, in effect, become a normal part of life for a significant percentage of young fathers. Few states keep an accurate count of the number of juvenile fathers in prison, but California Youth Authority has data suggesting that over 25% of its juvenile inmates are fathers (California Youth Authority, 1996). As the prison population continues to grow, we can assume that more and more young fathers will spend a part of their children's lives behind bars.

The rising number of young imprisoned fathers has far-reaching social implications for the men, their families, and their communities. Among these implications are the lack of male role models for children, economic strain for young mothers, and the estrangement of men from their own children. In addition to these effects, there is evidence to suggest that a man's fatherhood status can influence his risk of incarceration. Men who develop strong bonds with their children may have lower rates of recidivism (Sampson & Laub, 1993). These facts make it imperative that we begin to face the sociological challenges and policy issues presented by young incarcerated and paroled fathers.

In the pages that follow, I address the question of how father/child relationships are affected by the incarceration and homecoming of a father. My discussion of these issues is divided into the following sections: theory, previous research, methods, and analysis. The analysis section focuses on the impact of incarceration on paroled men's relationships with children and on their ability to provide monetary support.

PRISON, YOUNG FATHERHOOD, AND THE LIFE COURSE

The incarceration of a large number of young fathers has implications for a number of areas in sociology. One of the most interesting involves research on the life course. A sociological perspective on the life course focuses on the progression of people's lives, paying particular attention to how personal biography and prevailing social norms interact to affect the timing and experience of life transitions such as marriage, divorce, and childbearing (Elder & O'Rand, 1995). Life course researchers are also concerned with the results of life transitions and with their influence on the future of an individual. Any life transition profoundly affects other areas of a person's life. For example, getting married can have long-term repercussions on a person's career, lifestyle, education, and childbearing decisions. Much of the research done by life course researchers incorporates the idea of a "normative life course" in which particular transitions are presumed to precede others (Rindfuss, Swicegood & Rosenfeld, 1987). This normative life course assumes that people first finish school, then find employment, marry, and have children.

Young paroled fathers are an interesting group to investigate from a life course perspective. First, they are in clear violation of the normative life course; their transition to parenthood generally comes before they have married or completed their education. Second, time spent in prison is a potentially life-altering transition not taken into account by the normative life course perspective. While sociologists have studied the effects of an early transition to parenthood (for a review of the literature see Furstenberg, Brooks-Gunn & Morgan, 1987), there has been little research on the transition into or out of prison. Consequently, we know little about how juvenile incarceration affects later life events. This chapter begins to explore one specific issue: How does time spent in prison affect the experience of young fatherhood?

PREVIOUS RESEARCH

We can assume that there are paroled men in any study of young fathers. To date, however, researchers have not identified these men in their samples, nor have they considered the impact of incarceration on their respondents' fathering behaviors and attitudes. If prison does have a significant influence on father/child relationships, a failure to acknowledge its importance runs the risk of inaccurately portraying the lives and experiences of many young fathers.

Fatherhood research is not the only field that would benefit from an understanding of the link between incarceration and fathering attitudes and behaviors; the inner-city/underclass literature would also be strengthened. Researchers focusing on the inner city/underclass have viewed incarceration as a result of social problems, not a cause. William J. Wilson (1996), for example, argues that high levels of incarceration are one result of joblessness. Douglas Massey (1995) emphasizes incarceration as a direct result of the residential segregation of African Americans. While incarceration is sometimes mentioned in passing as a cause of the small number of marriageable black males, researchers have not yet explored how incarceration may contribute to the many other social problems in the inner city.

The lack of research directly focusing on the lives and experiences of newly released juvenile fathers requires that we look to other sources for guidance. There are two small bodies of literature that are particularly relevant. One literature, written mostly in Britain, discusses the lives and experiences of adult criminals who have been released from correctional institutions. The other literature focuses on returning prisoners of war. These men, like criminal fathers, experience a prolonged separation from their families. Comparisons, however, must be made with great care because many aspects of their experiences differ. Prisoners of war, for example, generally do not experience the stigma associated with criminal incarceration.

The literature on newly released prisoners of war and adult criminal fathers suggests that both groups of men face numerous problems adjusting to family life upon their release. These difficulties include marital problems and estrangement from children. In a seven-year study of 241 prisoners of war and their families, for example, Edna Hunter (1986) found that returning prisoners had a higher rate of divorce than did a matched control group of military families (30% as compared to 12%). Kathleen McDermott & Roy King's (1992) research with newly released prisoners in Britain indicates that the men experience high levels of stress in their relations with wives and children. A primary reason is that the children and women have undergone many changes during the men's absence. As a result, the men are no longer able to occupy their former role in the family, and they find themselves unsure of how to act.

While studies such as these provide intriguing glimpses of the experiences of newly released men, none examine in detail their relationships with their children, and none look at the specific experiences of juvenile prisoners. In this chapter, I provide some of the first data addressing the impact of prison on the postrelease relationships of young fathers and their children.

METHODS

The data for this chapter are drawn from three sources: participant observation in parenting classes provided to California Youth Authority (CYA) parolees; a survey administered to 258 paroled fathers in Northern California; and in-depth interviews conducted with a subset of 20 of the survey respondents. All of the men in this study were under the jurisdiction of the CYA, one of the largest juvenile correctional systems in the United States. It is the "last stop" for serious juvenile offenders in California. The northern region extends from Bakersfield, in the south, to the Oregon border. California, with its high juvenile incarceration rate and its heterogeneous population, is an ideal location for this study. Care must be taken, however, in making broad generalizations to the rest of the country, especially in light of the disproportionately high percentage of Latinos in the state.

The parenting classes I observed at the CYA were part of an ongoing pilot project to provide parental education and support services to parolees. Class topics included child development, child rearing, discipline, and legal issues associated with fatherhood. The curriculum also contained sections designed specifically for the parolee population; one session, for example, was devoted to a discussion of the transition from prison back into the lives of the young men's children. Based on a set of criteria provided by the CYA, parole agents selected 15 to 20 young men to attend each twelve-week parenting session. The agents were instructed to give priority to biological fathers or to young men whose wives or girlfriends were pregnant.

Between January and August of 1996, I attended approximately 40 parenting class sessions. My presence at the parenting classes was a unique opportunity to collect rich qualitative data about young parolees' concerns, problems, and experiences as fathers. Because most teachers made an effort to focus the classes on topics the young men requested, I was able to listen as the men discussed the issues they saw as most important in their lives. The classes also provided detailed data about the culture of fatherhood among young male parolees. The frequent group discussions, as well as the less-structured interaction of the men in the class, revealed much about their beliefs and feelings regarding fatherhood.

The data gathered from parenting classes cannot be used as a direct source for making inferences about the entire population of young paroled fathers. My observations indicated that there was some bias in class selection because parole agents either enrolled fathers who were having special problems with their children or, conversely, who were exceptionally involved with their kids. In addition, a number of the participants did not have children at all. In this chapter I use examples from parenting classes only when they are useful for illustrating some of the issues raised by the surveys and the in-depth interviews.

The second source of data for this chapter was a survey of all of the fathers on parole in Northern California between October 1996 and October 1997. The survey was designed to cover each man's current fathering practices, his experience as an incarcerated father, and his transition home to his children. For this reason, the questionnaire was fairly lengthy (about 45 minutes long). To keep it a reasonable length, I made the decision to limit the questions to one child. I asked each respondent about his oldest child, instead of younger children, because it increased the likelihood that he had been both incarcerated and on parole since the birth of that child. This decision was in accordance with the project's focus on the intersection of fatherhood and incarceration.

To compile a list of all fathers on parole, I requested names from every parole agent in Northern California. These agents are in a unique position to identify which men under their supervision are fathers. I supplemented the agent lists with Youth Authority records of prison-intake interviews. CYA staff members conducted these interviews when the young men first arrived at prison. One of the questions they were asked concerned fatherhood status. I obtained a record of the answer to this question given by all men who were on parole during the period of the study. Using the parole-agent lists combined with the intake records, I identified 380 young fathers. I believe this method produced a fairly complete list of those men willing to divulge their fatherhood status to authorities at the CYA. To locate men not willing to admit to being fathers, I spent time in the waiting rooms of parole offices talking with the parolees. Most of these men are required to report to their agents at least once a month, so I was able to explain the study to a large number of them. In this way I added 20 young fathers who were not on any list. All of the fathers were between the ages of 16 and 25.

In total, I identified 400 young fathers on parole in Northern California during the period of the study. Out of this group of men, seven interviewers and I contacted 275; of the 275 we administered surveys to 258. It should be noted, however, that for the purposes of this chapter, I only analyze data from 197 of the 258 survey respondents. Excluded are all respondents whose children were born after, rather than before, their prison stays. Men whose children were born postrelease obviously could not speak to the experience of being an incarcerated father who returns home to his children. In terms of the men I identified as fathers but who did not complete a survey, 125 were men I was unable to contact. About half of these men were absent without leave (AWOL) and could not be located by the parole office. About as many had been returned to prison before we could interview them. Because parolees must report to their agents on a regular basis, it was fairly easy to locate all those men who were not AWOL. There were, however, 17 young men whom we contacted but who did not participate in the study: five refused; the other twelve agreed to complete a survey, but logistical problems made contact impossible.

The survey's response rate of 64% (258 surveys administered out of 400 identified fathers) raises some concern about bias. As noted, the vast majority of the men we failed to interview were either AWOL or had been sent back to prison before we could contact them. Going AWOL and recidivism are related to delinquency and, for this reason, the survey respondents may underrepresent the more delinquent fathers on parole. The fact that we were in the field for a full year, however, helped to minimize the problem. We were in frequent contact with agents and were able to identify and administer surveys to many fathers who had just been released from prison. Many of these men later disappeared, or were sent back to prison, soon after the survey.

The administration of the surveys generally took place in private rooms at parole offices. Sometimes, however, a parolee lived far from the parole office and it was necessary to meet him at a public place closer to his residence: a library or a fast-food restaurant for example. The survey covered a large number of topics involving the men's experiences in prison and their current fathering practices. All of the surveys were read aloud and were conducted in English. The survey took about 45 minutes to complete. Although a large number of the Latino respondents spoke Spanish as their first language, all were proficient enough in English to be able to complete the questionnaires accurately.

After doing preliminary analysis of the survey data, I constructed an in-depth interview schedule. The purpose of the in-depth interviews was to explore areas of interest revealed during the survey stage of the project and to do further probing of the young men's attitudes about fatherhood. Because fathering behavior and attitudes are complex and emotionally charged, the in-depth interviews were especially important. The kind of rich and unique data obtained through in-depth interviewing are invaluable. They allowed me to explore attitudes, feelings, and reasons for behavior with a depth and sensitivity that are not available to an observer or to a survey interviewer.

To select candidates for the in-depth interviews, I used a purposive sampling strategy. First, I stratified the list of survey respondents by race/ethnicity (among all of the fathers in the survey, 14% were white, 31% were African American, and 55% were Latino). Once the list was stratified, I randomly selected five whites (25%), seven African Americans (35%), and eight Latinos (40%). I decided to include a disproportionately large number of whites in order to maximize racial/ethnic variation in the sample. After I selected the respondents, I checked to make sure that they were representative of the entire group. A few minor adjustments were necessary to make the sample geographically representative and to make sure that it contained a representative number of involved and noninvolved fathers. I made the necessary adjustments by adding randomly selected parolees from the underrepresented groups.

The in-depth interviews were conducted in a number of locations. Over a third were conducted in the young men's homes, or in a public place. Another third were conducted at parole offices. Four of the young men had been re-incarcerated, and their interviews took place in correctional facilities. All of the interviews lasted from forty-five minutes to an hour and were taped and directly transcribed.

THE TRANSITION HOME

Young incarcerated fathers leave prison and return home to find that many things have changed in their absence. Children have grown, friends have new lives, and jobs once held are no longer available. We know little about what happens to these young men as they try to adjust to their new lives on the outside, and our knowledge of their relationships with their children is particularly limited. How has the father/child relationship been changed during their incarceration?

In this chapter, I argue that newly released fathers leave prison with extremely high expectations for the relationship they will develop with their children. As time passes, however, they encounter challenges they do not expect. These challenges cause some fathers to decrease the amount of time they spend with their children. Before I begin to describe these challenges, a word of caution is necessary. While showing that the transition home from prison causes significant strain between fathers and their children, I do not mean to suggest that it is the sole source of relationship problems. Some inmates, for example, are estranged from their children before they are imprisoned. Incarceration, and a troubled homecoming experience, may have a further negative impact, but these things cannot be seen as the only cause of problems. In this chapter, I focus on the problems the parolees in my sample saw as directly related to their incarceration and parole.

High Expectations

In prison, young men are provided with some activities but a great deal of their time is empty. The in-depth interview respondents told me that juvenile inmates spend much of their free time imagining how their lives will be when they are released from prison. They make plans and think about changes they want to make. The literature on adult prisoners suggests that this kind of planning is not unique to the juvenile population (McDermott & King, 1992). Thinking about the future provides something to do and temporarily takes the prisoners away from the realities of imprisonment. The respondents in my study said that, while they thought about many aspects of their future, their prison daydreams frequently turned to the relationship with their children. They anticipated being very involved and spending a lot of time developing a close relationship. Marco described his image as follows:

> You picture it in your mind as good, great, you know, when you get out man, you're going to love her, love them, do this with her and do that. Doing – you're always telling yourself, do things my parents never did with me, and give her everything she wants, just be so good to my daughter. And love her, tuck her in, read to her, so you picture it in your mind, and you imagine it in your mind to be great, to be wonderful.

Alberto, who was back in prison when he spoke with me, also talked about the high expectations he had for a future relationship with his son. He said:

> I think about it and picture it every day. You know, soon as I get out, I'll see my son and I'll just be able to be at home with him, you know, watch him run around, watch him get into trouble . . . Just watch him grow up, just be there for him.

Marco and Alberto are not unusual; most of the men in the in-depth interviews told me that they spent hours in prison daydreaming and fantasizing about the relationship they were going to build with their children. They imagined activities they would engage in together and, even more importantly, they formed a picture in their minds of how their children would act. Central to this vision was the idea that their children would respond to them as "daddy," recognize them as father, and treat them with respect. In the following sections I discuss what happens when the men's expectations meet the reality of the outside world.

He Called My Brother "Daddy" – Unexpected Problems with Children

There is a fundamental problem with the expectations developed during time spent in prison: they tend to have an inadequate basis in reality. Most of the men have spent little or no time with their children and, as a result, have only a limited ability to imagine a real-world relationship. In this study, over 40% of the parolees' children were born while the men were in prison. Upon their release, some of these men had never met their children and the others had interacted with them only during visiting hours. As a result, few were capable of forming an accurate picture of their children. Fathers who spent time with their children before being arrested were able to develop somewhat more realistic expectations, but they too operated with limited knowledge. The children they remembered from the time before their incarceration changed during their absence and, by the time the men were released, had entered entirely new developmental stages.

Because of their high expectations, the paroled fathers were initially excited to see their children and made an effort to spend time with them. They quickly discovered, however, that their hopes of immediately assuming a daddy role were not to be realized. The first surprise many encountered was that children did not recognize them. Over 34% of the survey respondents agreed with the statement, "Being incarcerated made your child forget you." Such an experience was deeply upsetting for the parolees, especially when they had spent time in prison thinking about how much their children would love them. I talked with a young man named Miguel about how prison affected his relationship with his daughter:

Interviewer: How did incarceration change your relationship with your daughter?

Miguel: She knows who I am, she knows I'm her father but I see it in her sometimes she like . . . she feels I'm kind of a stranger in a way. She has that look in her eyes like she's kind of confused whether she should come and hug me. She treats me like a stranger sometimes cause she doesn't really know . . . that hurts me 'cause I look at her like I'm her dad you know.

Interviewer: Tell me about the first time you saw her when you got out?

Miguel: It was weird because when she'd see me in the institution she knew who I was – she was kind of running and giving me a hug, but when I got out that day I remember exactly – she looked at me like just stunned, like what are you doing here?

Another young man told me about the first time he saw his son after he was released from prison. He was disappointed to find that the child seemed confused by his homecoming:

Interviewer: Tell me about the first time you saw him when you were out.

Barry: It was kind of weird, he kind of like didn't know what to do. And I didn't really know what to do either 'cause he was so much bigger. It was weird.

Interviewer: Where were you?

Barry: Where was I when I first saw him?

Interviewer: Uhm hum.

Barry: I was at my grandma's house. She had to go pick him up because his mom still had the restraining order on me. He didn't really know what to say or do.

Interviewer: So what did you do?

Barry: Try to play with him and fix him some food, and try to be a father.

Like Barry, many of the men described being disappointed or angered by their children's reactions to them when they arrived home. Further compounding the problem, the men reported that their children sometimes expressed fear or refused to obey. About 15% of the survey respondents agreed that incarceration had made their child afraid of them. I asked Robert about his transition home to his son. He told me, "Well he'll sit there for a minute, he'll give me a hug and I'll try to talk to him the best I could. But then he'll start looking for his mom, start getting anxious, then start to run away and start crying or something." While some children are afraid, others simply refuse to do what their fathers say. I asked Marco how his three-year-old daughter had reacted the first few weeks he was out. He told me:

> Oh man, she didn't know me you know, it was like, if I told her anything, she didn't know whether to listen to me or run away from me. She still, she kind of knew who I was, so when I came, she greeted me, gave me a hug and a kiss, and whatever, but when it came to listening to me, or, you know, if I told her something, she didn't know how to act either. And I didn't know how to act whether, by telling her no, or you know, I wasn't really in that position to tell her no, because she didn't really know me.

I asked another man, B.J., if he had had any problems with his son since he was released from prison. He said, "Him talking back to me. Me trying to discipline him. He doesn't listen. Just if I tell him something, he'll right away say, 'No. I've never been there . . . . so he's just not used to me telling him what to do."

One of the reasons young men come home to find their children confused and disobedient is that other people have come to fill the father role in their absence. Jose commented on how frequently this occurs. He said:

> When a kid doesn't see their father, original or real father, they tend to latch on to whoever is next to them, you know what I mean? . . . If her mom has a new boyfriend, that's daddy . . . I mean, I see it everyday. When somebody gets locked up, what we call Sancho – you know, sidekick – that's the next dude in line – takes over and raises the kid with the kid calling him 'dad.'

There were many examples of "Sanchos" in the in-depth data. Robert, for example, reported that while he was incarcerated, his brother became like a father to his child. He said, "My brother, he called my brother dad, daddy, one time. And I was like, man – you know, cause he was there for him as much as possible or whatever." I asked him how he had felt when his child called his brother "daddy," he said, "I was hurt, bad, bad." In a similar story, another young man told me that, during his incarceration, his grandmother had taken over as the principal person in his child's life:

> It's kind of weird, but it's uncomfortable for me when my grandmother's around because I feel like my son forgets about me. Because he was so attached to her when I was gone. And she gave him more than 100% of her attention. It's just like she just dropped everything and did nothing but play with him all day, twenty-four hours a day, seven days a week, all she did was play.

Young men who come home expecting that they will be central to their children's lives and that their children will see them as "daddy" often face bitter disappointment. Alberto, who was out of prison for only twenty-one days before being rearrested, said that his time at home with his son had been upsetting to him:

> I felt kind of messed up cause when I had got out, it's like my little brother, it's like he was going to my little brother a lot too. I'd be carrying him and he'd see my little brother and it's like . . I felt like my little brother was a dad and I'm like, "I feel bad" you know ... My son probably think that he's his dad now.

Many of the young men in my study, like Alberto, told me that they felt they were replaced by other people who became a daddy for their children. While all these men expressed some degree of jealousy or anger, those who had been replaced by the boyfriend or husband of the child's mother reported particularly strong emotions. In addition to having to deal with feelings of being replaced, they also had to learn to negotiate with the new boyfriend for access to their children. Extreme jealousy and even violence can erupt between a newly released father and the mother's new boyfriend. Marco, who said he was having ongoing problems with the mother of his children, described what happened when he got out of prison and met her new boyfriend:

> I got out and I didn't know, just that day I found out she was with somebody else. So I was like, cool, I just want to see my kids I saw my kids, and told him, "Hey, whatever, you know, I ain't with her no more, you are ... come here (he gestures like he gestured to the new boyfriend) Don't ever touch my kids, I'll break your hands, don't ever try to get a hand on 'em, you know those are my kids."

Marco told me that, at first, his threats of violence against the boyfriend of his child's mother seemed to be an effective strategy because he was able to see his children when he wanted. Gradually, however, the tensions between the two men caused Marco to reduce his contact.

Many newly released parolees face problems with their children. The children do not recognize them, refuse to obey, or are afraid of them. Some of the children have found new people to assume a father role in their lives. The parolees, most of whom have build up high expectations for the future relationships with their children, are deeply disappointed by these reactions and many are jealous of the new father figure in their children's lives. The weight of these problems and tensions, combined with the disappointment of unfulfilled expectations, often encourages men to withdraw from their children.

Broken Promises and Other Problems with Children's Mothers

Research suggests that one of the factors most influencing the contact of non-resident fathers with their children is the mother of those children (DeLuccie,

1995; Furstenberg, 1995). As might be expected, when the relationship between a young man and the mother of his children is weak, he is less likely to see them. There are at least two important reasons for this. First, most mothers live with their children and act as a "gatekeeper" to control any interaction between the children and their father (Daly, 1993). They regulate the time, frequency, and duration of visits (Seltzer & Brandreth, 1994). When couples are not getting along, or when a mother feels the man is not living up to his commitments, she is able to restrict access to his child. In a study of families headed by teen mothers, Danziger (1987) found that over 18% of the mothers had, at one time or another, denied visitation to the father of their children.

Frank Furstenberg (1995) identifies a second way in which the relationship between men and their children's mother affects father/child contact. In his work with young parents, he finds that men do not view their relationship with their children apart from their relationship with the children's mother. Men see mothers and children as a "package deal." When the bond between men and their children's mother weakens, or is weak to begin with, the father separates himself from both the mother and the child. Furstenberg points out that weak parental relationships are particularly common among adolescents. Young couples often do not know each other well before they have a child and, in some cases, are "virtual strangers." The maintenance of a quality relationship in these cases is extremely difficult.

Given the unstable nature of most adolescent relationships, it is no surprise that newly paroled men return home to their children's mother and quickly encounter problems interacting with her. One of the first problems they discover is that the mother is frequently a quite different person from the one they knew before being incarcerated. The changes these women have undergone make it hard for the young men to negotiate a relationship with them. B.J., who had recently been released from prison after two years, told me that he was trying to maintain a good relationship with the mother of his child but that she had changed while he was gone:

> Now that I'm home, our relationship is not as good as it was before I went to jail. It's like we are trying to get to know each other again. Cause over the years people change, you know. You come home and they are not the same. It's kind of like meeting a new person again. It's hard.

When asked how incarceration affected the relationship he had with the mother of his child, Tyrell, after twenty-six months in prison, commented:

> It was like we wasn't really close friends anymore. It was like we didn't really know each other cause I was gone longer than I was with her. You know, so we had only been together maybe a year-and-a-half, two years and then I got locked up for over two years.

The literature on spouses of prison inmates suggests that the primary way that women change during their partner's incarceration is that they become much more independent and self-sufficient (Hunter, 1986; McDermott & King, 1992). When the men return, the women do not want to go back to the more dependent relationship they had before the man's incarceration. This new independence is hard for many men to accept, especially when they have spent their time in jail imagining how their wife or girlfriend used to act.

The data from this project suggest that many women not only become more independent in the men's absence, but that the parole system gives them additional power in their relationship with these men. During the in-depth interviews, a number of the young men told me that being on parole changed the power balance they had with their children's mother. These mothers were aware that parolees can be returned to the institution for any violation of their parole, and they understood that if they reported a violation it was likely that the young men would be rearrested. I heard about this problem in one of the first parenting classes I attended. The theme of the class was domestic violence and during the break a very bright young man named Jacob nervously approached the teacher of the class. He told the teacher that he was trying very hard not to hit his girlfriend, but felt as though she was always provoking him and then threatening to call his parole agent. He said he was at a loss as to how to handle the situation. Echoing similar sentiments, Charles told me:

> Like about two years ago, when my daughter was born and we split up, it was hard. We went through a lot because, like after she had the baby, she had an attitude I didn't know she had. So then she started jeopardizing my parole, calling my officer, telling my parole officers lies and stuff, so I had to just cut her off, because my freedom I felt was more important at the time.

The threat of being returned to the institution is obviously very serious and men will go to great lengths to avoid it. Jeremy, an African-American father of two children, told me that he took custody of his oldest son when the child's mother was arrested. He was concerned for the welfare of the child and did not want him to go back to the mother when she was released. He said:

> And then she got out, probably a week and a half later, and she called my grandma's house . . . talking about bringing my son home . . . "If you don't bring my son home, I'm going to call the police on you." Then my grandma's like, "Go ahead and take him home, take him home. You don't need more trouble, you're already on parole, just take him home."

These comments indicate that mothers of children can use the parole system to exert power over men and limit the access they have to their children.

Compounding all the relationship difficulties newly released young men face is the weight of unmet promises many have made to their children's mother from prison. As discussed above, incarcerated men spend time in prison building up their own expectations for their behavior after they are released. Once they develop this set of expectations, they naturally share them with their children and with their children's mother. By the time they are released, the high expectations they have for themselves are also shared by mothers and children. Marco described the promises he made to the mother of his child while he was in the institution:

> Like people always talk, you always write this and that to your girlfriend, "When I get out it's going to be like this, like that." This and that, and this and that. And you get out and it's like "you wrote it – do it – come on, I'm waiting for it you know."

Many incarcerated men share their hopes and plans with the mother of their children because they are excited and want to share them with someone. At the same time, some of the men also use their plans to encourage the mother of their children to stay in a romantic relationship with them. By making promises about how they will change, they raise the hopes of the mother. Tony described how he had kept his relationship with the mother of his child together by building up her expectations:

> I saw a lot (of breakups in prison), but for me, we got closer. The bad part was we got closer because of what I was telling her were lies. You know, I was telling her I was going to change when I get out, because I didn't want to be here. And I actually went home, and it was all a bunch of lies.

Mothers adopt the expectations the young men create and then begin to embellish them. Fernando told me that he got angry because the mother of his child had such high expectations for him. He commented, "She wanted to see what I was really made of. At times, she was like, she made me angry . . . cause she talked about how good it was going to be and I did too." When young men are finally released they are under great pressure, both self- and family-imposed, to live up to the promises they have made.

As young men fail to live up to the high expectations they have for themselves, they and their children's mother become disillusioned and disappointed. Fernando described what happened with the mother of his child when he was finally released:

> And when I got my job I was trying real hard to succeed and make everything right, as good as I could, cause she was talking about uh, not showing her nothing, I'm still the same person I was before I got locked up. And that kind of like really made me feel bad, you know what I'm saying? It kind of like downgraded me

Marco, like Fernando, found that he could not live up to the expectations his girlfriend had of him. He returned home from prison and was allowed to live with her and his child. Things quickly soured, however, and his girlfriend began complaining that he was a failure. He described what happened:

> I got out and I tried, you know, I got a job, you know, I was making a little bit of money and I tried. . . . but then after a while I think it just gets back into the same routine. You know, you get tired and you just work your way back into that same old crap.

The discouragement caused by the men's failure to fulfill their promises, combined with relationship difficulties with the mother of their children, decreases the desire of young fathers for involvement with either their children or the mother.

Allure and Danger of the Streets

In their work on a youth correctional facility in Arizona, Bortner and Williams (1997) describe the situation most young incarcerated men face as they leave prison and return to the community:

> For most youths, going home meant a return to poverty and unemployment, troubled homes, the allure of alcohol and drugs, the dominance of gangs, and daily hardships. Even when the prison's model program was functioning at its best, youths' changed attitudes and emergent hopes often were not sufficient to sustain them in the community. . . . Outside the prison, fundamental social conditions remained unchanged. When they were paroled, within a few days or a few hours youths confronted the same old pressures and many reverted to familiar responses and solutions (p. 148).

The pressures that Bortner & Williams describe for youth in Arizona are essentially the same as those faced by newly released men in California. Most of the parolees from this study returned to impoverished and, often, gang-controlled neighborhoods. The pressures of the streets in these neighborhoods have a strong impact on the men's relationship with their children. Tony told me about the plans he had for his homecoming to his son, and to what happened to those plans:

> It (prison) made me more, I was more, it made me different, not in a bad way. It made me different like I wanted to spend more time with him, I wanted to be around him all the time. That was, it's hard to explain, because I wanted to be around him all the time, but the street life was like more exciting. I could go around the streets and act wild, I know I can always go home and he'll be waiting for me. So really, he was like, in my heart he was first priority, in my mind I subconsciously made him second priority because I chose the streets and spent all my time there.

While many young people confront pressures from the streets, newly paroled men may be particularly susceptible because their time at the Youth Authority enhances their status in the eyes of their friends. The in-depth interview respondents reported that people they knew seemed to admire and fear them more when they were released. Simply spending time in prison was sufficient to evoke this response, but special admiration was reserved for those young men most adept at earning fear and respect from their fellow inmates. Stories of an incarcerated man's fighting prowess or his toughness leak out of the institution and build a reputation on the streets. Paul, a young white man from the suburbs, told me that he fought a lot in the institution. He described what it was like when he was first released:

> I had to hold up this image when I first got out because . . . like I find I'm highly respected around here . . . Then I got out and I was always running around. So, when I got out everyone was like, oh, you're crazy, you know, for this or that. And I never noticed it 'till, not 'till my friend woke me up to it and told me that you been trying to hold up an image for a long time and you know, I was like, what are you talking about? He said, "because you was known to get out of YA and you were known to fight and do this and that" and I was like "yeah, that's true" and I barely realized it but it took me a long time, I'd been out almost two years. And uh, I barely noticed that. I was trying to hold up an image all the time.

Men who are paroled back into their own neighborhoods often discover that they now have a higher status, and the streets become a place where they are validated and can feel good about themselves. Such attention is difficult to resist. It is almost impossible if it must be sacrificed in order to spend time with a child who is not giving the man the kind of respect and attention he would like.

A return to the street life does not necessarily mean that men forsake their children. Some continue to visit them and to take the children out with them, at least for a time. The problem, however, is that an active street life, combined with active fatherhood, can endanger the children. A few of the young men, in the in-depth interviews, talked about the difficulties of being with their children in their old neighborhoods where they feared gang violence. Jose told me that he didn't take his daughter outside his house. I asked him to explain and he said:

> I'm afraid, it's for my daughter, you know what I mean? I ain't worried about myself, what they could do to me, the same thing I could do back, I mean, it's only flesh, you know what I mean? We're destined to die, that's the way I look at it. So however I'm going to die, I'm going to die, I mean I can't avoid that, you know what I mean? So, if they shoot me, stab me, whatever, they go both ways. If I die, somebody is going to do back to them, so I'm not really worried about myself. I could care less. But my daughter, my family, and everybody else that's around me, they have nothing to do with that, that's a different story.

Tony told me about the first day he got out of prison:

> I was going to make up for lost time. I wanted him with me, every minute, (but) that changed because I, the first day I was out actually I was running into people I didn't get along with and I thought I can't have him with me when I'm out on the streets because I wouldn't be able to live if something happened to him. After that first day we were on the

> streets together, but after that, when I'd go out, I went by myself. I didn't take him like public places, I was so scared of what happened if I got caught by somebody I didn't get along with, somebody I did something wrong to. How would they react if I'm here with my baby?

Young fathers enjoy going out and doing things with their children and, when they are unable to do these activities because they fear for their child's safety, some cut back on the time they spend with them.

THE EFFECT OF PRISON ON MEN'S ABILITY TO SUPPORT CHILDREN

In the preceding sections I illustrated some of the ways that prison affects a father's contact with his children. While contact is an absolutely central component of fatherhood, it is, of course, not the sole component. Among other responsibilities, fathers are expected to provide monetary and in-kind support for their children. The data from this project suggest that prison alters men's employment prospects and changes their relationship with the legal child-support system. These changes, in turn, affect men's ability and desire to support their children.

Employment

Research indicates that there is a dearth of jobs available to young men, particularly for minority and poor youth in the inner city (Wilson, 1996). Parolees have an especially difficult time finding work. Although they do not have to volunteer information about their parole status to potential employers, they do need to answer truthfully if asked. Many simply end up telling employers they are on parole in order to explain where they completed high school or why they have no previous job experience. Sometimes a young man who has no employment record must list his parole agent as a character reference. Other young men must tell potential employers they are on parole because they have parole restrictions that limit the hours they can work. Scott had a great deal of trouble finding a job. He commented, "When you write 'parole' (on the application), most people . . . won't even give you no time of day at all. I mean, most places I've been to won't even talk to me." I asked him if he had to tell potential employers that he was on parole. He responded, "You're supposed to if they ask you, but I've been leaving, like the place I just got hired on, I left it blank, I didn't write yes or no, I just left it blank. And they hired me."

Not all of the problems parolees have finding a job can be traced to the job market, or even to employers' reluctance to hire men on parole. In truth, some of the young men harm their own chances of getting a job because they have tattoos or dress in a way that clearly gives the impression they are gang-affiliated. Obviously, men who appear to be in a gang, who have a bad attitude, or who dress poorly have difficulty finding employers willing to hire them. The fact remains, however, that many of the parolees do everything right and still fail to find work. I saw many cases of young men who were making every effort to find a job but were unable to obtain one. For example, I attended parenting classes with a young man who had been out on parole for a number of months and was actively seeking work. He spoke well, attended parenting classes faithfully, and dressed neatly. What follows is an excerpt from my field notes. Carl is the parolee attending the class and John is the teacher.

> Before class started the teacher was mostly talking to Carl (young African-American man who lives with girlfriend's child). He is trying to get a job – any kind of job seemingly but is not having any luck at all. John started lecturing about the proper way to dress but Carl said that he had been dressing up – with "dress shoes" and that he had even gotten a haircut which he had not wanted to do that cost him $10. He thinks that it is his parole record that is precluding him from getting a job. He applied to Vons and a number of other places. John argued that it was a tight job market in the city. Carl mentioned that he had robbery on his record and they wouldn't let him work the cash register

The inability to find a job may have a strong effect on a man's self-esteem. Unemployment also damages the relationship men have with their children because they are unable to make a material contribution to the children's support.

Child Support

The child-support system in the United States has not been very successful in establishing legal orders that require absent parents to pay child support. It is estimated that only 60% of households with absent fathers have a child-support order (Dinitto, 1995, p. 170). In households where a child has been born out of wedlock, the figure drops to 18% (Danziger, Kastner, & Nickel, 1993, p. 236). The solution is not simply issuing more child-support orders; compliance rates with established orders are extremely low. Full support is only collected in 30% of cases. Poor and minority households are significantly less likely to have an established child-support order, and are also less likely to receive full payment on an established order. Women of all races, who have children with adolescent fathers, also receive less child-support money (Pirog-Good & Good, 1995).

Little work has been done that enables us to get a good understanding of the impact of prison on child-support establishment and compliance. The data from this project suggest that the impact is mixed. On the one hand, men are easier to locate for the purposes of establishing orders, and they can receive help with the paperwork from YA staff. On the other hand, the structure of the system often results in men being saddled with child-support bills for the time spent in prison. The end result is that men become so discouraged and resentful that they give up trying to pay any support at all.

Many of the young fathers in the institution receive legal papers regarding their children. Over a quarter of the survey respondents reported that they had received such papers while they were in the institution. About 88% of these papers involved the establishment or adjustment of a child-support order. It is not a coincidence that so many young men in the survey received legal papers while incarcerated. As stated earlier, under current California law, a woman who applies for welfare must provide the name of the father of her child. Incarcerated men are unable to help the mothers of their children with expenses, and this may force some of the mothers to apply for welfare.

Young men who receive legal papers while in prison have a number of advantages over men who receive papers while they are living at home. The legal papers are usually difficult to understand and complete, and in prison the parolees have easy access to assistance. About 60% of the survey respondents who received papers reported that they had trouble understanding them. In an in-depth interview Paul told me that he had gotten papers in the mail several months previously. I asked him what the papers involved and he said:

> I don't know. I received one set of . . . so I read it but, I don't know what it meant. So, in truth I can't read good, I can do math, but I can't read good. I don't really know what those papers mean. They are just giving me papers and they think probably I'm like a lawyer. I don't know what you know, all these codes mean.

When men are in prison and find that they cannot understand legal papers, they are often able to obtain help from the staff. Close to 50% of the respondents who received legal papers said that they sought out and received help from a YA staff person. Only 20% said that there was no staff member available who could help them understand the legal papers. Men who are out in the community when they receive their papers are less likely to have the resources or the knowledge required to find help.

While current law makes easier the establishment of child-support orders, it does little to help young men, especially incarcerated young men, trying to comply with those orders. The situation is further complicated by institutional and bureaucratic problems. In California, fathers are not supposed to accrue child-support payment obligations while they are incarcerated. Unfortunately, there is no communication between the Youth Authority and the Division of Family Support, so Family Support has no way to know that the fathers are incarcerated unless the young men or their families inform them. The majority of the young men in the sample had low levels of education and few previous positive experiences with the law. As a result, many of them did not know their legal rights, nor did they feel that they had the power to protest their quickly mounting bills. The young men who did try to confront the Division of Family Support did so with mixed results. Some told the Division that they were incarcerated and received word back that their debts had been erased. Most, however, were not able to reduce their debt. I include a lengthy quote from Jeremy whose situation is fairly representative:

> The first time I got served with 'em, I was in the county jail. And, they called me out and they gave me some papers, they're like, "Here's some child support papers, you've been served" . . . and they're like "Are you the parent?" and I'm like "Yea." . . . So I signed it off. And then I went to YA and while I was there, they kept sending me bills for child support. $170 each month. And I would write to 'em and explain to 'em I'm incarcerated, I have no way to pay it, I'm not employed. They never responded, they just kept sending me more and more bills. Then I had my parole agent call and he called and left a message for them to contact him, but they never did . . . They just kept sending (the bills), right now it's like $6500.

Out of the 258 young men in the survey, 35 had legal child-support orders established before or during their incarceration. All were charged for the months they were incarcerated, but 11 took successful legal action to have their debt erased or reduced. Three men took unsuccessful legal action, and two cases were still pending at the time of the survey. The other 19 men did not take any action to have their debts reduced and most had incurred debts in the thousands of dollars for the time they spent in prison. When I spoke with them about their debts, many of the 19 men reported that they felt angry and helpless to do anything to change the amount they owed.

INCARCERATION AND YOUNG FATHERS: CONCLUSIONS, IMPLICATIONS, AND FUTURE RESEARCH

The data presented in this chapter indicate that when a young father spends time in prison, it significantly affects the long-term relationship he is able to build and maintain with his child. It is important that researchers investigating the attitudes and behaviors of young fathers begin to discuss the effect of prison, or risk an incomplete analysis of this large and growing group. It is also important for inner-city/underclass researchers to view prison as a possible cause of social problems, not simply as the result. Absent fatherhood may be a precursor of incarceration (Christmon & Lucky, 1994; Elster, Lamb & Tavare, 1987; Lerman, 1993), but prison further serves to alienate young fathers from their children. Similarly, while poverty often precedes incarceration (Bishop & Frazier, 1996; Bortner & Williams, 1997), it is also exacerbated by incarceration.

The high level of social and financial damage caused by the incarceration of young fathers indicates that one of the primary goals of public policy should be to reduce their numbers. While there are many policies that might accomplish this aim, I discuss only a few here. First, in order to provide alternatives to incarceration for young men, education and employment opportunities need to be expanded. William J. Wilson's work (1996) shows that the flight of unskilled jobs from the inner city has resulted in decreased employment opportunities for young men, particularly for men of color. There is evidence to suggest that juvenile criminal behavior can, in some cases, be reduced by the creation of strong ties to employment and school (Sampson & Laub, 1993).

Another strategy for reducing the number of fathers in prison is to reconsider the current guidelines increasing the length of juvenile sentences. New and more punitive laws are being enacted, resulting in longer periods of time spent in prison by juvenile offenders. This trend is seen most dramatically in the application of adult-sentencing laws, such as "Three Strikes," to some criminal young men. With increases in sentence length, we lose potentially productive citizens and fathers to the correctional system. We need to assess carefully our sentencing policies in order to protect the public but still limit the amount of time served by juveniles. We should also investigate and employ alternatives to incarceration including intensive probation, halfway houses, and electronic monitoring.

I believe that policies intended to decrease the number of young incarcerated fathers, such as those I have just described, make sense in light of the prohibitive costs of their imprisonment. Unfortunately, current public opinion regarding criminals indicates that such policies will be difficult to implement. Fear about crime, coupled with a desire to punish criminals, have resulted in an environment where incarceration is seen as the only way to achieve crime prevention. The public seems unwilling to consider the possibility that the long-term costs of imprisonment may outweigh the short-term costs of policies designed to reduce the number of fathers in prison. This political reality makes it clear that large numbers of young fathers will continue to enter our correctional system. For this reason, we must develop improved and enlightened policies for dealing with incarcerated and paroled men.

Before suggesting specific policies directed toward these men, it is important to acknowledge that the implementation of such policies raises a number of difficult questions. One of the most difficult involves the desirability of encouraging criminal men to be involved with their children. Some might argue that these men, given their criminal pasts, are unsuitable fathers and should be kept away from their children. This point is well taken. There is no question that some fathers commit such heinous crimes that allowing them to be in their children's lives could be harmful. On the other hand, it should be noted that a full 36% of the paroled fathers in this study were arrested for nonviolent crimes including crimes against property (auto theft, burglary, vandalism) and drug violations.

I believe that there are two groups of criminals whose potential danger clearly outweighs the benefits of involvement with their children. First, men who have been convicted of any type of child abuse should have restricted or no contact with their children. My data suggest that about 3% of inmate fathers in California are convicted of molesting or annoying children. Another small percentage mistreat or neglect children. The second group of men who should be allowed little or no access to their children are those who have committed acts of violence against the children's mother, or against other women. Studies suggest that children who are exposed to violence committed by their father against their mother have severe problems later in life. Females disproportionately become involved with men who abuse them, and males disproportionately batter their partners (Dobash & Dobash, 1979; Gelles & Straus, 1988). Given these facts, it is imperative that we act to protect children from men's violence.

Having acknowledged that there are categories of criminal men who should be denied involvement with their children, it would be a mistake to allow these men to dictate policy for all young incarcerated fathers. The majority of incarcerated young men do not pose a threat to their children, and they have the potential to be active and concerned fathers. For this reason, policies should be developed that help to mitigate the harmful effects of prison on father/child relationships. Space constraints keep me from discussing these policies at length, but it is important to mention several policies directly suggested by the findings reported in this chapter.

The first policy suggestion is that fathers in prison be given opportunities to maintain a relationship with their children. One way this can be accomplished is to encourage father/child visitation. Denise Johnston's work (1995) shows that prison visitation may ease the trauma associated with parent/child separation, and it may allow both parents and children to work out feelings of grief and loss that frequently accompany such separation. Her work also suggests that children who visit with their incarcerated parent are better able to develop realistic images of that parent. Children who do not visit with a parent are forced to rely on memory and imagination. When a child has a more realistic image of a parent, it makes it easier for the child to accept the parent back into their lives. Similarly, my data suggest that prison visitation might allow fathers to develop more realistic images of their children, thus easing their transition home.

Once men are released from prison, there are a number of measures that can be taken to help them readjust to life on the outside. First, as described above, one of the problems men face when they return home from prison is that they owe money for child support. In order to prevent debt from accruing while men are in prison, an information link between county child-support collection bodies and the prison system needs to be established. Such a system could work to the benefit of all involved. It would no longer be necessary for inmates to inform child support that they are incarcerated. Instead, child-support collection agencies would receive immediate notification, completely eliminating illegal charges. Once a man was released from prison, this information would be communicated to the proper authorities and child-support charges would begin to be levied. Child-support enforcement authorities could be encouraged to cooperate closely with parole agents who could help the young men develop a plan to begin or resume payment of child support.

Another recommendation suggested by this study is to develop public policies to help strengthen the relationship between paroled father and children's mother. I have shown that a young father's relationship with the mother of his child plays an important role in encouraging or discouraging his involvement with his children. Because personal relationships are complex and private, they are difficult to manipulate through policy. I believe, however, that there are some limited actions that are available to us. First, as a result of the recent upsurge in interest in young fatherhood, various states, localities, and nonprofit agencies are creating classes and support groups to aid young fathers. These programs should be designed to include the mother of the men's children. The California Youth Authority, in its parole parenting classes, has experienced success with its policy of including family members. In an evaluation of the classes, the research team observed that "the attendance of significant others in the class is extremely useful for helping parolees build strong relationships with their wives and girlfriends and in establishing a good classroom environment" (Cohen et al., 1996).

A second way to strengthen relationships between men and the mother of their children is to provide training in communication skills. This training could occur at parole offices, or at nonprofit agencies. My observations in parenting classes for parolees suggest that, like most adolescents, young fathers simply do not have the communication skills to be able to successfully negotiate their transition home to the mother of their children. I suspect that the mothers' communication skills are not any more advanced. Helping young men and the mothers of their children learn healthy ways to communicate may help them maintain amicable relationships. In turn, these amicable relationships can facilitate men's involvement with their children.

As the juvenile prison population continues to grow nationwide, more and more young fathers will spend a part of their children's lives behind bars. In addition to formulating a policy response to these men, it is important to make them a priority of our research agenda. One of the first projects that should be considered is a large longitudinal study that follows a group of young men from boyhood through adolescence. In this way, data could be obtained from men who end up in prison and from those who do not. This would allow us to measure more precisely the effect of prison on fathering attitudes and behaviors. While the data presented in this chapter suggest that prison and the transition home exert a strong influence on a man's relationship with his children, they do not allow us to analyze the strength of these effects. This could only be accomplished through a longitudinal comparison of men who go to prison with those who do not. Unfortunately, because of the time and expense required to do longitudinal studies, such data probably will not be available soon.

The life course perspective in sociology suggests a second important question for researchers interested in fathers and incarceration. This perspective calls on researchers to examine the "conditioning influences of prior life experiences" on new life transitions (Elder & O'Rand, 1995, p. 460). In the case of newly paroled fathers, we should direct our attention to the effect of prior life experiences on the relationship between prison and father/child relationships. Specifically, we might ask whether fathers with different life experiences and backgrounds (race, age, education) adjust their fathering behaviors differently in response to incarceration.

This chapter is an important step toward understanding the impact of prison on young men's relationship with their children. I have shown that prison can have potentially devastating long-term effects on the amount of contact parolees have with their children, and on how much support they provide. Given these findings, it is imperative that both sociologists and policymakers begin to consider prison as a primary force shaping men's ability and desire to be actively involved in raising their children.

REFERENCES

Austin, J., Krisberg, B., DeComo, R., Rudenstine, S., & Del Rosario, D. (1995). *Juveniles Taken into Custody Fiscal Year 1993 Statistics Report.* Washington D.C.: National Council on Crime and Delinquency.

Bishop, D. M., & Frazier, C. E. (1996). Race Effects in Juvenile Justice Decision-Making: Findings of a Statewide Analysis. *The Journal of Criminal Law and Criminology, 86,* 392–414.

Bortner, M. A. & Williams, L. M. (1997). *Youth In Prison.* New York: Routledge.

California Youth Authority. (1996). *Characteristics of CYA Population.* Sacramento: Research Division, Ward Information and Parole Research Bureau, Information Systems Unit. December.

Christmon, K., & Lucky, I. (1994). Is early fatherhood associated with alcohol and other drug use? *Journal of Substance Abuse, 6,* 337–343.

Cohen, L. E., Felmlee, D., Nurse, A. M., & Will, S. (1997). *Second Year Report to the California Youth Authority: Positive Parenting Program.* Sacramento. California Youth Authority.

Conley, D. J. (1994). Adding Color to a Black and White Picture. Using Qualitative Data to Explain Racial Disproportionality in the Juvenile Justice System. *Journal of Research in Crime and Delinquency, 31,* 135–148.

Daly, K. (1993). Reshaping Fatherhood: Finding the Models. *Journal of Family Issues, 14,* 510–530.

Danziger, S. K. (1987). *Father Involvement in Welfare Families Headed by Adolescent Mothers.* Discussion Paper No. 856–87, Institute for Research on Poverty, Madison, WI.

Danziger, S. K., Kastner, C. K., & Nickel, T. J. (1993). The Problems and Promise of Child Support Policies. In: R. I. Lerman & T. J. Ooms (Eds), *Young Unwed Fathers: Changing Roles and Emerging Policies,* (235–250). Philadelphia: Temple University Press.

DeLuccie, M. F. (1995). Mothers as Gatekeepers: A Model of Maternal Mediators of Father Involvement. *The Journal of Genetic Psychology, 156,* 115–131

Dinitto, D. D. (1995). *Social Welfare: Politics and Public Policy.* Boston: Allyn and Bacon.

Dobash, R. E., & Dobash, R. (1979). *Violence against Wives.* New York: The Free Press.

Elder, G. H. Jr., & O'Rand, A. M. (1995). Adult Lives in a Changing Society. In: K. S. Cook, G. A Fine & J. S. House (Eds), *Sociological Perspectives on Social Psychology,* (452–475). Boston: Allyn and Bacon.

Elster, A. B., Lamb, M. E., & Tavare, J. (1987). Association between Behavioral and School Problems and Fatherhood in a National Sample of Adolescent Youths. *Journal of Pediatrics, 111,* 932–936.

Furstenberg, F. F. Jr. (1995). Fathering in the Inner City: Paternal Participation and Public Policy. In: W. Marsiglio (Ed.), *Fatherhood: Contemporary Theory, Research, and Social Policy* , (119–147). Thousand Oaks, CA: Sage.

Furstenberg, F. F. Jr., Brooks-Gunn, J., & Morgan, S. P. (1987). *Adolescent Mothers in Later Life.* New York: Cambridge University Press.

Garwood, A. N., (Ed.) (1992). *Hispanic Americans: A Statistical Sourcebook.* Boulder: Numbers and Concepts.

Gelles, R. J., & Straus, M. A. (1988) *Intimate Violence.* New York: Simon and Schuster

Hagan, J. (1995). Rethinking Crime Theory and Policy: The New Sociology of Crime and Disrepute. In: H. D. Barlow (Ed.), *Crime and Public Policy: Putting Theory to Work,* (29–42). Boulder: Westview Press.

Hunter, E. J. (1986). Families of Prisoners of War Held in Vietnam: A Seven Year Study. *Evaluation and Program Planning, 9,* 243–251.

Hunzeker, D. (1995). Juvenile Crime, Grown-up Time. *State Legislatures,* May, 14–19.

Johnston, D. (1995). Jailed Mothers. In: D. Johnston & K. Gabel (Eds), *Children of Incarcerated Parents,* (3–20). New York: Lexington Books

Lerman, R. I. (1993). A National Profile of Young Unwed Fathers. In: R. I. Lerman, & T. J Ooms (Eds), *Young Unwed Fathers: Changing Roles and Emerging Policies,* (27–51). Philadelphia: Temple University Press.

Maguire, K., & Pastore, A. L. (Eds). (1994). *Sourcebook of Criminal Justice Statistics 1993.* Washington, D.C.: U. S. Government Printing Office.

Massey, D. S. (1995). Getting Away with Murder: Segregation and Violent Crime in Urban America. *University of Pennsylvania Law Review, 143,* 1203–1232.

McDermott, K., & King, R. D. (1992). Prison Rule 102: Stand by Your Man. In: R. Shaw (Ed.), *Prisoners' Children: What are the Issues?,* (50–73). London: Routledge.

Pirog-Good, M. A., & Good, D. H. (1995). Child Support Enforcement for Teenage Fathers: Problems and Prospects. *Journal of Policy Analysis and Management, 14,* 24–42.

Rindfuss, R. R., Rosenfeld, R. A., & Swicegood, C. G. (1987). Disorder in the Life Course: How Common and Does It Matter? *American Sociological Review, 52,* 785–801.

Sampson, R. J., & Laub, J. H. (1993). *Crime in the Making: Pathways and Turning Points through Life.* Cambridge, Massachusetts: Harvard University Press.

Seltzer, J. A., & Brandreth, Y. (1994). What Fathers Say about Involvement with Children after Separation. *Journal of Family Issues, 15,* 49–77.

Stouthamer-Loeber, M., & Wei, E. H. (1998). The Precursors of Young Fatherhood and its Effect on Delinquency of Teenage Males. *Journal of Adolescent Health, 22,* 56–65.

Tucker, M. B., & Mitchell-Kernan, C. (Eds). (1995). *The Decline in Marriage among African Americans.* New York: Russell Sage Foundation.

Wilson, W. J. (1996). *When Work Disappears: The World of the New Urban Poor.* New York: Vintage Books.

Wordes, M., Bynum, T. S., & Corley, C. J. (1994). Locking Up Youth: The Impact of Race on Detention Decisions. *Journal of Research in Crime and Delinquency, 31,* 149–165.

FATHERS BEHIND BARS: THE IMPACT OF INCARCERATION ON FAMILY FORMATION

Bruce Western and Sara McLanahan

Increases in divorce and non-marital childbearing have dramatically altered children's living arrangements and access to parental resources. Whereas in 1960, the vast majority of children lived with both of their biological parents from birth to adulthood, by 2000, less than half of all children were expected to grow up with both parents (Bumpass & Lu, 1999). Today, a third of all births in the U.S. are to unmarried parents, and many of these children will never live with their fathers (Ventura et al., 1995).

Not only have children experienced substantial declines in parental resources, their access to public resources have declined as well. Cash transfers to disadvantaged households have been discredited as an instrument of social welfare. In 1996, the Federal government abandoned it's national standard for public assistance by replacing Aid to Families with Dependent Children with Temporary Assistance for Needy Families. The new welfare law shifts responsibility for children from government to parents by limiting cash assistance to single mothers and by forcing non-resident fathers to pay child support. Although the shifting policy climate targets single mothers, the burden of welfare reform falls on children. Not surprisingly, child poverty rates have remained stubbornly high despite unprecedented prosperity throughout the economy as a whole. Although welfare policies seek to encourage marriage and

This chapter was written while Western was a visiting scholar at the Russell Sage Foundation, New York. Kristen Harknett provided invaluable research assistance.

Families, Crime and Criminal Justice, Volume 2, pages 309–324.

ISBN: 0-7623-0737-4

increase parental responsibility, parallel developments in our criminal justice policies have had the opposite effect. Criminal sentences have become more punitive, imposing long terms of incarceration on drug offenders and third-time felons. Under this new sentencing regime, the U.S. penal system grew by nearly 700,000 between 1990 and 1998 to include more than 1.8 million inmates (Gilliard, 1999). These aggregate figures conceal enormous racial disparity. Incarceration rates for African Americans are about seven times higher than those for whites. Estimates indicate that 12.1% of black men aged 20 to 35 were behind bars on an average day in 1996 (Western & Pettit, forthcoming). Figures based on 1991 incarceration rates indicate that 28.5% of black men will spend time in prison at some time in their lives compared to a lifetime risk of 4.4% for white men (Bonczar and Beck, 1997). Insofar as incarceration is concentrated among young poorly-educated minority men, the expansion of the penal system over the last two decades emerges as a key suspect in explaining the growing number of single-parent families in disadvantaged communities. Incarceration is likely to deter family formation both directly, by making it more difficult for fathers to live with their children, and indirectly, by reducing fathers' employment prospects and earnings capacity. Both outcomes discourage marriage and family formation.

In this chapter we use data from a new survey, the *Fragile Families and Child Wellbeing Study*, to examine the relationship between incarceration and family structure among new parents. The analysis looks at how incarceration influences cohabitation and marriage one year after the birth of a child. While other data sets provide information about marriage and cohabitation among new parents, the Fragile Families Survey is unique in offering a detailed picture of the criminal justice history of a sample of new fathers, including a large sample of unmarried fathers. An analysis of these data thus provides an important first step in assessing the impact of incarceration on family formation among low-income parents.

THE RISE IN INCARCERATION

Before discussing the relationship between incarceration and family structure, we provide some background by detailing the increase in incarceration in the United States. Throughout the first seven decades of the twentieth century, incarceration rates remained roughly constant. The proportion of people in prison remained at around 0.1% of the U.S. population. Rapid increase in the incarceration rate began in the early 1970s. By 1980, the incarceration rate had increased to 0.15% of the population. The growth of the penal system accelerated after this point and by 1998 nearly 0.45% of the population were housed in state or federal prisons. This understates the true extent of the penal system because it ignores inmates detained in local jails while awaiting trial or serving short sentences. The most recent figures indicate that the total incarceration rate is 0.67% (Gilliard, 1999). Prison and jail inmates are overwhelmingly male implying that well over 1% of the adult male population is currently behind bars. In a comparative survey of incarceration rates around the world, Mauer (1998, p. 23) found that only Russia exceeds the United States in the use of imprisonment. While easily surpassing the incarceration rates of Western Europe, the United States also imprisons a significantly larger proportion of its population than Singapore, China, Poland, and South Africa.

A striking feature of the expansion of the U.S. penal system is the extraordinary incarceration rates recorded by young minority men (Western & Pettit, forthcoming). In 1982, the incarceration rate for white men aged 20 to 35 was 0.83% compared 5.52% for young black men. By 1996, the same white incarceration rate had increased to 2.05% while 12.18% of young black men were in prison or jail. Among young male high school dropouts, the black incarceration rate was 36.30% in 1996 compared to 7.39% for whites. More young black low-education males were in prison or jail on average day in 1996 than were in paid employment. Indeed, prison and jail steadily eroded employment rates among young unskilled black men through the 1990s, despite rising employment rates throughout the labor market as a whole.

Why has incarceration in America increased so dramatically? Criminologists suggest that much of this growth is due to tougher sentences for repeat offenders and the increased likelihood of custodial sentences for drug offenders. Long sentences for repeat offenders have increased the average time served. Three strikes laws, truth in sentencing provisions, and the abandonment of parole boards have all increased the duration of prison sentences, explaining 36% of the rise in incarceration between 1980 and 1996 (Blumstein & Beck, 1999, p. 43). The intensified policing and prosecution of drug offenders has been particularly spectacular. The incarceration rate for drug offenses today is now comparable to the total rate of incarceration that prevailed in the United States over much of the past century. Although the link between drugs and criminal behavior is often claimed (Boyum & Kleiman, 1995), crime rates are not strongly related to trends in incarceration. The analysis of Blumstein & Beck (1999) attributes just 10% of the rise in state imprisonment between 1980 and 1996 to patterns of offending. By contrast, more the 50% of the rise in imprisonment is traced to the increasing use of custodial sentences, particularly for drug offenders.

The rise in incarceration thus involves a massive institutionalization significantly affecting young, poorly educated, minority males. This expansion of the penal system does not appear to be strongly related to crime rates but is instead rooted in policy shifts closely connected with Federal and state governments' 'war on crime' 'and war on drugs'.

INCARCERATION AND FAMILY STRUCTURE

The impact of male incarceration on parental relationships and child wellbeing has received little systematic empirical study (Hagan & Dinovitzer, 1999). However, other research on the post-release experiences of ex-inmates suggests how incarceration may influence family structure.

The most obvious consequence of incarceration is its incapacitative effect. Incapacitation commonly refers to criminals' inability to commit crimes while detained in prison (Zimring & Hawkins, 1995). We can expand this idea to consider a variety of positive social functions that inmates would perform, if released. If the incarceration rate were lower, many of those currently in prison might otherwise have held jobs and contributed to the support of families. Around 20% of male state prison inmates are married, so about 260,000 couples were separated by incarceration in 1998. The estimate is conservative because it ignores marital disruption through jail time or among female inmates (currently about 6% of the total prison population). The incapacitative effect of incarceration on children is even larger. According to the 1991 Survey of Inmates of State Correctional Facilities, 56% of state prison inmates have children under age 18. If this proportion remained constant, about 730,000 fathers were nonresident in 1998 due to their incarceration. Many of these fathers have more than one child. Thus recent studies find that 1.5 million children have a parent in prison (Hagan & Dinovitzer, 1999, p. 137). The total prison population grew at an annual rate of 6.7% in the 1990s, underlining a dramatic expansion of the number of families caught in the web of the penal system.

Incapacitation describes the immediate effect of prison detention, but the impact of incarceration also continues well after release. Incarceration is a watershed event that can disrupt key life course transitions setting in a motion a downward spiral of accumulating disadvantage (Sampson & Laub, 1993). The stigma of incarceration can produce strong feelings of shame and anger, both for inmates and their families (Hagan & Dinovitzer, 1999, pp. 126–127). Incarceration is thus likely to be a significant shock to family relationships, contributing to marital strain among ex-inmates. For those unmarried, the stigma of incarceration also shrinks the pool of possible partners. Thus counties with large populations of returning prisoners have large numbers of female-headed families (Sabol & Lynch, 1998). Ethnographic research also suggests that single mothers are reluctant to marry or live with fathers of their children if the father has a history of incarceration (Edin, 2000; Waller, 1997).

The impact of incarceration on family formation would be especially severe where prisons have a criminogenic effect on inmates. In this scenario, prisons serve as entry points for careers in crime. If the prison experience made men more prone to violence or abuse, incarceration would have large negative effects on family welfare. The criminogenic effects of the prison, however, are hotly contested. Evaluation research indicates that criminogenic effects – or least recidivism – can be reduced by well-designed treatment programs (Gaes et al., 1999). Resources for rehabilitation are scarce in the current policy context. For instance, while three-quarters or more of all state prison inmates require substance abuse treatment, only 13% of inmates will be in some kind of program during their prison term (McCaffrey, 1998). When prison overcrowding creates pressure to cut the costs of corrections, many state legislatures have been unwilling to expand resources for prisoner services. Deteriorating conditions inside prisons may thus be raising the social costs of incarceration after release.

Whatever the criminogenic effects of prison, there is reasonable evidence that incarceration reduces the employment and earnings of ex-inmates. Studies of administrative data and unemployment insurance records find relatively large effects of prison incarceration on employment and earnings (Grogger, 1992; Kling, 1999). While these studies usefully examine longitudinal data on legitimate earnings, the analyses provide no comparison with the un-convicted general population. Social survey data offer the opportunity of comparing ex-inmates to those that have never been incarcerated. Results from survey data tend to find larger economic penalties associated with prison time. Analysis of data from the National Longitudinal Study of Youth (NLSY) shows that juvenile detention can have extremely persistent negative effects on employment. Freeman (1991) reports that respondents serving time as juveniles experience an unemployment penalty of about 10 weeks in the year, up to eight years after juvenile detention. Using more recent data, Western & Beckett (1999) show that employment among NLSY respondents is almost 10% lower among juvenile detainees, 10 years after the original spell of incarceration, even controlling for accumulated work experience and adult incarceration. In this analysis the impact of juvenile incarceration was three times larger and more persistent than the employment effect of adult incarceration. Researchers argue that these negative economic effects may be due to employer discrimination against ex-inmates or the erosion of human capital during spells of prison time.

How does this research inform the study of family formation? The decision of low-income mothers to marry or remarry depends in part on the economic prospects, social respectability, and trustworthiness of their potential partners

(Edin, 2000). Incarceration undermines all these qualities. Ex-inmates tend to be low-earners with irregular employment records. The stigma of incarceration significantly reduces the social status of young men and signals their unreliability to possible marriage partners. In short, incarceration has potentially devastating consequences for the marriage markets of communities with high incarceration rates.

While the behavioral and economic effects of incarceration may disrupt family relationships, we must also consider the possibility that criminal conviction simply identifies those with few economic or social prospects. Prison time may be subject to endogeneity in which incarceration selects those unlikely to be married or at high risk of divorce or separation. The problem of endogeneity is central to much of the research on incarceration and employment. Thus analyses of panel data have tried to adjust for fixed or random effects, and instrumental variables have been used to identify the exogenous component of incarceration (Freeman, 1991; Western & Beckett, 1999; Kling, 1999).

Although endogeneity certainly poses a threat to causal inference we caution that this threat may be overstated in the current policy climate. The incarceration rate has increased threefold in the two decades since 1980. There is little support for the idea that criminal behavior is driving the rise in incarceration. Instead, it appears that the exogenous influence of criminal justice policy is now capturing new kinds of offenders who would not have been detained earlier under a less punitive sentencing regime. The endogeneity of imprisonment to criminal and other anti-social behavior is thus likely to be weaker now in the period of mass incarceration.

Support for this idea can be seen from Table 1 comparing male state prison inmates to noninstitutional males in the years 1979 and 1991. The noninstitutional men come from the March Current Population Survey (CPS). We restrict analysis to CPS men aged 18–45 because 95% of all prison inmates are aged in this range. The table shows that the inmate pool has become somewhat more mainstream in several respects. The educational attainment of prisoners increased between 1979 and 1991. In 1979, inmates were only 20% as likely as noninstitutional men to have attended at least some college. By 1991 the gap in college attendance had closed as the chances of college attendance among inmates had increased to 30% of the college attendance rates of their noninstitutional counterparts. The difference in high school dropout rates also shrunk slightly between 1979 and 1991. Other analysis, not reported here, shows that gap in schooling shrinks even more if inmates are compared only to young noninstitutional black and Hispanic men. The family and marital status of prisoners also improved in relation to the noninstitutional population through the 1980s. While inmates in 1979 were 3.5 times more likely to be divorced

Table 1. Characteristics of Male State Prison Inmates, All Noninstitutional Men Aged 18–45 (CPS Men), 1979 and 1991.

	1979			1991		
	Inmates (1)	CPS Men (2)	Ratio (1)/(2)	Inmates (1)	CPS Men (2)	Ratio (1)/(2)
Mean age (years)	29.1	29.8	–	31.9	31.5	–
Race or ethnicity						
White	41.5%	83.0%	0.50	35.4%	78.0%	0.45
Black	46.6	10.7	4.36	45.5	11.9	3.82
Hispanic	9.5	6.2	1.53	16.8	10.1	1.66
Schooling						
Less than HS	52.6	19.8	2.66	40.3	15.8	2.55
HS or equivalent	39.1	38.4	1.02	46.0	38.9	1.18
Some college	8.4	41.8	0.20	13.7	45.3	0.30
Family Status						
Never married	52.5	34.0	1.54	55.9	37.5	1.49
Married	22.4	59.2	0.38	18.2	53.2	0.34
Divorced or Separated	23.1	6.5	3.55	24.3	9.1	2.67
Children under 18	50.8	56.9	0.89	56.0	48.8	1.15
Sample size	9142	30074	5213	11031	29943	7244

Source: Survey of Inmates in State Correctional Facilities (1979, 1991), March Current Population Survey (1979, 1991)

or separated than noninmate men, the relative separation rate of state prison inmates had fallen to 2.67 by 1991. The relative rate of nonmarriage fell slightly. The average age of prisoners increased and they were also more likely to have young children in 1991 than in 1979. In contrast to these trends however, the proportion of non-Hispanic whites has declined while the proportion of Hispanics has increased by nearly three-quarters. In short, the average inmate is increasingly coming to resemble the modal young male of the noninstitutional minority population.

DATA AND METHODS

To study the impact of incarceration on marriage and cohabitation, we analyze data from the Fragile Families survey which contains information both on the relationship status of new parents and the fathers' history of contact with the criminal justice system. Data for the Fragile Families and Child Wellbeing Study is being collected in twenty U.S. cities, stratified by different labor market conditions and varying welfare and child support policy regimes. The sample will be representative of nonmarital births to parents residing in each of the twenty cities. When all the data are collected, the sample will be representative of nonmarital births to parents residing in cities with populations over 200,000. A comparison group of married parents will also be followed. The total sample size will be 4700 families, including 3600 unwed couples and 1100 married couples. New mothers will be interviewed at the hospital within 24 hours after they have given birth. Follow-up interviews with both parents will be conducted when the child is 1, 2½, and 4 years old. At the time of writing, baseline and follow-up data were available for 656 families in Oakland and Austin.

Table 2 describes the characteristics of our samples of married parents and unmarried parents taken from the baseline interviews in Oakland and Austin in 1998. Although the two samples share similar age distributions, the unmarried parents are substantially more likely to be African American and less educated than married parents. Among fathers, African Americans make up nearly half the unmarried sample (48.0%) while they account for fewer than one in six (14.9%) of the married sample. The racial composition of the mothers' sample is similar. Nearly half of the unmarried mothers are African American compared to just 16% of the married mothers. Ethnic differences between married and unmarried parents are less pronounced. Around 40% of unmarried mothers and fathers are Hispanic compared to around 50% of married parents. The sample of married parents is also characterized by relatively high rates of college attendance. Almost half of the married mothers and fathers have completed at least some college education while only about a quarter of the unmarried sample (24% for fathers and 22% for mothers) has attended college. The unmarried sample shows higher rates of high school graduation than the married sample, and similar proportions of couples with less than twelve years of schooling. The high rate of high school dropouts among the married sample is due to the large proportion of Hispanics in the married sample. We investigate the relationship between marital status and education more systematically below.

Table 3 provides a preliminary idea of the accuracy of the Fragile Families measure of incarceration status compared to other data sources. The incarceration status of the Fragile Families fathers is recorded in two ways. First, we note whether the father was in jail at the time of the baseline interview, and second, we note whether the father had ever been incarcerated by the time of the follow-up survey. Whether respondents have even been incarcerated at the time of the follow-up interview reflects their cumulative risk of incarceration. This information about fathers is obtained in both the fathers' and mothers' surveys.

The table reports the cumulative risk of incarceration and the point-in-time incarceration rate for the samples of new fathers. Data on the cumulative risk

Table 2. Characteristics of Mothers and Fathers in the Fragile Families Survey at Baseline Interview, 1998.

	Unmarried	Married
Father's age (median)	26.0	26.0
Mother's age (median)	24.0	24.0
Father's race		
White	9.2%	32.8%
Black	48.0	14.9
Hispanic	42.8	52.2
Total	100.0	100.0
Mother's race		
White	11.5%	32.1%
Black	44.8	16.4
Hispanic	43.8	51.5
Total	100.0	100.0
Father's education		
Less than H.S.	39.5%	39.5%
H.S. graduate	36.5	14.2
Some college	24.0	46.3
Total	100.0	100.0
Mother's education		
Less than H.S.	48.0%	37.3%
H.S. graduate	39.8	15.7
Some college	22.2	47.0
Total	100.0	100.0
Sample size	400	134

of incarceration matches the variation by race found in other data sources. Black men have a much higher incarceration rate than whites. The survey data tend to understate the cumulative incarceration risk for Hispanics, which falls between incarceration risks for blacks and whites calculated by the Bureau of Justice Statistics (BJS). Still, the proportions of black and Hispanic men in the Fragile Families data who report they have been incarcerated at some time approximately equals BJS figures and cumulative incarceration risks reported by respondents from the National Longitudinal Sample of Youth.

The mother's reports of the father's status indicate a much higher risk of imprisonment. Mother's reports are particularly informative about fathers who

Table 3. Percentage of Males Incarcerated in Fragile Families Survey Compared to Other Data Sources. (Counts in parentheses.)

	Race or Ethnicity		
	Black	White	Hispanic
Cumulative risk			
Ever incarcerated (FF Fathers)	28.1 (39)	18.2 (12)	12.4 (23)
Ever incarcerated (FF Mothers)	55.6 (110)	23.1 (39)	27.7 (39)
Lifetime incarceration risk (BJS)	28.5	4.4	16.0
Ever incarcerated, 1981–1994 (NLSY)	17.2 (242)	4.4 (144)	9.9 (89)
One-day rate			
Incarcerated at baby's birth (FF)	5.6 (60)	1.1 (4)	2.0 (5)
Incarceration rate, 1996 (BJS)	7.5	1.1	

Note: Lifetime risk is reported for state prisoners using 1991 incarceration rates by Bonczar & Beck (1997). The NLSY cumulative risk is measured proportion of respondents interviewed in correctional facilities between 1982 and 1996. All Fragile Families results use the follow-up survey, except for data on incarceration status at interview which uses the full baseline survey. Cell entries for fathers use father's reports of incarceration status. Cell entries for mothers use mother's reports of father's incarceration status The 1996 rate using BJS data are calculated by Western and Pettit (forthcoming). Hispanics are includes among blacks and whites for the BJS rates

are hard to find and whose incarceration risk is likely to be high. Because the fathers in the Fragile Families survey are disproportionately young with little education, it seems likely that the father's reports understate the true incidence of incarceration. Mother's reports of cumulative incarceration risk may be more accurate. In the analysis below, we contrast incarceration data from mother's and father's reports in an effort to minimize the effect of nonrandom sample selection in the re-interview of fathers. The final two lines of Table 2 compare the BJS incarceration rates for 1996 to the percentage of fathers known to be in prison or jail at the time of the baby's birth. The BJS incarceration rates are calculated for the adult male population aged 18 to 65. The relatively young and uneducated Fragile Families sample is thus likely to have a higher incarceration rate than is reflected in most survey data.

RESULTS

The relationship status of non-inmates and ex-inmates is described in Table 4. If incarceration status is measured by the father's cumulative risk of imprisonment, father's reports indicate that men who have never been incarcerated are about twice as likely to be married as ex-inmates. Ex-inmates are also 50% more likely

Table 4. Percentage Distribution of Relationships by Incarceration Status, Fragile Families Follow-up Surveys, Oakland and Austin.

	Father's Report Ever Incarcerated?		Mother's Report Ever Incarcerated?	
	No	Yes	No	Yes
Married	41.1%	21.5%	42.8%	16.3%
Cohabiting	34.4	31.6	34.1	23.5
Visiting	5.8	17.7	6.9	13.8
Other	18.7	29.1	16.2	46.4
Total	100.0	100.0	100.0	100.0
Sample size	343	79	334	196

to have no relationship with their baby's mother a year after the birth. Similar but stronger patterns can be found in the mother's reports of father's incarceration status. The mothers' reports indicate that nearly half of all ex-inmate fathers are neither living with nor romantically involved with the mother at the time of the follow-up interview. Finally, data from the full baseline survey show a similar pattern for the men who were incarcerated at the time of the baby's birth. Virtually none of the inmates in the sample were married and over 15% were no longer in a relationship with the mother at the time of birth.

Regression analysis offers a more systematic picture of the impact of incarceration on relationship status. The regression analysis proceeds in two stages. We first estimate the probability that a couple is living together (either cohabiting or married) at the time of the follow-up interview. Then conditional on the couple living together, we estimate the likelihood of marriage. The key predictor is a dummy variable that indicates whether the father was ever incarcerated. The main control variables are the age, race, ethnicity, and education of the parents. To minimize collinearity among the predictors we code the race and ethnicity of the couples rather than the race or ethnicity of individual partners. We also add a measure of relationship quality that reflects how often the couple socializes and engages in other activities together. This relationship index is intended to capture the father's commitment to the mother. Father's that are strongly committed to their partners are more likely to be interviewed and more likely to be married or cohabiting. Because the father's commitment to the mother influences the probability of observing the father and the parents' relationship status, the relationship index can be understood as a control for sample selectivity in the analysis of the father's data. Sample selectivity can also be investigated by contrasting father's reports and mother's reports of father's status. Mother's reports are more complete since they provide information about those fathers that have been hard to find and interview. By including a larger proportion of the high-risk fathers, the mother's reports are likely to yield relatively large incarceration effects.

Table 5 reports the logistic regression results. The results show that father's incarceration status has a large negative effect on the likelihood that the parents will be living together one year after the birth of their child. Using fathers' reported history of criminal justice system contact we find that prior incarceration status lowers the odds of cohabitation by about 50% ($e^{-0.66} = 0.51$). As expected, mother's reports yield even larger incarceration effects. Mother's data suggest that prior incarceration lowers the odds of living together by about 70% ($e^{-1.01} = 0.36$). Results for the other independent variables indicate that couples' relationships are strongly patterned by race and age. Mothers' and fathers' data both suggest that black couples are less than half as likely as white couples to be living together a year after the birth of their child. Hispanics, however, are not strongly different from whites. Mother's age is positively and significantly related to the probability of living together. Each additional year of age is estimated to raise the odds of living together by about 10%. By contrast, the effect of father's age is negligible. Finally, the results also suggest that the relationship index – measuring the degree of the partner's commitment to each other at baseline – provides some control for sample selectivity. We suspect that nonresponse among fathers creates bias because it is systematically related to relationships status. Consistent with this idea, the relationship index is significantly and positively related to the probability of living together in the fathers' data, but not the mothers'.

Table 5 also reports the effects of incarceration and other independent variables on the odds of marriage, given that couples are living together at the time of the follow-up interview. In this case, both data sources indicate that father's prior incarceration status is unrelated to the likelihood of being married. This suggests that incarceration works nonlinearly, to destabilize couples with relatively weak relationships. The results also show that race and education are significantly related to the chances of marriage among couples that are living together. African American couples are highly unlikely to be married compared to whites. According to the estimates, the odds of a black couple being married a year after their child's birth is only 10% ($e^{-2.27} = 0.10$) of that for a white couple with similar age and education. Father's education is not strongly related to the probability of living together, but college attendance among mothers nearly triples the odds ($e^{1.05} = 2.86$) of marriage among couples that are living together. Finally, mixed race couples that live together are much less likely to

Table 5. Logistic Regression of Relationship Status on Mother's and Father's Characteristics, Fragile Families Survey, Oakland and Austin.

	Father's Reports		Mother's Reports	
	Living Together	Married	Living Together	Married
Intercept	-2.40	-1.23	-1.26	-0.45
	(2.38)	(1.08)	(1.30)	(0.43)
Couple characteristics				
Black	-0.85	-2.26	-1.00	-2.27
	(1.76)	(3.83)	(2.09)	(4.16)
Hispanic	0 80	-0.75	0 35	-0.69
	(1.45)	(1.19)	(0.67)	(1.19)
Mixed	-0.23	-1.41	-0.35	-1.32
	(.41)	(2.13)	(0.64)	(2.11)
Relationship index	0 32	0.06	0.15	-0.07
	(2.43)	(0.39)	(1.25)	(0.49)
Mother's characteristics				
Age	0 10	0.05	0.10	0.06
	(2.84)	(1.56)	(2.73)	(1.58)
Less than HS	-0.09	-0.46	-0.22	-0.21
	(.27)	(1.12)	(0.69)	(0.55)
Some college	0.25	1.13	0.18	1.05
	(64)	(2.57)	(0.48)	(2.49)
Father's characteristics				
Age	0.00	0 01	0.00	0.01
	(0.14)	(0.22)	(0.03)	(0.16)
Less than HS	0.36	0.61	0.49	-0.11
	(1.00)	(1.44)	(1.52)	(0.28)
Some college	0.01	0.70	0.10	0 16
	(0.03)	(1.63)	(0.28)	(0.37)
Incarcerated	-0.66	0.03	-1.01	-0.10
	(2.02)	(0.06)	(3.72)	(0.30)
N	355	264	387	272

be married than same race couples. In sum, the evidence suggests that incarceration operates at the fringes of the marriage market, on the weakest relationships. However, mother's reports show that incarceration affects more than half of the follow-up sample and the estimated magnitudes of the incarceration effects are very large so the effects of incarceration on family formation are likely to be extensive.

DISCUSSION

Results from this chapter point to the large destabilizing effects of the penal system on low-income families. The evidence indicates that incarceration narrowly influences whether a father is present in the household; it does not affect the likelihood of marriage among couples that live together. If – controlling for age, race, and education – the causal effect of incarceration on father absence were just one-fifth as large as the estimated effect, the rate of father absence would be 20% lower among ex-inmates. Given estimates of lifetime risks of incarceration and 1990 population figures from the Bureau of Labor Statistics, the regression results suggest that around 425,000 African American fathers are absent due to incarceration. This totals about 15% of the black nonresident father population.[1] Similar calculations for whites suggest that about 8% or 520,000 nonresident fathers are absent because of the post-release effect of incarceration on family formation. Combine these admittedly rough estimates with a prison population that that grows at more than 6.7% each year, and it is clear that the penal system is eroding the fabric of family life in poor minority communities.

From a methodological viewpoint, the analysis also demonstrates the difficulties of applying standard survey research methods to young economically disadvantaged men. They are hard to interview and high rates of survey non-response contribute large biases to survey data. The Fragile Families survey provides data on young high-risk men collected from their partners. These data collected from mothers were substantially more complete and yielded estimates of cumulative incarceration rates nearly double, in some cases, those obtained from fathers. The men's data also provided regression coefficients that were significantly smaller than those obtained from the data provided by women in the sample. If only the men's data were available, as is usually the case, both the incidence and the effect of incarceration may have been substantially under-estimated. Although most methodological attention in the study of incarceration effects has focused on the problem of nonrandom selection into the penal system, the fundamental problem of accurately counting ex-inmates may prove to be an even larger obstacle to accurately assessing the social impact of the penal system. Results from the Fragile Families data are certainly encouraging in suggesting a method – interviewing accessible and nearby informants – about groups that are generally under-represented in social surveys.

These findings can also be placed in the broader context of other research on the social impact of incarceration. First, if we combine the current findings with research on the employment opportunities of ex-inmates, it is clear that the resources for supporting children are significantly depleted by incarceration. Men released from prison do poorly in the labor market. In addition to their grim economic prospects, the stigma of incarceration also makes ex-convicts highly unattractive partners for marriage or cohabitation. Consequently, they are poorly-equipped to contribute financially to raising children and they are relatively uninvolved in parenting. While prisons remove men from families, the echo of incarceration continues well after release from prison.

Second, much research on family formation, crime, and incarceration has focused on how strong family relationships direct young men away from crime and the risk of serving time in prison. In this work, the social bonds of marriage reinforce social control, reducing deviant behavior. Our analysis turns this relationship around. In our approach, the experience of incarceration can undermine social bonds, straining martial and other family relationships. The formal social control of the prison may thus undermine the informal social control of family ties, contributing further to crime and deviance in high incarceration communities.

Finally, the possibility of a self-reinforcing relationship between marriage and incarceration also has important policy significance. We have tried to emphasize that the link between crime and incarceration is institutionally variable. Changes in sentencing policy opened the prison gates to new categories of people. Although changes in sentencing policy represent a significant effort at crime control, the effects of this policy may create the conditions for higher crime rates in the future. In contrast to old arguments about the prison as a school for criminals, the criminogenic effect of the prison may work through its impact on the pattern of social relationships in high incarceration communities. The American experiment with mass incarceration may thus be a self-defeating strategy for crime control. Without assistance for families disrupted by incarceration, the negative social effects of the penal system may aggravate the problems it was designed to solve.

NOTES

1. This figure is calculated by first estimating the number of African American adult males who have ever been incarcerated. We calculate this by multiplying the African American adult male population (8.479 million) by the lifetime risk of incarceration (0.285). We then multiply the ever-incarcerated population by the estimated difference in rates of father absence between ex-inmates and non-inmates (0.20) to obtain the number of nonresident fathers due to incarceration.

REFERENCES

Blumstein, A., & Beck, A J. (1999). Population Growth in U.S. Prisons, 1980–1996. In: M. Tonry, & J. Petersilia (Eds), *Crime and Justice: Prisons, 26*, (17–62). Chicago. University of Chicago Press.

Bonczar, T. P., & Beck, A. J. (1997). Lifetime Likelihood of Going to State or Federal Prison. Bureau of Justice Statistics Special Report NCJ-160092. Washington DC: U.S. Department of Justice.

Boyum, D., & Kleiman, M. A. R. (1995). Alcohol and Other Drugs. In: J. Q. Wilson, & J. Petersilia (Eds), *Crime*, (295–326). San Francisco, CA: ICS Press.

Bumpass, L., & Lu, H.-H. (1999). Trends in Cohabitation and Implications for Children's Family Context in the U.S. Center for Demography and Ecology, University of Wisconsin-Madison.

Edin, K. (2000). Few Good Men: Why Poor Mothers Don't Marry or Remarry. *American Prospect, 11*, 26–31.

Freeman, R. B. (1991). Crime and the Employment of Disadvantaged Youths. Working paper no. 3875. National Bureau of Economic Research Working Paper.

Gaes, G. G., Flanagan, T. J., Motiuk, L. L., & Stewart, L. (1999). Adult Correctional Treatment. In: M. Tonry & J. Petersilia (Eds), *Crime and Justice: Prisons, 26*, (361–426). Chicago: University of Chicago Press.

Gilliard, D. K. (1999). Prison and Jail Inmates at Midyear 1998. Bureau of Justice Statistics Bulletin. NCJ-173414. Washington, DC: U.S. Department of Justice.

Grogger, J. (1992). Arrests, Persistent Youth Joblessness, and Black/White Employment Differentials. *Review of Economic and Statistics, 74*, 100–106.

Hagan, J. & R. Dinovitzer. (1999). Collateral Consequences of Imprisonment for Children, Communities, and Prisoners. In: M. Tonry and J. Petersilia (Eds), *Crime and Justice: Prisons*, Vol 26, (121–162). Chicago: University of Chicago Press.

Kling, J. (1999). The Effect of Prison Sentence Length on the Subsequent Employment and Earnings of Criminal Defendants. Discussion Paper no. 208. Woodrow Wilson School, Princeton University.

Mauer, M. (1999). *Race to Incarcerate*. New York: New Press.

McCaffrey, B. (1998). Drug Treatment in the Criminal Justice System. Washington, DC: Office of National Drug Control Policy.

Sabol, W. J., & Lynch, J. P. (1998). Assessing the Longer-run Consequences of Incarceration: Effects on Families and Employment. Paper presented at the 20th Annual Research Conference of the Association for Pulbic Policy Analysis and Management, New York.

Sampson, R. J., & Laub, J. H. (1993). *Crime in the Making: Pathwys and Turning Points through Life*. Cambridge, MA: Harvard University Press.

Ventura, S. J. Bachrach, C., Kaye, K., Holcomb, P., & Koff, E. (1995). The Demography of Out-of-Wedlock Childbearing, In Report to Congress on Out-of-Wedlock Childbearing, DHHS pub. No (PHS) (95–1257). Washington D.C.

Waller, M. R. (1996). Redefining Fatherhood: Paternal Involvement, Masculinity, and Responsibility in the 'Other America'. Doctoral Dissertation, Princeton University.

Western, B., & Beckett, K. (1999). How Unregulated is the U.S. Labor Market? The Penal System as a Labor Market Institution. *American Journal of Sociology, 104*, 1030–1060.

Western & Pettit. Forthcoming. Incarceration and Racial Inequality in Men's Employment. *Industrial and Labor Relations Review*.

Zimring, F. E., & Hawkins, G. (1995). *Incapacitation: Penal Confinement and the Restraint of Crime*. New York: Oxford University Press.

THEY DO TIME TOO: THE EFFECTS OF IMPRISONMENT ON THE FAMILIES OF WHITE-COLLAR OFFENDERS

Karen A. Mason

Few events are more stressful for families than having a member incarcerated. Families of inmates often are referred to as the "hidden victims of crime." They suffer from emotional, social, and financial hardships (Bakker, Morris & Janus, 1978; Brodsky, 1975; Fishman, 1990). Previous research has helped to identify the problems that families face and the adjustments they must make when a father, mother, son, or daughter is imprisoned. However, most of this research is based on common street offenders and their families. Little is known about the impact of imprisonment on the families of white-collar offenders.

Most researchers agree that the impact of imprisonment varies depending on the family's adjustment prior to incarceration, whether incarceration is a recurrent pattern, and the material and social resources available to the family (Carlson & Cervera, 1992; Fishman, 1990; Flanagan, 1981). Because of their higher social status and extensive social reputations, families of white-collar offenders are assumed to avoid or neutralize many of the problems created by conviction and imprisonment. On this latter point there is some dispute. Some have argued that, relatively speaking, white-collar offenders and their families are more likely to be socially and economically damaged by conviction and imprisonment because of their high social status (Mann, Wheeler & Sarat, 1980).

Families, Crime and Criminal Justice, Volume 2, pages 325–337.

ISBN: 0-7623-0737-4

Because families of white-collar offenders have not been systematically included in research, we cannot assess the validity of either perspective.

To fill the gap in the literature, this chapter describes the impact of incarceration on the families of white-collar offenders. Information is presented on the amount of familial awareness of criminal activity before arrest, stigmatization throughout adjudication, and adjustment strategies used by family members during incarceration. I argue that families of both types of offender's undergo similar stresses and strains, and that criminal justice processes may provoke rather than deter negative sentiments about the legitimacy of the law.

WHO IS THE WHITE-COLLAR CRIMINAL?

There are two answers to this question. The more traditional approach limits the offender pool to those individuals of an upper or elite social class. Based on Sutherland's work, the offender-based approach focuses on professional offenders such as doctors, lawyers, the managerial ranks of public and private corporations, public officials with significant discretionary powers, and the owners of substantial capital. The second approach includes anyone who has committed a white-collar offense, regardless of his or her social or occupational status. Rather than emphasizing the offender, this perspective considers how a crime is committed. Using the offense-based approach, white-collar crime is defined as, "an illegal act or series of illegal acts committed by non-physical means and by concealment and guile to obtain money or property – to avoid payment or loss of money or property or to obtain business or personal advantage" (Edelhertz, 1970, p. 3).

For this study, white-collar crime is defined using the offense-based perspective. One reason for selecting this approach was to create a broad sample of offenders. Since little research has focused on the ordinary, garden-variety white-collar offenders, using an offense driven approach increases the likelihood that such individuals will be identified. Also, a more inclusive definition may more accurately reflect the types of crimes that are prosecuted in most federal districts as indicated by the small number of high level white-collar offenders in previous quantitative studies (see Weisburd, Waring, Wheeler & Bode, 1991).

Second, an offense-based approach is more likely to include women offenders. Previous studies focusing on major forms or organizational crime and workplace crime committed by professional workers rarely include a woman. One reason is that, relative to men, few women are in powerful, professional positions. But there are other reasons for the absence of women in these studies: few scholars have thought that it is important to interview female professional workers to investigate how gender relations may influence occupational or organizational crime (Daly, 1989). Whether the absence of women in research is a product of researcher neglect or a reflection of the sex distribution within occupations, arrest data confirm that a conceptually broad definition of white-collar crime is more likely to identify female offenders.[1]

THE STUDY

The study was conducted under the auspices of the United States Probation Office of the United States District Court for the Eastern District of Tennessee. Interviews with current and former probationers were conducted from February through October 1998.[2] All probation officers were provided a memo describing the study that included a list of federal statutes to be used to identify potential subjects. The list of offenses was adapted from the Weisburd et al. (1991) study of middle-class, white-collar offenders. The list was used to guide the probation officers' review of cases and to familiarize them with the types of crimes that are the focus of the study.

The offenses selected are included in most conceptions of white-collar crime. Some of the choices were those that virtually any American would regard as white-collar crimes, such as securities and antitrust violations. Bribery and embezzlement offenses were chosen because they specifically included the abuse of public and financial trust. Mail and wire fraud statues were chosen since many of the most important financial transactions require the use of mail or wire communications. Mail and wire fraud can be used to prosecute any crimes in which the postal service or other federally regulated communication systems are used for fraudulent purposes. Three other violations involving fraud or misrepresentation were also selected including: tax fraud, false claims and statements, and credit fraud. Because people in the highest income categories have the most to gain from tax frauds, such offenses are regarded as especially prevalent in the white-collar population. The final offense category is credit or lending institutions fraud, in other words knowingly making a false statement on loan and credit applications to federally insured financial institutions. Since the United States economy operates on credit, it is reasonable that the submission of fraudulent applications for loans is an important form of nonviolent economic crime (Weisburd et al., 1991, p. 9–11).

A total of 182 current and former probationers were identified.[3] Potential subjects were mailed a form letter with a self-addressed, stamped envelope that asked how the investigator could contact individuals interested in participating in the study. Of those receiving the letter, 35 offenders agreed to be interviewed.

The interviews were conducted at either the offender's home, work, or at a local restaurant. The interviews ranged in length from two to four hours.

The case files of all participants were reviewed to gather data on the offenders and their offenses. Twenty-two interviewees, seventeen men and five women, were sentenced to minimum-security federal prison camps. This chapter focuses only on those offenders sentenced to serve a term in prison.

Most of the men (10) were sentenced to serve time at a federal prison camp in Montgomery, Alabama. The other male offenders served at federal prison camps in Atlanta, Georgia and Manchester, Kentucky. The women were sentenced to either the federal prison camp in Alderson, West Virginia or the federal prison camp in Lexington, Kentucky. The length of imprisonment ranged from 14 months to 6 years.

Conducting research on individuals convicted of crimes is difficult. Among other problems, there are few incentives that can be offered for participation and little reason for offenders to want to revive a stressful and negative life event. Thus, it is not surprising that few people volunteered to share their experiences. Another concern is the use of information on the family's reactions from only one member of the family, the offender. While discussing the effects of incarceration on the family, offenders may have consciously or unconsciously omitted or neglected information describing the depth of the stress and strain suffered by their families. For these reasons, the information presented here must be viewed with caution.

Considering the criteria used in selecting the sample, the demographic characteristics of the subjects are not surprising. Unlike common offenders, the offenders in this study are predominantly white, over 40 years of age, and educated. Most were married with children at the time of their conviction. Eleven offenders had school-aged children and eight offenders had adult children. The remaining three offenders were divorced (2) or single (1). However, these figures do not convey the degree to which the majority of the subjects of this study do not fit the stereotype of the white-collar offender. Images of white-collar offenders presented in the media and in a significant amount of research, portray these offenders as members of the upper echelon of the corporate world. Although this image is true of some cases, it is a misconception to think that the stereotypical white-collar offender represents the majority of cases.

Perhaps the most striking characteristic of the sample is the ordinariness of the offenders. With the exception of the vice president of one of Tennessee's well-known banks most of the offenders were far removed from the corporate elite. Most of the offenders worked a 40–hour a week job and were scraping to carve out their piece of the American Dream. Some of the offenders were barely making enough to pay the bills, while others had a comfortable life where they owned their home, had a stable career, and took yearly vacations. The typical offender in this sample was lower middle to middle-class and led a moderately comfortable lifestyle.

RESULTS

Finding Out: Awareness and Uncertainty

Rarely are the criminal activities of the typical street offender a family secret (Irwin, 1970). Often working class wives of prisoners consider crime as something the men do to get money and do not view it as unusual or disturbing behavior (Fishman, 1990). While working class families may have some familiarity with crime, jails and prisons, the same is not true of most middle-class families. Several studies have described the criminal behavior of conventional people, or "Square Johns," and have stressed their lack of criminal commitment and connections to other aspects of the criminal identity (Irwin, 1970; Gibbons, 1968). For instance, Donald Cressey, in his classic study of embezzlers, described how these "conventional and trusted offenders" rationalize hiding their "unshareable problem" from family and friends (1963, p. 30).

The finding in previous research that most middle-class families are not aware of any criminal activity was supported by this study. Most family members (18 of 22) were not aware or suspicious of any criminal behavior until either an investigation had started or the offenders feared that arrest was eminent. The absence of awareness is not surprising considering the conventional and law-abiding lifestyle of many white-collar offenders and their families. The initial reactions of family members as reported by the offenders included shock and anxiety concerning the uncertainty of the future.

> The worst part of it for my wife was the unknown. Not knowing what was going to happen to me, and if I was going to have to spend years in jail. We went through a really hard time, but as time went a long she was better able to handle the stress. The kids [*two adult children*] handled it a little better. When I told them about it they couldn't believe that I had done it. But, I told them I did and that I was looking at some serious legal trouble. My wife and kids both just worried about what was going to happen. As soon as the sentencing was over and we all knew that I was going to prison for such and such a time, well, we all figured we could handle that. I mean the part of not knowing was the killer part of the whole deal (Male comptroller convicted of embezzlement).

> My wife couldn't handle it. She told me that she felt like her life was spinning out of control. So, I guess she thought she could gain some control by getting a separation and divorce from me. She couldn't understand why I had jeopardized so much and ruined things for her. She told me that if she had to live alone she would rather be divorced. So, that's what she did (Male business owner convicted of a false statement of a loan application).

The lack of suspicion by family members may be a related to the nature of white-collar crime. Since white-collar offenses are often committed during routine business activities and involve little or no change in the lifestyle and day-to-day activities of the offenders, signs that the offender is 'up to no good' may not be as evident.

Although the routine nature of the many white-collar crimes may facilitate secrecy on the part of the offender, emotional and psychological signs of stress were not as easily hidden. Four of the offenders reported that their families were suspicious that 'something was wrong'. In these instances the families did not know, but suspected with varying degrees of certainty, that something was going significantly wrong in the life of the offender.

> Both my wife and mother knew or thought that something was going on. My wife would ask me all the time about what had me so down, or what was going on. But, I would put her off and tell her it was a bad day at work. My mom said that she knew something was going on, and she was hoping that I would correct it. I think that was her exact remark (Male insurance agent convicted of embezzlement).

> I was at home and a girl from the bank called. She said that the examiners were there and that they wanted to talk to me. Well, I knew what was going to happen. I thought, I've got to find [*my husband*] and tell him what's going on. I found him at a house he was working at, and told him the whole story right there. I told him, if you want a divorce right now we'll go get one. But the first thing he said was that he loved me. He said, I'm glad to know that's your problem. That's why you're so nervous. And he said do what you have to do. Since the whole thing, you know, we've talked about things. He said that he knew that something was eating at me. Like, when we would go on vacation I would always have to call the bank 'cause I had to cover myself. I never went anywhere where I couldn't at least call the bank and find out what was going on. I was a nervous wreck. At one point I pulled out my eyelashes from nerves. I think he was just relieved that I was really going crazy. He just said that now that he knew what it was that we could handle it together and do what we needed to get through it (Female bank manager convicted of bank fraud).

While few white-collar offender families suspected wrongdoing by the offenders, previous research on the families street offenders finds that often these families know at least partial information about criminal activity (Fishman, 1990; Lesieur, 1977). In addition to knowing about the criminal activities, families of street offenders often tolerate illegitimate activities as long as they are not involved in them. The disruption caused by possible conviction and imprisonment may not be considered a source of difficulty or stress in street offender families. In the case of middle-class families learning about potential family disruption caused by probable arrest and conviction represents a significant change in their expected life experiences and opportunities. None of the offenders' families in the present sample had prior encounters with law enforcement agents and often considered themselves to be upstanding, law-abiding citizens. As evident in the examples above, the discovery that a member of their family was involved in a crime initially caused great strain on all family relationships.

Stigma: Acceptance and Rejection

Although differences in the level of awareness exist between the families of white-collar offenders and families of street offenders, most families of individuals convicted of crimes experience stigmatization. As Goffman (1963) noted, when the offender's stigma and moral culpability is shared by family members, both the offenders and their families become discreditable. The official labeling process that is applied to the offender maybe accompanied by unofficial labeling processes in the community that extends beyond the offender to ensnare his or her family.

The amount of stigma experienced by a family may depend on the type of community in which they live. In a study of prisoner's wives Fishman (1990) explained community variation along class lines by stating, "the extent to which wives reported positive or hostile community reaction is partially explained by the kinds of communities in which they lived. Arrests are not extraordinary events in most working-class communities. People are in and out of jail. In some communities, being officially labeled as a criminal's wife is so common that it becomes personally irrelevant" (p. 121). If this characterization is correct then we would expect the process of stigmatization to be more extreme in middle-class communities.

Another factor that may affect the intensity of stigma experienced by families is the presence of media accounts of the offender's crimes. Certainly, exposure by the media is embarrassing to both the offender and the family, but the process of being exposed publicly may be especially damaging for those families that have a stake in maintaining a legitimate reputation. Benson (1990) explains that in the case of white-collar crime, media accounts focus on the exceptional status and background of offenders to create a newsworthy story. Benson further argues that the emphasis on the offender's social status and income "magnify" the crime and add to the stigmatization processes. Over half of the offender's (63%) in the present study accused the media of furthering feelings of shame and embarrassment experienced by themselves and their families.

When families are faced with the possibility of stigmatization they must decide on the appropriate response. According to Page (1984) there are two options, either to reject or accept the stigma. The acceptance of a stigma entails a consensus between an individual and the community about the moral culpability of the offender. If the stigma is rejected then an individual may:

(1) passively reject, whereby the individual holds sentiments of hostility towards the community and refuses to participate in the rejection of the offender, or (2) flaunt the stigma in a effort to expose the community's erroneous label (Page, 1984). In this study, expressions of both acceptance and rejection characterize the offenders' accounts of familial stigma management processes.

In the cases where the family members accepted the stigma, most distanced themselves from the offender in an effort to minimize shame. Separating from the offender meant that spouses would file for separation or divorce and other family members would stop contact with the offender.

> When you go through this sort of thing you find out real fast who really loves you and who don't. Those family members like my one brother and my mother have always been there for me and they've stood by me. The rest, and I'm talking about sisters, aunts, uncles, well, they all act like you're a dirty piece of nothing. I haven't seen or spoken to most of them in years (Male business owner convicted of mail fraud).

Nearly all of the offenders in the sample reported at least some degree of rejection and separation from family members and friends. However, the rejection reported by these offenders was never as extreme as the complete abandonment found in studies of other incarcerated offenders (Carlson & Cervera, 1992; Chesney-Lind & Rodriquez, 1983; Owen, 1998). This finding could be related to differences between the types of offenders. Often street crimes are linked to other kinds of unconventional behaviors that keep the offender outside the home, such as partying, selling drugs or fencing stolen property, These types of behaviors limit contact and interaction with family and friends outside the criminal lifestyle, and may make it easier for the family to abandon the offender after arrest.

Overall, rejection of the stigma was most common among family members of the white-collar offenders in this sample. This finding is surprising since we expect individuals who have more invested in the maintenance of dominant social values to be more likely to accept and endure the stigma. Rejection of the stigma by families did not include a desire to conceal the offender's label or any re-invention of the situation. In some cases family members passively retreated from individuals and community groups who may have ridiculed the offender or family.

> We changed our lifestyle. [My wife] She definitely changed what she did and who she spent time with while I was gone. She told me that it was just to hard to be around people who didn't know the whole story and who just thought either you were a criminal or you were not. Basically, we don't hang out in the same groups of people like the ones at church and so forth. See, people have no idea and they just assume they know what really went on. We don't need that. We have plenty of friends and family and people that do, and they don't think anything about what I did, and surely not anything about my wife and family (Male business owner convicted of theft of government property involving a kickback).

Families also rejected stigma by criticizing the social controls that embody the stigma. In these instances offenders reported that they and members of their families loss faith in the legitimacy of the criminal justice system.[4]

> Look, it's like this. If it can happen to me it can happen to anyone. That's the way it is now. You know, love your country but fear your government. That saying that, I don't know, maybe you have seen that saying before. There is a question that everybody has been asking me. Hey, I mean yes I did wrong and I paid for it, but there was no use in putting me in jail and separating my family. That was the thing that my daughter and my wife wanted to know. Why did they have to suffer. They were the ones that were really punished. It makes no sense (Male business executive convicted of fraud).

> When I went for sentencing the case that was up before mine was about these two men that had shot a deer in Smokies [*Great Smoky Mountain National Park*]. We were all there, my whole family and a lot of people showed up to support me. Anyway, that judge sentenced those two boys to 14 months, or something like that for shooting a deer. When that happened I knew I was doomed. My son turned around, I'll never forget this. He looked at me and said well momma, I'm glad you didn't shoot a deer. I mean that is just how ridiculous we think the whole system is. My sentence was just as ridiculous as the one those fellas that shot that deer got (Female convicted of signing a false statement.).

The rejection of a commitment to social order is particularly troubling because we assume that families who hold positions of respect in the community to willingly accept the law's legitimacy. In a study of the emotional effects of adjudication on white-collar offenders, Benson (1990) argued that the stigmatizing nature of the criminal justice system acts as a form of "disintegrative shaming."[5] Rather than making offenders feel repentant, it provokes dysfunctional sentiments of anger and resentment. As the examples presented here suggest, those same dysfunctional processes may also involve members of the family other than the offender. By stigmatizing offender's and their families, the American justice system emphasizes punishment and deterrence. But, if these stigmatizing processes cause more individuals to reject rather than accept the legitimacy of the law, then we must question whether deterrence and punishment approaches really work or merely exacerbate the problem they were intended to solve.

Adjustment: Resources and Roles

After the offender goes to prison the family is also stuck 'doing time'. The family must find ways to adjust financially and emotionally to the separation. Previous research on inmates and their families presents different techniques of familial coping and adjustment including mobilizing for support from extended family and community, seeking spiritual or psychological counseling, and assuming new roles within the family (Kepford, 1994). Research on families of inmates consistently find that those families that have the most emotional and financial resources before incarceration also report higher levels of adjustment after the inmate returns home.

The families of offenders in the current sample report similar means of adjustment as families of other kinds of offenders in previous research. Most of the white-collar offenders had some financial collateral such as stocks, savings, vacation homes, and motor vehicles. Many families also had social capital in the form of positive and supportive relationships, with a number or extended family members and friends, and civic and religious ties in the community. Families used these resources while the offender was incarcerated.

> The whole ordeal put a financial strain on us. It hasn't been easy. We don't have a condo at the beach. We had to sell that, and we don't live like we used to. I guess I was lucky in that I had the money put away so that I could afford a good attorney. See those people that have the court appointed ones usually end up in bad shape (Male business owner convicted of mail/wire fraud).

> My mother was able to help us out some. She paid the phone bill so I was able to call home almost every night, and mother helped my husband be able to afford to come and see me. The visits and the phone calls is how I made it through. A lot of those women in there never got any visitors. It really is a sad, sad thing. I think the visits helped me and my husband become closer on an emotional level. I know this might sound strange, but it was almost like we were forced to talk about things like we hadn't taken the time to do in a long time. I mean, what else could you do in that stuffy visitation room surrounded by strangers and guards (Female bank officer convicted of bank fraud).

In some cases the spouse of the offender had to take on new responsibilities and roles, such as becoming the primary caregiver and provider for the family. The transition into new roles was made easier by the education, previous work experience and support from family and friends possessed by the offender's spouse.

> My wife had to go back to work. I think it was real hard on her and the kids. Most of all she had to become like a single parent. She had to fill my place. People at the church were real good about checking on her while I was away. She had a couple of good friends that would call and see how she was doing, and offer to do things that would give her a break. I mean even it was just going out for ice cream or something (Male lawyer convicted of embezzlement).

While a decline in financial security was common among the families of offenders in this sample, most eventually recovered. The ability of middle-class families to regain most if not all of their financial status has been documented in previous research on white-collar offenders. In an examination of the loss of one's occupational status as a consequence of white-collar crime, Benson

(1984) found that most offenders, even those regulated by state licensing agencies, were able to regain their occupational and financial standing. An indication of the ability of families and offenders in the present study to maintain a significant portion of their financial status is the fact that all offenders who owned homes were able to retain ownership after incarceration. The ability to remain in the same neighborhood and same household suggests that the most basic element of the families' lifestyle was preserved.

CONCLUSION

With few exceptions, the stresses and strains caused by arrest and conviction are similar for families of street offenders and white-collar offenders. Although, initially families of white-collar offenders may experience more emotional stress when first learning of the crime the families social, psychological and financial resources allow them to regain and maintain a lifestyle similar to that before arrest. What white-collar families do not seem to recover is their acceptance of our laws and legal system. As a result of the stigmatization and public ridicule these offenders and their families report feelings of alienation from justice processes. Sentiments of alienation are particularly troubling since we assume that these families have a significant investment in community standing with higher stakes in conformity. As a result of their experiences, most of what these families learn about the criminal justice system is that the system does not care about them, and that their previous achievements and conventional lives do not mean much.

In conclusion, by focusing on the experiences related to adjustment and stigmatization among white-collar offenders and their families, I do not wish to suggest that these offenders deserve any special consideration. The struggles and difficulties suffered by white-collar offenders pale in comparison to those of many street offenders. For example, white-collar offenders, particularly those that charged in federal courts, typically serve their time in minimum to medium security facilities where contact with violent offenders is less likely. However, the data presented here suggests that families of both types of offenders encounter similar pains and problems as a result of the offender's imprisonment.

NOTES

1. The percentage of arrests for forgery, fraud, and embezzlement that were women were 36%, 41% and 44% respectively (Maguire & Pastore, 1997).
2. Only those individuals whose cases were closed within the last five years were included on the closed case list. This cutoff point was used because of the unreliability

of older addresses. It is not uncommon for probationers to relocate after completing their terms under court supervision.

3. Originally 302 offenders were identified but 120 of those did not have accurate address information and the letters were returned to the researcher.

4. In effect, the offenders are using the neutralization technique of condemning the condemners (Sykes & Matza, 1957). Benson (1985) suggests that white-collar offenders use this technique in the expression of outrage and anger toward the prosecutor and other criminal justice officials.

5. Disintegrative shaming is a stigmatization process at the societal level in which a deviant characteristic is assigned to an offender as a master status. Braithwaite (1989) argues that disintegrative shaming is counterproductive because it pushes offenders further outside mainstream society.

REFERENCES

Bakker, L. J., Morris, B., & Janus, L. M. (1978). Hidden Victims of Crime. *Social Work, 23*, 143–148.

Benson, M. L. (1990). Emotions and Adjudication: Status Degradation Among White-Collar Criminals. *Justice Quarterly, 7*, 515–528.

Benson, M. L. (1985). Denying the Guilty Mind: Accounting for Involvement in White-Collar Crime. *Criminology, 23*, 583–604.

Benson, M. L. (1984). The Fall From Grace: Loss of Occupational Status as a Consequence of Conviction for a White Collar Crime. *Criminology, 22*, 573–593.

Braithwaite, J. (1989). *Crime, Shame and Reintegration.* Sydney: Cambridge University Press.

Brodsky, S. L. (1975). *Families and Friends of Men in Prison.* Lexington, MA: D.C. Heath Company.

Carlson, B. E., & Cervera, N. (1992). *Inmates and Their Wives: Incarceration and Family Life.* Westport, CN: Greenwood.

Chesney-Lind, M., & Rodriquez, N. (1983). Women Under Lock and Key: A View from the Inside. *The Prison Journal, 63*, 47–65.

Cressey, D. (1963). *Other People's Money.* New York: The Free Press.

Daly, K. (1989). Gender and Varieties of White-Collar Crime. *Criminology, 27*, 769–794.

Edelhertz, H. (1970). *The Nature, Impact and Prosecution of White-Collar Crime.* Washington, D.C.: USGPO.

Fishman, L. T. (1990). *Women at the Wall: A Study of Prisoners' Wives Doing Time on the Outside.* Albany, NY: SUNY Press.

Flanagan, T. J. (1981). Dealing with Long-Term Confinement: Adaptive Strategies and Perspectives Among Long-Term Prisoners. *Criminal Justice and Behavior, 8*, 201–222.

Gibbons, D. (1968). *Society, Crime, and Criminal Careers.* Englewood Cliffs, NJ: Prentice-Hall.

Goffman, E. (1963). *Stigma: Notes on the Management of Spoiled Identity.* Englewood Cliffs, NJ: Prentice-Hall.

Irwin, J. (1970). *The Felon.* Englewood Cliffs, NJ: Prentice Hall.

Kepford, L. (1994). The Familial Effects of Incarceration. *International Journal of Sociology and Social Policy 14*, 54–90.

Lesieur, H. R. (1977). *The Chase: Career of the Compulsive Gambler.* Garden City, NY: Doubleday Anchor.

Maguire, K., & Pastore, A. L. (1997). *Sourcebook of Criminal Justice Statistics, 1996.* Washington, D.C.: USGPO.

Mann, K., Wheeler, S., & Sarat, A. (1980). Sentencing the White-Collar Offender. *American Criminal Law Review, 14*, 479–500.

Owen, B. (1998). *In the Mix: Struggle and Survival in a Women's Prison.* Albany, NY: SUNY Press.

Page, R. (1984). *Stigma: Concepts in Social Policy.* Boston: Routledge.

Shapiro, S. P. (1990). Collaring the Crime, Not the Criminal: Reconsidering the Concept of White-Collar Crime. *American Sociological Review, 55*, 346–365.

Sykes, G., & Matza. D. (1957). Techniques of Neutralization: A Theory of Delinquency. *American Sociological Review, 22*, 664–670.

Weisburd, D. S., Wheeler, S., Waring, E., & Bode, N. (1991). *Crimes of the Middle-Classes: White-Collar Offenders in the Federal Courts.* New Haven, CT: Yale University Press.

'IT'S LIKE DYING EVERY DAY': THE FAMILIES OF CONDEMNED PRISONERS

Margaret Vandiver and Felix M. Berardo

AN OVERVIEW OF THE DEATH PENALTY IN THE UNITED STATES

This chapter considers the effect of death sentences and executions on the families of condemned prisoners. Where possible, we draw upon the evolving loss literature for a framework or set of concepts that helps elucidate the reactions and adaptations of these families. Within that context, we review the scholarly studies and memoirs that address this subject, and summarize related literature on families of prison inmates and families that have experienced violent bereavement. We conclude with suggestions for future research.

The number of condemned prisoners in the United States is at an all time high, with 3,625 people currently on death row. This number increases each year, because more people are sentenced to death than leave death row by execution, natural death, suicide, clemency, and reversal of sentence. The population of death row is overwhelmingly male, with the fifty condemned women comprising only 1.38% of all death sentenced inmates. Sixty-three condemned prisoners, 1.74% of the total, were juveniles at the time of the crimes for which they were convicted. White inmates are 46.51% of death row prisoners, African

The authors wish to thank Erica Owens and the editors of this volume for their helpful comments and suggestions. The title quotation is from Crawford (1997, p. 173).

Families, Crime and Criminal Justice, Volume 2, pages 339–358.

ISBN: 0-7623-0737-4

Americans 42.59%, Latinos 8.58%, Native Americans 1.27%, and Asians 0.86% (NAACP Legal Defense Fund, 1999).

Between the resumption of executions in early 1977 and June 30th, 2000 650 people were executed in American states. In the first half of the 1990's, an average of 27.4 executions occurred each year, increasing to an average of 68.2 from 1995 through 1999 (Death Penalty Information Center, 2000a). There were more executions in 1999 than in any year since 1951 and it is likely that number will continue to slowly increase for the foreseeable future. To put these figures in perspective, however, it is important to note that the death penalty remains a rare punishment. Hugo Bedau (1997, p. 32), working with figures from the mid-1990s, estimated that one person was "executed each year for every 1,000 criminal homicides."

A striking characteristic of the modern death penalty in America is the long delay between sentencing and execution (Crocker, 1998). The Bureau of Justice Statistics (1998) recently reported that the average time spent on death row for those executed between 1977 and 1997 was 111 months, or more than nine years. Time spent on death row has reached extraordinary lengths for some inmates, especially those whose sentences have been reversed and reimposed one or more times. There are condemned prisoners who are approaching a quarter century on death row, and 15 to 20 year stays are not unusual. Some inmates have spent more than half their lives under sentence of death.

Although death is a rare punishment, its impact is much broader than its direct effect on the persons immediately involved. Capital punishment has great symbolic importance, and whether it should be used or abolished is one of the most contentious moral questions of our day. In the debate over the death penalty, its effects on offenders, victims' families, potential victims, and society at large are frequently discussed and disputed. Almost always absent from the discussion is any consideration of the families of condemned prisoners. These families suffer much from the sentence of their condemned relatives, and their distress is not ended by the execution. Neither popular debate nor scholarly literature has paid sufficient attention to these families.

It is important to learn more about the families of condemned inmates for several reasons. First, their experiences, along with those of the families of homicide victims, should be considered in any assessment of the benefits and costs of the death penalty. Without knowing how capital punishment affects these families, we lack information necessary to fully evaluate its use. More broadly, the study of these families is important because what they experience is so extreme. All legal punishment affects families as well as offenders; this is an unfortunate but inevitable aspect of our criminal justice system. But just as capital punishment is the ultimate punishment for offenders, it has more severe effects upon their families than do other types of punishment. A better understanding of the particular pressures they face and how they deal with their loss and its associated grief may suggest the kinds of assistance required to move them toward recovery and a more normal mode of daily living.

The concept of *loss* may be useful here. It has been defined as "produced by an event which is perceived to be negative by the individuals involved and results in long-term changes in one's social situations, relationships, or cognitions" (Miller & Omarzu, 1998, p. 12). Losses, of course, are part of the human experience. They can range from the trivial to the traumatic, from temporary to permanent, from anticipated to unexpected, from short-term to extended, from merely upsetting to catastrophic, from symbolic to physical, with singular or multiple consequences. Moreover, losses will differ in their impact on individual lives as well as their behavior as members of a group, such as a family. Hence, loss is an event that affects people in very idiosyncratic ways.

People confront loss many times during their lives, and each will define them differently. Some losses are considered ephemeral or minor and don't have deep or lasting effects on the person/group. Other losses are considered major, like losing a spouse or sibling. Major losses give rise to grief, while other loss events may provoke some degree of distress (Weiss, 1998, p. 343). The more significant or important a relationship is to us, the more we grieve the loss (James & Friedman, 1998, p. 4)

For most people, recovering from a significant emotional loss is a difficult task. For families of death row inmates, as well as those of homicide victims, this task is particularly onerous. In both cases, but especially that of families of death row inmates, the problem of unresolved grief may have to be dealt with. *Unresolved grief*, that grief which cannot easily be brought to closure, is often associated with ambiguous loss (Boss, 1999). *Ambiguous loss* includes situations in which an incomplete or uncertain loss has the effect of freezing the grief process and thereby preventing its completion. For some, ambiguous losses in personal relationships can be more stressful or devastating than ordinary, clear-cut losses. The uncertainty inherent in such situations can lead to considerable distress and impede the recovery process. Incomplete recovery can have a lifelong negative impact on a person's well-being and his/her capacity for happiness.

The death of a significant other, in the case of the condemned prisoner, is generally anticipated, though the thought of a last minute reprieve often lingers to the very end. Everyone knows the person eventually is likely to be executed, but delays can drag the process on over great lengths of time. In such circumstances, members of the family may begin the grief process before the person is actually put to death, rehearsing the role of survivor, imagining what their life would be like once this significant other is finally gone. Though the evidence is inconsistent, *anticipatory grief*, or forewarning of the impending death of another, is generally viewed as useful for future adjustment. However, thoughts of what life would be like once the loved one is gone also may bring some guilt.

Generally speaking, the death of a family member requires the development of alternative patterns of behavior if the survivors are to maintain satisfactory relations within their family, the kin group, and with the community, and if they are to sustain personal equilibrium. Sociologically, death precipitates a reorganization of the family as a social system. Roles must be reassigned, status positions shifted, and values reoriented. Families exhibit considerable diversity in the ways they react to and are affected by death, and in their efforts to accomplish such transitions. As they move toward recovery, they often receive social and psychological as well as practical support from friends, neighbors, and the community at large. In the case of the families of death row inmates, however, such support systems may either be unavailable, minimal, or withheld. Thus, these families may find themselves dealing with their loss in somewhat of a vacuum.

REVIEW OF THE RESEARCH

Scholarly research on families of condemned prisoners

We know almost nothing about how executions affected prisoners' families prior to the U.S. Supreme Court's 1972 ruling in *Furman v. Georgia*, which invalidated all then-existing death penalty statutes. No systematic research has been done on families of those executed before 1972. Glimpses of their experience can be gained by looking through the case files preserved in archives, which often contain letters begging the governor to spare a relative's life. These desperate, hand scrawled letters are almost the only evidence remaining of the anguish felt by hundreds of families facing the execution of a family member.

Two studies addressed the issues presented by having a family member on death row during a time when executions occurred very infrequently. Radelet, Vandiver & Berardo (1983) interviewed and observed condemned inmates' family members in Florida in the early 1980s. The authors' sample consisted entirely of family members who did visit their relatives and thus does not provide any information about families who could not or chose not to visit. At the time, only one Florida inmate had been executed in the previous 20 years, although death warrants were frequently signed, and the threat of execution was quite real. The authors found that the prisoners' families faced tremendous social and emotional difficulties as they tried to cope with their relatives' death sentences. Efforts to maintain contact with condemned relatives through visits were a further source of strain. The families lived in chronic dread of the eventual execution of their relatives, and experienced a prolonged and unresolved period of anticipatory grief, complicated by their inability to affect the outcome of the case.

Smykla (1986, 1987) interviewed families of eight inmates on Alabama's death row during a period when executions were rare. Although he did not have a random sample, his sampling techniques were more systematic than those used in other studies of condemned prisoners' families. Smykla wrote to the sixty five persons on death row in Alabama in 1983 explaining that he wanted to study the impact of the death penalty on their families, and requesting the names and addresses of relatives; forty three inmates responded. He interviewed 40 members of eight families, chosen by location and the length of time their relatives had been on death row.

Smykla found prolonged and distorted grief reactions in each family. Prolonged grieving for the condemned relative occurred during the period of appeals, and took the form of self-accusation, powerlessness, and social isolation (1987, p. 338). Family members often blamed themselves for their relative's situation. They felt powerless to prevent the execution or even to make their relative's time on death row easier, and described the situation as hopeless. The families' social isolation was such that it 'actually made families prisoners in their own homes. [It] was accompanied by intense anger, rage, frustration, the fear of losing one's mind, as well as an unbelievable mixture of other physical, emotional, and social problems' (1987, p. 343).

Smykla identified a number of distorted grief reactions among these families (1987, pp. 343–45). Nearly all respondents reported having medical problems they believed were the result of the stress caused by their relative's death sentence. Loss of social relationships was frequent, with some families moving often to try to escape the stigma of the sentence. Depression was common, and some family members considered suicide. Disturbingly, some respondents reported that they had been treated with mockery, contempt, and deliberate cruelty because of their relatives' sentence.

Vallejo (1995) interviewed ten family members of death row inmates, eight from California and two in other states. Her particular interest was how social workers could assist the families with problems resulting from their relatives' sentence. Vallejo's interviews revealed many of the same stresses and reactions noted in earlier studies, including rejection and fear of rejection, loss of friendships, lack of support, and emotional difficulties. Interestingly, she also found that several respondents felt that they had become stronger as a result of their experiences.

King & Norgard (1999) approached the issue from a legal perspective, arguing that the law should provide an opportunity for defendants' families to make a statement to the sentencing jury about the impact the death sentence would have on them. They note that the U.S. Supreme Court's ruling in *Payne v. Tennessee* (1991) allows the relatives of homicide victims to give statements to the jury describing the impact of the crime on their families, but equivalent testimony by defendants' families is usually not permitted. In support of their argument, King & Norgard presented the results of their interviews with 28 family members of death row inmates from around the country. Their findings are strikingly consistent with those of other studies, although they cited none of the previous studies and seem to have been unaware of them, noting that they "found no books or articles" on the topic (p. 1141).

King & Norgard found that the family members they interviewed suffered from stigmatization, social isolation, depression, and chronic grief. Family members reported that they were viewed by others only as relatives of the condemned, and that they lost "their individual identities" (p. 1138). As was true in Smykla's sample, King & Norgard found that many family members had experienced some type of health problem related to the stress of their relative's sentence (p. 1127). The authors found much evidence of chronic grief among the subjects they interviewed. They compared the families of condemned inmates to families of children with chronic illnesses, noting many similarities, but also noting that as terrible as it is to wait for a sick child's death, these families "are not awaiting the homicide of their child to be committed by the state in the name of the community" (p. 1142).

It is important to note that families of condemned prisoners are not a representative cross section of American families. They are multiply disadvantaged, and their problems began long before their relatives were sentenced to death. Overwhelmingly, they are poor. Those who had any financial resources before their relative's trial very likely exhausted them in his/her defense. Like other poor families, those of condemned prisoners lack education, have few skills that could lead to steady and well paying employment, and face many problems in their daily lives concerning housing, social services, and medical care. The few studies on the backgrounds of condemned prisoners indicate that their childhoods were spent in families with extraordinarily high levels of violence, alcoholism, drug use, mental illness, and other dysfunctions (Haney, 1995, 1998; Lewis et al., 1986, 1988). The families are likely to have few financial and emotional resources for coping with the stresses of their relatives' death sentence.

Memoirs and essays on families of condemned prisoners

Several essays and memoirs mentioning the experiences of families of executed inmates have been written in recent years by persons involved in one way or another with the prisoners' cases. These memoirs are especially valuable since none of the scholarly literature has dealt with samples of families whose relatives have been executed. Most important among these writings are the memoirs of family members.

Shirley Dicks, whose son was on death row in Tennessee for many years, has written about her experiences, and interviewed other death row family members (1990, 1992), giving important information on how they perceive and are affected by the death sentencing system. Mikal Gilmore, brother of Gary Gilmore, the first person to be executed under current statutes, wrote a book exploring the roots of violence in his family, and the effects of his brother's death sentence and execution on surviving family members. He wrote of the pain and alienation felt by his family after the execution:

> For the rest of your life, you will have to move around in a world that wanted this death to happen. . . . You will have to try to find a way to live with the sorrow that will now always be at the heart of your heart. You will have to try to find a way to live in this world, in this life, and not hate it – and you will have to try despite the impossibility of such a task (Gilmore, 1994, p. 349–50).

Pamela Crawford, the sister of a man executed in Alabama in 1996, wrote a chapter for a book on capital punishment, expressing how her brother's execution affected her and her family. The excerpts quoted below indicate the unresolved grief felt by her family and the mental and emotional strain placed upon them by the execution.

> It's like dying every day. I feel somewhat like death row inmates who feel that they are already dead. I can't get any sense out of it. I'm still trying to understand what happened and I have a problem dealing with that and then I have to try to explain to my children. I don't have an answer for them. . . . There was not just one murder. There are two dead people now. There is not just one family hurt – there are two. There are children hurting that had nothing to do with it. There are people that have died with broken hearts. People have forced themselves into an early death because of the pain. It's a long chain. It goes on and on and on. I don't see a stopping point. I don't see who benefits (Crawford, 1997, p. 173–4).

Descriptions of final visits between families and their relatives can be found in books by a minister (Ingle, 1990), a nun (Prejean, 1993), and a prison warden (Cabana, 1996). A journalist's report of the execution of Ricky Ray Rector in Arkansas contains a sensitive portrait of the condemned man's sister (Frady, 1993). Vandiver (1989, 1992, 1998) wrote of the effects of death watch and executions on prisoners' families, based upon her experiences as a legal assistant to lawyers representing condemned inmates. She observed that the families of condemned prisoners have to cope with deaths that are both lingering and violent, and that the extent of the damage this inflicts on families, especially upon children, is unknown.

OTHER RESEARCH RELEVANT TO FAMILIES OF CONDEMNED PRISONERS

Families of prisoners

More people are incarcerated in the U.S. than ever before. Both the numbers and rate of incarceration have increased tremendously over the past two decades. Sentence lengths have increased as well, and parole has been cut back or eliminated in many jurisdictions, meaning that more people are incarcerated and they are staying in prison longer. Thus, more families are confronted with a relative's incarceration, often for many years. No agency of the criminal justice system gathers information on these families, or even attempts to count them, so the precise number of people affected by a relative's incarceration is not known. A recent estimate by the Bureau of Justice Statistics is that 1.96 million children have a parent or close relative incarcerated on any given day (Butterfield, 1999).

There is a substantial body of research on prisoners' families. Several studies give general demographic characteristics of prisoners and their families and provide an overview of the issues facing them (Bloom, 1995; Dallao, 1997; Fox, 1981; Gabel & Johnston, 1995; Gabel & Shindledecker, 1993; Hairston, 1995; Johnston & Gabel, 1995; Kiser, 1991; Koban, 1983; Riggs & Kilpatrick, 1990). Collins (1997), King (1993), and Swan (1981) focused their research on the effects of incarceration on the families of African American inmates. Generally stated, this literature concludes that inmates' families undergo a number of serious hardships connected with their relatives' incarceration. Commonly mentioned hardships are financial strains, emotional and behavioral problems, especially among children, difficulties with maintaining the relationship between the inmate and family, social stigma and shame, and problems concerning the inmate's release and readjustment after prison.

Several articles have focused on the particular difficulties facing families with children (Boudin, 1998; Gabel & Girard, 1995; Johnston, 1995a; Weilerstein, 1995); legal issues such as the termination of parental rights and regaining custody after release (Barry, 1995; Gentry, 1995; Schoenbauer, 1986; Smith, 1995; Zealand, 1998); and issues of care of children, including foster care

(Norman, 1995) and relatives serving as caregivers (Phillips & Bloom, 1998) during the parent's incarceration. Gabel (1992) gives a methodological critique of the research on prisoners' families.

The effect of parental incarceration on children is an area of special concern. The emotional, psychological, and social strains placed on children by the incarceration of a parent, especially if that parent is the primary caretaker, have been the focus of a number of studies (Beatty, 1997; Henriques, 1982; Johnston, 1995b; Kampfner, 1995). Reed & Reed found that children "respond [to a parent's imprisonment] in various ways, including sadness, withdrawal, low self-esteem, excessive crying, depression, diminished school performance, truancy, disciplinary problems, alcohol and other drug use, running away, and aggressive behavior . . ." (1997, p. 159).

Families of condemned prisoners share many of the experiences of other inmates' families but the negative aspects of their situation are more serious, and they have less reason for hope. The literature on inmates' families makes frequent mention of the stigma and shame they feel because of their relatives' imprisonment. To some extent this feeling may be attenuated as imprisonment becomes more common; it is difficult to feel particularly ashamed of an experience many people of your acquaintance have shared. The stigma of the death penalty, on the other hand, has not decreased. Death remains a rare and extreme form of punishment, the ultimate rejection a society can inflict. The death penalty continues to attract a highly disproportionate amount of media attention, which contributes to the families' sense of public exposure and humiliation. And the many newsworthy developments as the case moves through the system ensure that their humiliation will recur many times before the case is finally resolved.

Confinement on death row is more severe than confinement in a maximum security prison (Jackson & Christian, 1980; Johnson, 1981; Magee, 1980). The emotional effect of such harsh confinement, in combination with prolonged dread of execution, constitutes a condition referred to by the European Court of Human Rights as the "death row phenomenon" (Lillich, 1991). The impact of the death row phenomenon is also felt by the prisoners' families, who face many years of uncertainty and emotional strain as their relatives' cases work their way through the courts.

Families of inmates have not been successful in organizing themselves on a large scale to bring pressure for beneficial changes, though there are many small and local organizations of prisoners' family members attempting to assist themselves and each other. Prisoners' families generally lack economic and social resources, and they are a group many politicians are unwilling to recognize as part of their constituency. There are many fewer family members of death row inmates than of prisoners in general, and thus they tend to be more widely scattered and less able to organize effectively. And the stigma attaching to the death sentence is such that these families are even less able to generate support for reforms that might ease their problems.

While most prisoners' families can look forward to being reunited after the inmates' release, death row families have little more to hope for than that their relatives will remain alive and incarcerated. What for the other families is the worst outcome – life in prison – is the best the death row families can reasonably expect. This does not mean that death row prisoners are never released. Since 1973, 87 condemned prisoners have been exonerated and set free (Death Penalty Information Center, 2000b); these inmates were able to overcome formidable practical and legal obstacles to proving their innocence. These exonerations are frequent enough to torment families with hope, but so rare as to make release in any particular case unlikely indeed.

Families of homicide victims

Early accounts of the experiences of the families of homicide victims took the form of case studies, anecdotal accounts, interviews with a few survivors, or autobiographical accounts. Since the mid 1970s, research on homicide victims' families has increased substantially. Several researchers take a clinical approach (Amick-McMullan et al., 1989; Amick-McMullan, Kilpatrick & Veronen, 1989; Burgess, 1975; Johnson & Young, 1992; Rynearson, 1984; Rynearson & McCreery, 1993; Sprang, McNeil & Wright, 1989). Several studies focus on specific populations, such as children who witness the homicide of one parent by another, or the young siblings of homicide victims (Burman & Allen-Meares, 1994; Freeman, Shaffer & Smith, 1996; Malmquist, 1986; Rinear, 1988), while others are concerned specifically with therapeutic interventions and practical responses (Craig, 1999; Getzel & Masters, 1984; Jackson, 1979; Lyon, Moore & Lexius, 1992; Masters, Friedman & Getzel, 1988; Parkes, 1993; Redmond, 1989; Rock, 1998; Rynearson, 1995; Spungen, 1998; Thompson & Vardaman, 1997), or the effects of the criminal justice system upon survivors (Day & Weddington, 1996). A few recent studies have attempted to use sophisticated sampling and methods of analysis (Amick-McMullan, Kilpatrick & Resnick, 1991; Sprang, McNeil & Wright, 1992–1993; Vargas, Loya & Hodde-Vargas, 1989).

Research on families of homicide victims indicates that the emotional trauma people experience after the murder of a family member is prolonged and severe. The reaction frequently takes the form of symptoms typical of Post Traumatic Stress Disorder. The violent death of their relative can shatter survivors' sense of justice and safety, and make them lose any sense of purpose and meaning in their lives. Feelings of guilt for not protecting the murdered relative are common, and people often experience intrusive fantasies of revenge against the murderer. The impact of the homicide is so great that survivors experience it in many areas of their lives, encountering difficulties in their other relationships and in their work. Both formal and informal supports for homicide victims' survivors are inadequate, and they often feel profoundly alone and isolated. The criminal justice system often does little to help survivors, and sometimes actually victimizes them further. Recovery from a family member's homicide takes years, if not decades. Outsiders' expectations of quick grieving and recovery are unrealistic and damaging. Even after survivors resolve their grief, it is unlikely that their emotional lives will ever be the same as before the crime (Vandiver, 1998, p. 480).

The families of executed inmates are also violently bereaved, but their experiences are significantly different from those of families of homicide victims. Vandiver (1998, p. 486) summarizes these differences. Death by homicide is sudden and unexpected, while executions occur after years, if not decades, of expectation. Families of homicide victims will generally encounter expressions of sympathy and regret for their loss, while families of executed inmates know very well that their relatives' deaths were desired by many and were deliberately inflicted by respected and powerful persons and institutions. The isolation that homicide victims' survivors often feel is compounded for these families by a pervasive sense of alienation from their own society.

SUGGESTIONS FOR FURTHER RESEARCH

Research on the death penalty has provided us with a wealth of information on how the current system functions. We have extensive information about racial disparities and geographical patterns in sentencing, about the death penalty's failure to deter more than long prison sentences, about the appellate process and the cost of criminal justice systems with the death penalty, and about the ways juries function in capital cases (Bohm, 1999). We have some knowledge, though not enough, of how death row confinement affects inmates, and how executions affect correctional personnel. But we still know very little of the effects of death sentencing and almost nothing of the long term effects of executions on the families of the prisoners. Much basic research is still required in this area. In what follows, we outline suggestions for directions this research might take.

The existing scholarly research on families of condemned prisoners suffers from serious methodological flaws. (For a discussion of similar flaws in the research on prisoners' families, see Hagan (1996); many of the same weaknesses also exist in the research on families of homicide victims). The first problem concerns the samples. No study of condemned prisoners' families has used a random sample, and only Smykla used systematic sampling techniques. None of the studies has attempted to identify and interview family members who were not in contact with their condemned relatives. We do not even know whether it is more common for families to maintain or to drop contact with their condemned relative – or indeed, how many families had already lost contact before their relative was sentenced to die.

Another major weakness of existing research is the lack of comparison groups, such as families of life sentenced prisoners. By comparing families of inmates sentenced to life and to death, it would be possible to sort out the effects of the death sentence on the families from the effects of imprisonment. Comparisons could also be made between families who do maintain contact with their relatives and those who do not. A third limitation of existing studies is that they all do interviews at one time only, and thus do not capture changes over time (Smykla's study is the only one even to try to address this issue). Interviews with families at certain intervals over a period of time would constitute a significant improvement in design. Finally, none of the existing studies attempted to collect baseline information about the families before the death sentence. This makes it very difficult to determine which problems existed previous to the relative's death sentence and which are the result of that judgment.

In sum, existing studies of condemned prisoners' families are descriptive accounts of availability samples of respondents, interviewed at one time, without comparison groups and without baseline data. With these limitations in mind, however, the similarity of the studies' findings, and their consonance with the memoirs of families and observers, indicate that researchers have probably identified real patterns of emotional and behavioral responses in the families of death row inmates. In what follows, we suggest a number of directions that future research could take.

Who and how many are affected, and for how long?

The demographics of prisoners' families are presumably similar to those of the prisoners themselves. While we know the ethnicity and age of condemned prisoners, and something about their education level and prior criminal records, it would be useful to know more about their families of origin (family of orientation) and those they subsequently formed as adults (family of procreation). What are the sizes of condemned prisoners' families? How many people are affected by an execution? Research such as that done to estimate the number of people affected by homicide (Redmond, 1989; Amick-McMullan, Kilpatrick & Resnick, 1991) should be done to allow better estimates of these numbers. Another question concerns the length of time families are affected by the execution of a relative. Do the effects end with the generation living at the time of the execution, or does the execution continue to have an influence on following generations? How do families tell new generations about the execution?

What are short and long-term effects of an execution on the adult members of the prisoner's family?

Researchers have begun to learn about the effects of losing a family member to homicide. It seems reasonable to assume that families of executed prisoners would experience some of the same responses. But the differences between the sudden loss of a family member to an individual's violence and the loss of a family member through a state's execution after years of waiting may be greater than any similarities. These differences need to be systematically explored to gain a greater understanding of the experiences of both sets of bereaved relatives.

How does the experience of a relative's execution differ by relationship with the executed person?

Parents of the condemned inmate may blame themselves for his crime and feel guilty for any of their own behavior that may have contributed to his criminality. Wives, girlfriends, and friends may feel ashamed of their voluntary association with and affection for a person who could commit such a crime. Children probably bear the heaviest burden, however, and do so with less ability to understand and cope with their situation. How can they reconcile respect for their parent with the fact of society's condemnation not just of their parent's act, but of his/her existence?

What are the short and long-term effects of a parent's execution on his children?

How do the children of condemned prisoners face the planned killing of their father or mother? How do they understand what has happened to them and their family? What are the long term effects of such radical trauma on their development and subsequent adjustment? What difference does age of child at time of execution make? Do children who were close to and visited their parent on death row react differently from children who had less contact?

What are the effects of a parent's execution on children when the homicide is domestic?

Although it is unusual for domestic homicides to result in death sentences (Rapaport, 1990), this does sometimes occur. Families in this situation face all the problems of victims' families and of defendants' families. This may be the absolute worst situation for children. Sam Reese Sheppard, whose mother was murdered when he was seven, and whose father was convicted (and later exonerated) of the murder, has said, 'If the state of Ohio had executed my father, I would not have lived to grow up' (Sheppard interview, 1999). The loss of a parent to murder is one of the most damaging and traumatic experiences a child can face; the subsequent loss of the other parent to execution can compound the effects on the child. The situation may be even more damaging if the child's testimony was used in the case against the condemned parent.

What effect, if any, do media attention and public support for the execution have on families?

Some cases of condemned inmates receive extraordinarily high levels of media attention, while others are almost ignored. In cases with high exposure, there are often public calls for the execution of the inmate, and occasional public celebrations of his death. Do these variables exacerbate the pain of the inmate's family and make their recovery after the execution more difficult?

Does the length of time spent on death row make any difference?

Is it more difficult for relatives to cope with the execution of a family member who has been on death row for 20 years rather than two years? Or does the extremely long period of anticipatory grieving actually ease the pain of families once the execution does occur? Is there a length of time beyond which most family members drift away from the condemned inmate? Does this length of time differ by relationship to the inmate?

Implications of reversal and reapplication of death sentence

Work is needed focusing on those families who prepare for the death of a relative who then receives a reprieve, which subsequently is lost, meaning the family must prepare once again for his/her demise. In such cases, having the sentence temporarily suspended and then reinstituted would no doubt aggravate the feelings of ambiguous loss the family has already experienced.

This may be analogous to cancer patients who undergo remissions and the like, or families with a member who has AIDS and whose health may go from nearly fine to death's door and back. Is there a point where proper grief work is no longer possible because of too many 'cry wolf' episodes? Can a person become burned out on anticipatory or unresolved grief?

Do executions benefit the families of victims?

Although this chapter focuses on the families of defendants, we believe that their experience should not be considered in isolation from the experiences of the families of victims. Policy makers should weigh the harm that executions do to defendants' families against any good they may do for victims' families. Although over 600 people have been executed under current laws, there has been no research that attempts to study whether the annihilation of the offender offers any benefits to the victim's family. Enough executions have occurred in the past 20 years to permit systematic analyses of their impact on victims' families over long periods of time. Future research should address these unexplored yet important issues in the use of capital punishment.

Several years ago Lincoln & Straus (1985) noted that social scientists seemed to be divided into two independent research groups – one studied crime, the other the family. What was needed, they argued, was a conceptual framework that encompassed both areas and which would explain how the two are linked together. The framework they developed allowed them to grasp the research findings and the essential issues surrounding crime within, by, and against the family. Studies of families of death row inmates similarly could benefit from a perspective that organized thinking about the issues raised in this chapter, and also suggested new research questions and insights.

REFERENCES

Amick-McMullan, A., Kilpatrick, D. G., & Resnick, H. S. (1991). Homicide as a Risk Factor for PTSD among Surviving Family Members. *Behavior Modification, 15*, 545–559.

Amick-McMullan, A., Kilpatrick, D. G., & Veronen, L. J. (1989). Family Survivors of Homicide Victims: A Behavioral Analysis. *The Behavior Therapist, 12*, 75–79.

Amick-McMullan, A., Kilpatrick, D. G., Veronen, L. J., & Smith, S. (1989). Family Survivors of Homicide Victims: Theoretical Perspectives and an Exploratory Study. *Journal of Traumatic Stress, 2*, 21–35.

Barry, E., Ginchild, R., & Lee, D. (1995). Legal Issues for Prisoners with Children. In: K. Gabel & D. Johnston (Eds.), *Children of Incarcerated Parents*, (147–166). New York: Lexington Books.

Beatty, C. (1997). *Parents in Prison: Children in Crisis*. Washington: Child Welfare League of America.

Bedau, H. A. (Ed.). (1997). *The Death Penalty in America: Current Controversies.* New York and Oxford: Oxford University Press.

Bloom, B. (1995). Public Policy and the Children of Incarcerated Parents. In: K. Gabel & D. Johnston (Eds.), *Children of Incarcerated Parents*, (271–284). New York: Lexington Books.

Bohm, R. M. (1999). *Deathquest: An Introduction to the Theory and Practice of Capital Punishment in the United States.* Cincinnati: Anderson Publishing Co.

Boss, P. (1999). *Ambiguous Loss: Learning to Live With Unresolved Grief.* Cambridge, MA: Harvard University Press.

Boudin, K. (1998). Lessons from a Mother's Program in Prison: A Psychosocial Approach Supports Women and Their Children. *Women and Therapy, 21*, 103–125.

Bureau of Justice Statistics. (1998). Capital Punishment 1997. (December). U.S. Department of Justice, Bureau of Justice Statistics. Washington, DC: U.S. Government Publishing Office.

Burgess, A. W. (1975). Family Reaction to Homicide. *American Journal of Orthopsychiatry, 45*, 391–398.

Burman, S., & Allen-Meares, P. (1994). Neglected Victims of Murder: Children's Witness to Parental Homicide. *Social Work, 39*, 28–34.

Butterfield, F. (1999). Parents in Prison. *The New York Times* (April 7), A1.

Cabana, D. A. (1996). *Death at Midnight: The Confession of an Executioner.* Boston: Northeastern University Press.

Collins, C. F. (1997). *The Imprisonment of African American Women: Causes, Conditions, and Future Implications.* Jefferson, NC and London: McFarland & Company.

Craig, R. J. (1999). Healing the Wounds of Sibling Survivors of Lethal Violence. In: H. V. Hall (Ed.), *Lethal Violence: A Sourcebook on Fatal Domestic, Acquaintance, and Stranger Violence*, (527–542). Boca Raton, FL: CRC Press.

Crawford, P. (1997). My Brother, Ed Horsley. In: C. Whitman & J. Zimmerman (Eds.), *Frontiers of Justice: The Death Penalty*, (165–177). Brunswick, ME: Biddle Publishing Co.

Crocker, D. (1998). Extended Stays: Does Lengthy Imprisonment on Death Row Undermine the Goals of Capital Punishment? *The Journal of Gender, Race & Justice, 1*, 555–574.

Dallao, M. (1997). Coping with Incarceration – From the Other Side of the Bars. *Corrections Today, 59*, 96–98.

Day, L. E., & Weddington, M. M. (1996). Grief and Justice in Families of Homicide Victims: Initial Results from a Study of the Impact of the Criminal Justice System. Paper presented at the meeting of the American Society of Criminology, Chicago, IL.

Death Penalty Information Center. (2000a). Executions: Executions Since 1976 by Year. (July 20). Available at http://www.essential.org/dpic/facts.htm l.

Death Penalty Information Center. (2000b). Innocence: Freed from Death Row. (July 20). Available at http://www.essential.org/dpic/Innocentlist.html.

Dicks, S. (1992). *They're Going to Kill My Son.* Far Hills, NJ: New Horizon Press.

Dicks, S. (1990). *Death Row: Interviews with Inmates, Their Families and Opponents of Capital Punishment.* Jefferson, NC: McFarland & Company, Inc.

Fox, S. S. (1981). Families in Crisis: Reflections on the Children and Families of the Offender and the Offended. *International Journal of Offender Therapy and Comparative Criminology, 25*, 254–264.

Frady, M. (1993), February 22. Death in Arkansas. *The New Yorker:* 105–133.

Freeman, L. N., Shaffer, D., & Smith, H. (1996). Neglected Victims of Homicide: The Needs of Young Siblings of Murder Victims. *American Journal of Orthopsychiatry, 66*, 337–345.

Furman v. Georgia. (1972). 408 U.S. 238.

Gabel, K., & Girard, K. (1995). Long-Term Care Nurseries in Prisons: A Descriptive Study. In: K. Gabel & D. Johnston (Eds.), *Children of Incarcerated Parents,* (237–254). New York: Lexington Books.

Gabel, K., & Johnston, D. (Eds). (1995). *Children of Incarcerated Parents.* New York: Lexington Books.

Gabel, S. (1992). Behavioral Problems in Sons of Incarcerated or Otherwise Absent Fathers: The Issue of Separation. *Family Process, 31,* 303–314.

Gabel, S., & Shindledecker, R. (1993). Characteristics of Children Whose Parents Have Been Incarcerated. *Hospital and Community Psychiatry, 44,* 656–660.

Gentry, P. M. (1995). Termination of Parental Rights among Prisoners: A National Perspective. In: K. Gabel & D. Johnston (Eds.), *Children of Incarcerated Parents,* (167–182). New York: Lexington Books.

Getzel, G. S., & Masters, R. (1984). Serving Families who Survive Homicide Victims. *Social Casework: The Journal of Contemporary Social Work, 65,* 138–144.

Gilmore, M. (1994). *Shot in the Heart.* New York: Doubleday.

Hagan, J. (1996). The Next Generation: Children of Prisoners. In: *The Unintended Consequences of Incarceration,* (21–39). Vera Institute of Justice. Available from the Vera Institute of Justice, 377 Broadway, New York, New York, 10013.

Hairston, C. F F. (1995). Fathers in Prison. In: K. Gabel & Denise Johnston (Eds.), *Children of Incarcerated Parents,* (31–40). New York: Lexington Books.

Haney, C. (1995). The Social Context of Capital Murder: Social Histories and the Logic of Mitigation. *Santa Clara Law Review, 35,* 547–609.

Haney, C. (1998). Miitigation and the Study of Lives: On the Roots of Violent Criminality and the Nature of Capital Justice. In: J. R. Acker, R. M. Bohm, & C. S. Lanier (Eds.), *America's Experiment with Capital Punishment,* (351–384). Durham, NC: Carolina Academic Press.

Henriques, Z. W. (1982). *Imprisoned Mothers and Their Children: A Descriptive and Analytical Study.* Washington, D.C.: University Press of America, Inc.

Ingle, J. B. (1990). *Last Rights: Thirteen Fatal Encounters with the State's Justice.* Nashville: Abingdon Press.

Jackson, A. M. (1979). The Availability of Mental Health Services for Dependents of Homicide Victims. In: H. M. Rose (Ed.), *Lethal Aspects of Urban Violence,* (91–100). Lexington, MA: Lexington Books.

Jackson, B., & Christian, D. (1980). *Death Row.* Boston: Beacon Press.

James, J. W., & Friedman, R. (1998). *The Grief Recovery Handbook.* NY: HarperCollins.

Johnson, N. C., & Young, S. D. (1992). Survivors' Response to Gang Violence. In: R. C. Cervantes (Ed.), *Substance Abuse and Gang Violence,* (136–146). Newbury Park: Sage Publications.

Johnson, R. (1981). *Condemned to Die: Life under Sentence of Death.* New York: Elsevier.

Johnston, D. (1995a). The Care and Placement of Prisoners' Children. In: K. Gabel & D. Johnston (Eds.), *Children of Incarcerated Parents,* (103–123). New York: Lexington Books.

Johnston, D. (1995b). Effects of Parental Incarceration. In: K. Gabel and D. Johnston (Eds.), *Children of Incarcerated Parents,* (59–88). New York: Lexington Books.

Johnston, D., & Gabel, K. (1995). Incarcerated Parents. In: K. Gabel & D. Johnston (Eds.), *Children of Incarcerated Parents,* (3–20). New York: Lexington Books.

Kampfner, C. J. (1995). Post-traumatic Stress Reactions in Children of Imprisoned Mothers.' In: K. Gabel & D. Johnston (Eds.), *Children of Incarcerated Parents,* (89–100). New York: Lexington Books.

King, A. E. O. (1993). The Impact of Incarceration on African American Families: Implications for Practice. *The Journal of Contemporary Human Services, 74*, 145–153.

King, R., & Norgard, K. (1999). What About Our Families? Using the Impact on Death Row Defendants' Family Members as a Mitigating Factor in Death Penalty Sentencing Hearings. *Florida State University Law Review, 26*, 1119–1173.

Kiser, G. C. (1991). Female Inmates and Their Families. *Federal Probation, LV*, 56–63.

Koban, L. A. (1983). Parents in Prison: A Comparative Analysis of the Effects of Incarceration on the Families of Men and Women. *Research in Law, Deviance and Social Control, 5*, 171–183.

Lewis, D. O., Pincus, J. H., Feldman, M., Jackson, L., & Bard, B. (1986). Psychiatric, Neurological, and Psychoeducational Characteristics of 15 Death Row Inmates in the United States. *American Journal of Psychiatry, 143*, 838–845.

Lewis, D. O., Pincus, J. H., Bard, B., Richardson, E., Prichep, L. S., Feldman, M., & Yeager, C. (1988). Neuropsychiatric, Psychoeducational, and Family Characteristics of 14 Juveniles Condemned to Death in the United States. *American Journal of Psychiatry, 145*, 584–589.

Lillich, R. B. (1991). Notes and Comments: The Soering Case. *The American Journal of International Law, 85*, 128–149.

Lincoln, A. J., & Straus, M. (1985). *Crime and the Family*. Springfield, IL: Charles C. Thomas.

Lyon, E., Moore, N., & Lexius, C. (1992). Group Work with Families of Homicide Victims. *Social Work with Groups, 15*, 19–33.

Magee, D. (1980). *Slow Coming Dark: Interviews on Death Row*. New York: Pilgrim Press.

Malmquist, C. P. (1986). Children Who Witness Parental Murder: Posttraumatic Aspects. *Journal of American Academy of Child Psychiatry, 25*, 320–325.

Masters, R., Friedman, L. N., & Getzel, G. (1988). Helping Families of Homicide Victims: A Multidimensional Approach. *Journal of Traumatic Stress, 1*, 109–125.

Miller, E. D., & Omarzu, J. (1998). New Directions in Loss Research. In: J. H. Harvey (Ed.), *Perspectives on Loss: A Sourcebook*, (1–20). Philadelphia, PA: Brunner/Mazel.

NAACP Legal Defense and Educational Fund (1999). Death Row U.S.A. (Fall). Available from NAACP Legal Defense Fund, 99 Hudson Street, Suite 1600, New York, NY 10013.

Norman, J. A. (1995). Children of Prisoners in Foster Care. In: K. Gabel & D. Johnston (Eds.), *Children of Incarcerated Parents*, (124–134). New York: Lexington Books.

Parkes, C. M. (1993). Psychiatric Problems Following Bereavement by Murder or Manslaughter. *British Journal of Psychiatry, 162*, (49–54).

Payne v. Tennessee. (1991). 501 U.S. 808.

Phillips, S., & Bloom, B. (1998). In Whose Best Interest? The Impact of Changing Public Policy on Relatives Caring for Children with Incarcerated Parents. *Child Welfare, LXXVII*, 531–541.

Prejean, H. (1993). *Dead Man Walking: An Eyewitness Account of the Death Penalty in the United States*. New York: Random House.

Radelet, M. L., Vandiver, M., & Berardo, F. M. (1983). Families, Prisons, and Men with Death Sentences: The Human Impact of Structured Uncertainty. *Journal of Family Issues, 4*, 593–612.

Rapaport, E. (1990). Some Questions about Gender and the Death Penalty. *Golden Gate University Law Review, 20*, 501–565.

Redmond, L. M. (1989). *Surviving: When Someone You Love Was Murdered*. Clearwater, FL: Psychological Consultation and Educational Services, Inc.

Reed, D. F., & Reed, E. L. (1997). Children of Incarcerated Parents. *Social Justice, 24*, 152–169.

Riggs, D. S., Kilpatrick, D. G. (1990). Families and Friends: Indirect Victimization by Crime. In: A. J. Lurigio, W. G. Skogan, & R.C. Dairs (Eds.), *Victims of Crime: Problems, Policies, and Programs,* (120–138). Newbury Park: Sage Publications.

Rinear, E. E. (1988). Psychosocial Aspects of Parental Response Patterns to the Death of a Child by Homicide. *Journal of Traumatic Stress, 1,* 305–322.

Rock, P. (1998). *After Homicide: Practical and Political Responses to Bereavement.* Oxford and New York: Clarendon Press.

Rynearson, E. K. (1984). Bereavement after Homicide: A Descriptive Study. *American Journal of Psychiatry, 141,* 1452–1454.

Rynearson, E. K. (1995). Bereavement after Homicide: A Comparison of Treatment Seekers and Refusers. *The British Journal of Psychiatry, 166,* 507–510.

Rynearson, E. K., & McCreery, J. M (1993). Bereavement after Homicide: A Synergism of Trauma and Loss. *American Journal of Psychiatry, 150,* 258–261.

Schoenbauer, L. J. (1986). Incarcerated Parents and Their Children – Forgotten Families. *Law and Inequality, 4,* 579–601.

Sheppard, S. R. (1999), February 18. Interview with M. Vandiver.

Smith, G. (1995). Practical Considerations Regarding Termination of Incarcerated Parents' Rights. In: K. Gabel & D. Johnston (Eds.), *Children of Incarcerated Parents,* (183–195). New York: Lexington Books.

Smykla, J. O. (1986). The Impact of Capital Punishment on Death Row Inmates' Families. *USA Today* (January), 86–90.

Smykla, J. O. (1987). The Human Impact of Capital Punishment: Interviews with Families of Persons on Death Row. *Journal of Criminal Justice, 15,* 331–347

Sprang, M. V., McNeil, J. S., & Wright, R. (1989). Psychological Changes after the Murder of a Significant Other. *Social Casework: The Journal of Contemporary Social Work, 70,* 159–164.

Sprang, M. V., McNeil, J. S., & Wright, R. (1992–1993). Grief among Surviving Family Members of Homicide Victims: A Causal Approach. *Omega, 26,* 145–160.

Spungen, D. (1998). *Homicide: The Hidden Victims: A Guide for Professionals.* Thousand Oaks, CA: Sage Publications.

Swan, L. A. (1981). *Families of Black Prisoners: Survival and Progress.* Boston: G.K. Hall and Company.

Thompson, M. P., & Vardaman, P. J. (1997). The Role of Religion in Coping with the Loss of a Family Member to Homicide. *Journal for the Scientific Study of Religion, 36,* 44–51.

Vallejo, C. A. (1995). Death Sentence Experience: The Impact on Family Members of Condemned Inmates. Masters thesis, Social Work Department, California State University, San Bernandino, CA.

Vandiver, M. (1989). Coping with Death: Families of the Terminally Ill, Homicide Victims, and Condemned Prisoners. In: M. L. Radelet (Ed.), *Facing the Death Penalty: Essays on a Cruel and Unusual Punishment,* (123–138). Philadelphia: Temple University Press.

Vandiver, M. (1992). Beyond All Justice: The Death Penalty and Families of the Condemned. *Odyssey* (Spring), 56–63.

Vandiver, M. (1998). The Impact of the Death Penalty on the Families of Homicide Victims and of Condemned Prisoners. In: J. R. Acker, R. M. Bohm & C. S. Lanier (Eds.), *America's Experiment with Capital Punishment,* (477–505). Durham, NC: Carolina Academic Press.

Vargas, L. A., Loya, F., & Hodde-Vargas, J. (1989). Exploring the Multidimensional Aspects of Grief Reactions. *American Journal of Psychiatry, 146,* 1484–1488.

Weilerstein, R. (1995). The Prison MATCH Program. In: K. Gabel and D. Johnston (Eds.), *Children of Incarcerated Parents,* (255–264). New York: Lexington Books.

Weiss, R. S. (1998). Issues in the Study of Loss and Grief. In: J. H. Harvey (Ed.), *Perspectives on Loss: A Sourcebook*, (343–352). Philadelphia, PA: Brunner/Mazel.

Zealand, E. (1998). Protecting the Ties that Bind from Behind Bars: A Call for Equal Opportunities for Incarcerated Fathers and Their Children to Maintain the Parent-Child Relationship. *Columbia Journal of Law and Social Problems 31*, 247–281.

PART IV

OPPORTUNITIES FOR SYSTEM REFORM

THE INVISIBLE CHILDREN OF INCARCERATED MOTHERS

Jacquelyn Sandifer and Suzanne Kurth

> My son was in prison and I am now in prison again. I wasn't in a position to help him. There are a lot of mothers in here with kids in prison. A lot. (Owen, 1998 p. 125).

Within American society the expectation persists that mothers will be their children's primary caregivers. As the rates of single parenthood increase, more and more women are both the primary caregiver and economic provider for their children. Substantial numbers of single mothers never obtain the education or employment skills necessary to legitimately provide for themselves and their dependent children. A subset of these women may resolve this dilemma through illegitimate means or escape in the world of drugs. For these women the stage is set for them to enter the criminal justice system, and for their children to become invisible victims of their mothers' incarceration. These women and their children are the focus of this chapter.

The contemporary response to crime is increasingly to incarcerate offenders without regard for their personal characteristics (e.g. being a single mother) or the relative severity of their offenses. Embedded in the overall pattern of increased incarceration is a dramatic increase in female imprisonment and concomitantly of the number of children deprived of contact with the parent who has both nurtured and provided for them within the limits of her ability.

Imprisonment produces many unanticipated consequences that are often unacknowledged and negative for children, their families, and society. Spending on corrections is outpacing spending for social programs that might reduce criminogenic conditions. In conjunction with changing policies governing child

Families, Crime and Criminal Justice, Volume 2, pages 361–379.

ISBN: 0-7623-0737-4

welfare, the consequences of increasingly imprisoning mothers are ones which policy makers and the public have inadequately considered.

The needs of prisoners' children are easily overlooked by the overburdened criminal justice and child welfare systems. These children are the unacknowledged victims of their mothers' behaviors, current sentencing practices, and marginal economic circumstances. Additional problems confront children of incarcerated parents; they experience more difficulties and hardship than children still living with a parent in similar socioeconomic circumstances (Gabel, 1992; Hostetter & Jinnah, 1993). Laws and programs intended to aid children by keeping them from a childhood of multiple foster home placements, foster care drift, may have the unintended consequence of making family reunification more difficult.

We address a series of questions in this chapter. Who are the women incarcerated in our jails and prisons and why are they there? What are the costs for the children and society of having mothers incarcerated? Which programs or policies hinder or aid caring for the invisible children of incarcerated mothers?

WHO ARE THE WOMEN OFFENDERS?

Until the 1970s rates of female incarceration were quite low. During the last decades of the 20th century three policy shifts occurred which increased those rates: reduced flexibility in sentencing, a general 'get tough' approach emphasizing incarceration of offenders, and a "war on drugs" (Chesney-Lind, 1991).

Almost one million women were under correctional supervision (on probation or parole, in jail, or in prison), as the 1990s drew to a close (U.S. Department of Justice [USDJ], 1999b). The majority of these women were on probation or parole. Women have had low rates of incarceration for several years. However, recently the number of women in jail or prison has been rapidly increasing. The number of women incarcerated (in jail or prison) in 1998 (146,507) was 13 times greater than it was in 1970 (11,154). Both the numbers of incarcerated women and men have risen dramatically, but the number of men has increased at a slower rate (about a fivefold increase from 1970 to 1998) than the number of women (U.S. Department of Justice [USDJ], 1971; U.S. Department of Justice [USDJ], 1999a).

Personal Characteristics

The personal characteristics of women inmates vary to some extent depending on the level of facility in which they are located. Examination of their race, age, educational, economic, and family histories provides a profile of women who are members of minority groups, equipped with minimum levels of education, lacking in work skills and experience, and disproportionately single parents who were raised by single parents.

Although the majority (62%) of women under probation supervision are white, the majority of incarcerated women are black, Hispanic, or of another minority group (64% of female jail population; 67% of female state prison population, and 71% of the female federal prison population are minorities) (USDJ, 1999b).

About three quarters of incarcerated women are of child bearing and rearing age, that is, between 25 and 44 years of age. Yet, there are differences in age associated with type of incarceration. About a quarter (21%) of female jail inmates are 24 or under, while nearly a quarter of Federal prison inmates are at least 45 years of age (USDJ, 1999b).

In recent decades the educational level of women prisoners rose as did that of the general population, but it is still low. A substantial minority (about 45% of women in jail, 44% of those in state prison, and 37% of those in Federal prison) have not completed high school or earned a GED (USDJ, 1999b). Their reasons for dropping out of school are associated with a lack of commitment to education. For example, in a 1987 survey 59% of the women reported boredom as their reason for dropping out, while 34% gave "pregnancy" as the reason (American Correctional Association, 1990).

Incarcerated women are substantially less likely to have ever married than women in general. Nearly one half of the women in jails and state prisons have never married. Of the somewhat older female in federal prisons, slightly over one third have never married (USDJ, 1999b).

Research on incarcerated women for the past twenty-five years has consistently reported that at least three quarters of them are mothers (American Correctional Association, 1990; Baunach, 1982; USDJ, 1999b; Johnston, 1995; McGowan & Blumenthal, 1978). A Department of Justice special report indicated over two thirds of the women (about 70%) under correctional care, custody, or control in 1998 were mothers of minor children. Almost three quarters (73%) of incarcerated mothers reported they had custody of their children at the time of their arrest and the majority reported living with some or all of their minor children before incarceration (USDJ, 1999b).

Female offenders generally were in an economically marginal position before their arrest. Almost two thirds (60%) of the women in state prison in 1998 reported they were not employed full-time before their arrest. Their low levels of full-time employment are paralleled with low incomes (about 37% reported monthly income of less than 600 dollars) and dependence on welfare (almost 30% received welfare assistance before arrest) (USDJ, 1999b). A 1991 survey of women state prison inmates reported a similar low level of full-time employment with about a third (35.7%) working full-time before their arrest (USDJ, 1994). When employed, women inmates earned little more than the minimum wage, less than they needed to support themselves and their children (American Correctional Association, 1990). This employment and income profile has led some to conclude that women who become involved in illegal activity "tend to commit survival crimes, most often petty economic crimes to earn money (or) feed a drug-dependent life" (Owen, 1998: p. 11).

Female inmates were twice as likely to have grown up in a single parent household as members of the general population. In a 1991 survey of females in state prisons, 42% grew up in single parent households (39% with mothers; 3% with fathers) and 16% in households where neither parent was present (United States Department of Justice [USDJ], 1994).

The women's families of origin often had other members in the criminal justice system. Almost half of the female state prison inmates (47%) reported that at least one member of their immediate family had been incarcerated with a higher percentage of black than white women reporting they had a family member who had been in prison or jail. The immediate relative most likely to have been incarcerated was a brother (USDJ, 1994).

Drug use and abuse characterize the lives of many female offenders. About 5 in 10 females incarcerated in state prisons in 1998 described themselves as a daily user of drugs and 6 in 10 described themselves as using drugs in the month before their offense. Nearly 4 in 10 were under the influence of drugs when committing their offense (USDJ, 1999b).

Types of Offenses

For every category of major crime for the time period 1990–1996 – violent, property, drugs, and other felonies – the rate of increase in the number of convicted female defendants was greater than the changes in the number of convicted male defendants (USDJ, 1999b). The increased conviction and incarceration of women are a consequence of shifts in how certain types of offenses (e.g. drug related crimes) are categorized and changes in sentencing patterns for females brought about by mandatory sentencing guidelines, not a major female crime wave. Researchers argue increases in drug related arrests are propelling the explosive growth of women in prison (Owen, 1998). Chesney-Lind (1991) further asserts that the combined effects of the 'war' approach to reducing drug use and new sentencing guidelines have produced a de facto war against women evidenced by the rapid growth of female incarceration for drug related offenses USDJ, 1994).

More than one-half of all women in prison in 1991 were serving a sentence for a non-violent offense (USDJ, 1994). Of those females who were incarcerated in 1998 for committing violent offenses, three out of four had committed simple assaults. The rate of serious violent offenses (e.g. murder) has been declining since 1980. For female defendants convicted in state courts, nearly 90% of the increase in the "number of violent felons was accounted for by aggravated assault, perhaps reflecting increased prosecution of women for domestic violence" (USDJ, 1999b:6). Thus, the women filling our prisons in increasing numbers are not serious violent offenders.

More than one-half of the increase in the number of women defendants convicted of property crimes in 1996 was attributable to rising numbers of individuals convicted of forgery, fraud, or embezzlement – less serious than some forms of property crimes. The proportion of women incarcerated for property offenses rapidly declined (41.2% in 1986; 28.7% in 1991) as women increasingly were imprisoned for drug offenses (12% in 1986; 32.8% in 1991) (USDJ, 1994).

From 1990 through 1996, the number of convictions for drug trafficking grew by 34% while the number of convictions for drug possession increased 41% (USDJ, 1999b). Of women incarcerated in 1998, 72% of female inmates in Federal prisons, 34% of inmates in State prisons, and 34% of female jail inmates are incarcerated for drug offenses. The pattern of offenses among females in state prisons has been changing. The proportion of women in prison for violent and property crimes has decreased steadily since 1979, while the proportion of women in prison for drug and public-order offenses grew during the same time period – almost tripling between 1986 and 1991 (USDJ, 1999b).

In addition to incarceration for drug offenses, drugs and/or alcohol increasingly play a part in the commission of all types of crimes committed by females. Victimization surveys in 1998 revealed that an estimated four in ten females engaging in violence were perceived as being under the influence of drugs or alcohol at the time of the crime (USDJ, 1999b). Over one-fourth of females incarcerated in late 1987 indicated that their primary reasons for committing any type of crime was to pay for drugs (American Correctional Association, 1990). Of inmates incarcerated in State prisons in 1998, nearly 1 in 3 reported "they had committed the offense which brought them to prison in order to obtain money to support their need for drugs" (USDJ, 1999b:9).

The combination of personal characteristics and offense patterns of all female offenders suggests that the typical incarcerated female is of child rearing age, unmarried, a minority group member, a mother of minor children, under-educated, economically marginal, and has considerable experience with or is dependent on drugs or alcohol. She was incarcerated for a less serious non-violent offense, directly or indirectly related to the use or abuse of drugs or alcohol. That the typical female offender is a danger to the safety of others or that the personal and social problems that may have encouraged her criminality will be lessened by incarceration is subject to question (Fletcher, Shaver & Moon, 1993).

WHERE ARE THEIR CHILDREN?

The children of prisoners are the largely invisible victims of their single mothers' poverty and 'tough on crime' laws which are blind to defendants' gender and parental status. Systematic data regarding these children are not routinely gathered and maintained as women move through the criminal justice system, for they are not the focus. Researchers focused on the children's welfare often must estimate the numbers, ages, living situations, and problems the children face (Bloom & Steinhart, 1993). Self-report data from mothers are the best available, but they may under represent the number of children. Mothers may be reluctant to provide information about their children that may jeopardize retention of child custody or access to public assistance (Baunach, 1982; Hairston, 1991; McGowan & Blumenthal, 1978).

The majority of incarcerated females have minor children they were living with before they were arrested, although they may not be living with all of their children at the time of their arrest (Hungerford, 1996). Self-report studies of incarcerated mothers have consistently indicated they have between 2 and 3 children (American Correctional Association, 1990; Baunach, 1982; Bloom & Steinhart, 1993; USDJ, 1999b; Johnston, 1995; Koban, 1983, Stanton, 1980).

Conservatively, there are "more than 1.3 million minor children who are the offspring of women under correctional sanction; (with) more than a quarter million of these children hav(ing) mothers who are serving time in prison or jail" (USDJ, 1999b:8). The National Council on Crime and Delinquency projects that as many as ten million children have a parent who at some point in their lives has spent time in prison or jail, or has been on probation or parole (Gabel & Johnston, 1995). The number of children affected by the imprisonment of a parent is rapidly growing in tandem with the rising prevalence of imprisonment as the preferred legal sanction

The Impact of Maternal Incarceration on Children

Children of incarcerated mothers typically are among the disadvantaged of society prior to parental imprisonment. Their mothers may have had difficulties with the law before the event that led to their incarceration. They did not have the culturally mythologized stay-at-home mother (Johnston, 1995); however, this does not mean that their mothers were not trying to parent the children as best they could based on their knowledge and resources or that the children are better off without them.

The role that penal sanctioning of parents plays in children's lives, alone or in combination with other negative childhood experiences and events is often depicted in descriptive accounts (Gabel, 1992). The effects of parental incarceration on children are most often derived from surveying their parents; no large-scale study of prisoners' children exists (Johnston, 1995). Few researchers are able to directly assess children and those that do, obtain clinical samples of children limiting the generalizability of their findings.

"Many children are angry that they're separated from their mothers, and they're guilty – sometimes they feel responsible that she was taken away" (Watterson, 1996:355–6). At the very least children of incarcerated mothers experience the negative effects associated with parental separation, problems with bonds of attachment and associated ability to trust (Johnston & Carlin, 1996). Typically, whatever special needs they had before her imprisonment and any generated by her imprisonment are unmet. The limited data available suggest that other negative outcomes, such as pursuing pathways which may later result in deviant or criminal behavior may result from imprisoning parents, particularly caregiver mothers (Bloom & Steinhart, 1993).

WHAT HAPPENS AFTER A MOTHER IS DETAINED?

Arrest: Finding Caregivers

Police and social service officials face a placement and care dilemma each time they are aware that a mother of dependent children is arrested. Police contact social service agencies if children are present at the time of the arrest or they know the children are alone somewhere. Officers often encourage arrested mothers to find care for their children with relatives or friends. State child protection workers may respond to the site of arrest to assist with and supervise children's transition either to a parent designated caregiver or to formal state custody. Mothers are advised about the risks of non-kin foster care and of the shortage of quality foster homes, particularly for sibling groups, adolescents, and children of color (American Bar Association, 1993). When placement with a relative is not immediately possible, children are transported to temporary shelters where efforts are continued to secure care with a relative. If kin care cannot be arranged, children are assigned to foster families or a group home.

Time: Sentences and Threats to Custody

Criminologists Irwin & Austin assert that children of incarcerated mothers routinely do not receive the "supports guaranteed to other children in America by the 1980 Federal Adoption and Child Welfare Act (PL 96–272), a federal law mandating services primarily to promote the preservation of families" (1997, 163). Other researchers (Barry, 1995; Beckerman, 1994; Bloom, 1995) expand on the nature of the problems, namely maintenance of contact with children, participation in permanency planning, and availability of the mother to resume care within the specified time frame due to the length of her sentence. The chances of mothers and children becoming beneficiaries or victims of protective legislation are linked to how the children are cared for while their mother is in prison. Arrangements range from informal private ones with relatives to formal public arrangements (group homes, foster parents) with hybrid arrangements such as kin foster parents in between.

Most children of incarcerated mothers are in the care of relatives, other than their fathers who often had limited involvement in their lives before their mothers' incarceration. In one sample 70% of the children of incarcerated mothers resided with grandparents or other relatives (Johnston, 1995). A survey of women in the Ohio prison system revealed about two thirds of the women's children resided with a maternal grandparent (50%) or a maternal sibling (15%) and about another 20% with their natural fathers or paternal kin (Hungerford, 1996). A small percentage of children may be in the care of quasi-kin, such as godparents. Grandparents, particularly maternal grandmothers, often play a key role in caring for incarcerated women's children.

Research on kin care focuses on the effects of adding these children to typically already financially and emotionally strained family living units (Gabel & Johnston, 1995). Many grandmother caregivers report that caring for their incarcerated daughters' children is physically taxing (Dressel & Barnhill, 1994). Relative caregivers report being unprepared to deal with the children's psychological reactions to the emotional trauma of arrest and separation from their mothers, in addition to their own.

Inmates' children may not accept the authority and discipline of the caregiver. Discipline and control of prisoners' children may be complicated by the incarcerated mothers' efforts to parent. Attempts to maintain a decision-making role over her children may generate confusion and hard feelings between caregivers, children, and inmates (Hungerford, 1996; Minkler & Roe, 1993). Major problems caregiving grandmothers face include: adjusting their lifestyles to accommodate children, dealing with their own emotional reactions (anger, grief, shame) to their daughters' behaviors and incarceration, finding sources of support/relief for the stresses they experience, and stretching their limited financial resources (Hungerford, 1996).

Before 'welfare reform' and the change from Aid to Families with Dependent Children (AFDC) to Temporary Assistance to Needy Families (TANF), a substantial minority of these caregivers (over 40%) relied on AFDC to care for their incarcerated relatives' children. The 1996 Personal Responsibility and Work Opportunity Reconciliation Act (P.L. 104–193) may make some relatives ineligible to receive payments or food stamps because persons who have violated probation or parole or who were convicted of specific drug felonies are ineligible for benefits. Other kin caregivers, for example those with poor job skills and work histories, may encounter difficulties with the work requirement provisions of the Act and the time limit (a lifetime total of 60 months) for receipt of assistance (Phillips & Bloom, 1998).

Foster Care

Only 10% of the children of inmate mothers participating in an Ohio prison survey were in the formal custody of the state (Hungerford, 1996). Representatives of the criminal justice system from police to judges through both action and inaction facilitate relatives taking over care of children before questions of state custody can arise. The low percentage of children of incarcerated mothers in state custody reflects the preferences of the mother and her kin (Phillips & Bloom, 1998), as well as social service agencies. The belief that families should care for their relatives, as well as the limitations of the child welfare system (dearth of good placements and other resources) and the lack of interest in the criminal justice system, account for the low percentages of children placed with non-relatives.

Non-kin foster care placement of children of incarcerated parents typically begins with being abruptly removed from their residence (McGowan & Blumental, 1978). Children may move through a sequence of caregivers while at the same time they are deprived of interaction with their siblings and other relatives. While experiencing disruptions in caregivers, residence, and schools, they receive limited help in coping with their experiences. Typically, these children have had difficult lives (e.g. poor housing, parental absences) before they entered foster care, so one potential benefit of state custody is that they might be more likely to receive any special assistance they may need than they would otherwise.

Increased demands on the foster care system in the late 1980s and early 1990s led to more and more relatives being formally recognized as foster parents. The Personal Responsibility and Work Opportunity Reconciliation Act encouraged all states to give preference to relatives as foster care providers, a practice followed in two thirds of the states before this legislation was enacted (Geen, Boots & Tumlin, 1999). Benefits for foster parents are typically higher than those kin can receive otherwise, but foster care payments can only be received if children are formally in state custody. Relatives may have to state that the mother was abusive or neglectful for foster care status to be granted.

In whatever setting the children may be, three common themes run through the literature on the care of prisoners' children. One, both foster and kin care givers are untrained and ill-equipped to address the special problems and circumstances of children of incarcerated parents. And, both foster and kin caregivers are, for the most part, financially and emotionally strained. Three, contact between imprisoned mothers and their children is difficult to maintain regardless of whether a child is cared for by relatives or by the state.

REBUILDING MOTHER-CHILD RELATIONSHIPS

Parental absence disrupts parent-child bonding. When fathers and mothers are incarcerated they do not relinquish their parental roles, obligations, or concerns (Bloom & Steinhart, 1993), although about one-half of them do not receive any visits from their children while incarcerated (USDJ, 1994). After a mother is released from prison, underdeveloped bonds of attachment with her child could contribute to parenting and disciplining problems for her and attachment problems for the child (Gabel & Girard, 1995).

Barriers to Contact

Female offenders may be seen as engaging in behaviors which harm their children and strain others' limited resources. Caregivers may be reluctant to expend scarce resources (money, time, and energy) for children to visit their mothers. Caregivers' perceptions of the incarcerated woman may be negative; she may be seen as having chosen to follow a path that harmed her children and other family members. One imprisoned addict reported "my mother only came once to jail to tell me that she always knew that I would turn out like this, and that I was a disgrace to the family and that she didn't want to see me" (Moore & Devitt, 1989:65). Even when caregivers and inmates are not in conflict, logistical constraints (transportation to rural prisons, visitation regulations, and lack of funds) may serve as barriers to mother-child contact.

Transportation can be a substantial expense for people living on the margin. Over 60% of mothers in prison are incarcerated over one hundred miles from their children. Women prisoners are usually incarcerated further distances from their families than are men due to the relative scarcity of facilities for women. Transporting children for maternal visits may be prohibitively expensive and the logistics nearly impossible when incarceration is in prisons located in rural areas that lack public transportation. Local or state welfare policies often set geographical limits beyond which parent child visitation is not required for foster parents (Chicago Legal Aid to Incarcerated Mothers [CLAIM], 1997).

Maintenance of limited contact is possible by telephone. Again, costs are substantial if prisons are located far from where the children live with kin. Mothers cannot make calls based on their perceptions of their children's needs because they are dependent on others' resources and approval to make calls. Even telephone contact between mothers and children may be hampered by where facilities are located. State funds usually are not provided for long distance telephone calls to or collect calls from mothers in prison (Bloom, 1995).

Prison visitation policies and facilities are not designed to accommodate children. Mothers and alternative caregivers may be reluctant to have children see their mothers in such settings. For maternal visitation to occur, caregivers would have to be highly motivated for they would have to make arrangements and use typically scarce resources for brief visits (Barry, 1995).

Bureaucratic Obstacles: Policies and Practices

Some bureaucratic policies and practices undermine family reunification efforts while the absence of others block family reunification. An example of an undermining practice occurs when sibling groups are taken into state custody, for they are most often separated with each child assigned a different caseworker. Caseworkers rarely have the time or resources to coordinate their efforts for these children. Visitation between siblings is difficult to arrange and visitation between all the siblings and the mother simultaneously is rare.

Judicial processes create other stumbling blocks. When children are placed in foster care, months may elapse before mothers who cannot place their children with relatives discover where their children are, and even more time before the court grants visiting privileges for the children. Visiting privileges can be revoked if case workers or foster parents ask that visits be suspended. If visits are denied and children have no contact with the mother while she is incarcerated, resumption of life as a family after release is very difficult (Gabel & Johnston, 1995), particularly with the demands of permanency planning.

Policies governing child welfare may undermine incarcerated mothers' efforts to retain custody of children placed in foster care. Their sentences are longer than the maximum time prescribed for foster care (Barry, 1995; Bloom, 1995), under the Adoption and Safe Families Act of 1997 (P.L. 105–89). Reunification policy requirements of mother-child contact may be impossible to fulfill because of 'barriers' beyond their control. In most instances a caseworker may have few cases involving a parent in prison, and neither the time nor the experience necessary to help prisoners meet all standards. Mandated counseling or training may not be available to the prisoner.

These barriers may account for why incarcerated mothers with children in foster care are less likely than other mothers to be reunited with children supervised by the child welfare system (McGowan & Blumenthal 1978; Hairston, 1998). The National Black Child Development Institute's (1989) research suggests that termination of parental rights may occur disproportionately for imprisoned parents, particularly if their children entered foster care because mothers could not provide materially for them while incarcerated.

THE CHILDREN'S INTERESTS

Criminologists recognize that parental deviance and criminality are risk factors for children's antisocial behavior, delinquency, and eventual adult criminality (Rowe & Farrington, 1997; Gottfredson & Hirschi, 1983; Sampson & Laub, 1993). The relative level of risk cannot be measured but some researchers estimate that children of incarcerated parents may be as much as six times more likely as other similarly situated children to enter the criminal justice system at some point in their lives (Barnhill & Dressel, 1991).

The question is what *is* in the children's best interest. Family preservation advocates generally argue that children are better off with family members regardless of the problems in the family system (CLAIM, 1997). Children remain with familiar others, are less likely than children in non-kin foster care to repeatedly have new caregivers, and their mothers do not lose custody of them. Skeptics argue that children may fare better in foster care because the state then has the mandate to provide services for these children and their mothers that are not available with informal custodial arrangements (Gaudin & Sutphen, 1993). Imprisoned parents whose children are in foster care are entitled to assistance in maintaining contact with their children through letters, telephone calls, and visits, as well as to counseling, parenting education, and caseworker reunification services. The hybrid kin foster care model provides some of the benefits identified with kin care and with state care, but the mother does not retain custody. Kin foster placements are not as closely monitored as non-kin foster care so children's special needs still may not be addressed. The hybrid model may primarily address the problem of providing a somewhat higher level of financial assistance to kin caring for children when the number of foster parents needed far exceeds the supply (Phillips & Bloom, 1998).

A mother may move her children into informal kin care to protect or conceal them from the state, so that she not lose legal custody. The children's welfare and special needs may not be known and addressed by appropriate social service agencies. Even if there were programs for their and their caregivers' special needs, the children's whereabouts are unknown and therefore, services cannot be designed for or provided for them. In addition, many relative caregivers may live at or near the poverty level and may be unable to provide the children with positive role models.

Children may be instructed to keep their parents' incarceration a secret, so that the family caregiver will not be embarrassed, official agencies will not be alerted to the children's status, and the children will not be stigmatized. Secrecy further isolates these children and may inhibit the formation of social bonds with teachers or other non-familial adults (Hostsetter & Jinnah, 1993), as the interests of others are assigned higher priority than the children's.

A mother's (or father's) imprisonment may be a turning point, an event that combined with other adverse life experiences sets a child on a negative pathway that culminates in adult criminality (Hagan & Palloni, 1990). The impact of a particular event in childhood may lie less in its immediate consequences than its setting in motion a chain reaction in which one negative event leads to another, propelling a non conforming life course trajectory (Caspi, Bem & Elder, 1989). Children may be traumatized by the arrest event, the subsequent separation, prison visits, as well as circumstances in their alternative care setting. Any or all of these events increase the risk of emotional stress and negative behavior

PROGRAMS FOR INCARCERATED WOMEN AND THEIR CHILDREN

Policies emphasizing imprisonment stimulate the spending of money on physical facilities to the detriment of spending money on programs aimed at decreasing the risk that prisoners will recidivate or that their children will follow in their footsteps. In response to a recent National Institute of Justice (NIJ) survey few states identified more than a handful of innovative corrections programs and 17 states identified none (Morash, Bynum & Koons, 1998). Two types of prison programming are needed. One set addresses the need for continuing contact between mothers and children both to meet social and emotional needs and to fulfill legal requirements for resuming custody of children in foster care. Another set addresses improving the mother's ability to be an effective nurturer and provider through expanding and modifying her life skills.

Mother-Child Contact

Mothers and children need assistance in having the contact necessary for relationship maintenance, as well as to meet the family reunification plan requirements of foster care (Hairston, 1991). Yet, some correctional facilities treat visits with children, as rewards for good behavior, rather than as prisoner rights. Women in the Nebraska Correctional Center may have up to five visits every month with their children one to nine years of age as a reward. The needs of children are not addressed when visitation is used as a prisoner reward.

A model prison program is the Children's Visitation Program (CVP) in Michigan that offers a child-centered environment for interaction (Bloom & Steinhart, 1993) in contast to the child unfriendly visiting areas in most prisons. Other innovative programs such as Motheread in North Carolina work to increase contact and improve parenting attitudes at the same time as they improve literacy skills, by having mothers compose stories and letters and make audio tapes to send to their children (Martin & Cotton, 1995). Such facilitative programs are scarce. During a period of rapid growth in the female prison population, a nationwide survey in 1992 found virtually no change from 1985 in the availability of parental programs, such as supportive facilities for mothers and children (Collins, 1997).

Even when programs exist, opportunities for visits do not materialize into visits unless mothers and children receive assistance with logistics. Various programs depend on non-profit organizations and volunteer workers. The Bedford Hills Correctional Facility in New York has a summer program and a weekend which permits more visits than are available in most prisons by relying on host families to house visiting children, as well as volunteer drivers. The successful Girl Scouts Beyond Bars program which increases mother-child interaction also relies on volunteers (Zaplin & Dougherty, 1998). Volunteers may serve in additional capacities; organizations such as the Women's Prison Association in New York work to establish a broad range of services necessary to reunite and support families (United States Department of Justice, 1998).

Life Skills Programs

Most prisons have some programs to help prisoners improve themselves, e.g. by earning a GED or combating addictions. Most do not have the variety of programs (vocational, substance abuse, relationship skill) or enough spaces in the programs they do have to meet the needs (mandated or not) of incarcerated mothers who want to resume custody of their children when released.

Most women in prison were not working full-time before arrest and had a low level of income. With their records of imprisonment added on to their poor employment histories and limited job skills, their opportunities for employment after release are generally poorer than they were when they entered. Income from welfare programs may be unavailable. Mothers are permanently ineligible for TANF payments or food stamps, if they were convicted of drug-related felonies (Devine, 1999). Therefore, developing their work skills before they leave prison is essential.

The work training programs for women in prison continue to be limited in number and scope. They tend to emphasize a narrow range of traditional female occupations, such as clerical worker or cosmetologist (Schram, 1998). The emphasis on a limited set of female occupations may reflect traditional gender role attitudes held by those operating the criminal justice system (Zaplin & Daugherty, 1998). In addition, women in prison often hold very traditional gender role attitudes that lead them to expect someone will provide for them (Owen, 1998). Work training programs need to do more than impart skills; they need to educate women prisoners about the need for them to work, particularly in better paying non-traditional occupations.

As the female correctional population rapidly increased, the percentage of that population with alcohol or drug problems rose to about 70% (National Women's Law Center, 1995). Despite the influx of women into prisons and jails for drug-related crimes, the number of drug counselors and programs available for them has not increased proportionately. The war on drugs needs to lead to more than the temporary warehousing of women whose primary problem is drug abuse. Prison programs are not designed to meet the needs (e.g. visitation) of mothers of minor children (Prendergast, Wellisch & Falkin, 1995).

Counseling programs are available for only a fraction of the women who could benefit from them. For a mother to effectively reassume care of her children and guide them along a different path than she followed, she typically needs to acquire new skills. Fewer women than men in prison have access to counseling, anger management/conflict resolution, and parenting classes (CLAIM, 1997). Initiatives such as the Children's Visitation Program mentioned above offer parenting classes and support groups (Bloom & Steinhart, 1993). Despite the relative success of programs such as these, the costs of providing beds for the increasing number of women prisoners has been the focus in most states.

POLICY IMPLICATIONS

Criminal justice and child welfare policies unintentionally support family separation – children from mothers and siblings from each other (Dressel, Porterfield & Barnhill, 1998). In the short term the costs of current sentencing and drug policies are born primarily by the children and kin of incarcerated women. A broader perspective is needed.

Pro-child policies would facilitate keeping mothers out of prison and development of programs to aid mothers in making changes that would allow them to legitimately support their children. One policy change would be to use alternative sanctions (e.g. house arrest, electronic monitoring) for mothers found guilty of various non-violent crimes. The proactive use of alternative sanctions has to be done in conjunction with increased provision of job preparation and drug treatment programs. Such social programs will be more cost effective than sending mothers to prison not only because of the costs of imprisonment, but also because of the short-term and long-term costs of caring for their invisible children who are at risk of pursuing deviant pathways.

When imprisonment occurs, more support for families is necessary. Family preservation advocates such as Chicago Legal Aid to Incarcerated Mothers (CLAIM) suggest that continued family connections throughout the mother's incarceration offer greater consistency and fewer traumas for children than drifting through the social services system. Continuing parent-child contact may reduce the children's sense of alienation and is crucial for their mothers' maintenance of custody. Another potential benefit for children and society is that maintaining ties with her children may contribute to a female offender's rehabilitation. If she is involved with her children, her sense of responsibility may be maintained or heightened. Therefore, supporting families' ties potentially reduces costs by contributing to the current and future welfare of the children and to prisoners' rehabilitation.

Additionally, some changes need to occur in social service policies. Children of incarcerated parents need to be identified. Child welfare agencies with substantial numbers of children with prisoner mothers could redesign case load assignments, so that a few caseworkers would handle all the prisoner cases and be better able to manage the logistics of prison visits.

CONCLUSIONS

The costs of increasingly incarcerating mothers are high for everyone. The financial costs of building new prisons and caring for the rapidly increasing numbers of women in prison are well known. Our prison building and spending spree cannot keep pace with the increasing numbers of women being incarcerated. Money being spent on building prisons is money not being spent on social programs that might help poor mothers and their children pursue non-criminal pathways.

One unintended consequence of getting tough on drugs is that increasing numbers of children's lives are seriously disrupted by their parents' incarceration. A parent's incarceration severely reduces a child's chances to accumulate the emotional and social resources important to following a pro-social pathway. Without such social capital children may follow paths that lead to criminality and thus the intergenerational perpetuation of crime (Hagan & Polloni, 1990). Our criminal justice system may be spawning a next generation of criminals as an unintended consequence of imprisoning women on its economic margins (Bloom & Steinhart, 1993; Hagan, 1997).

Various types of programs can make a difference. Fostering of parental ties may improve behavior in prison, reduce recidivism after release, and contribute to children's welfare. We need to modify both criminal justice and social service policies and practices that undermine mother-child relationships using successful programs and practices as models. We need to increase the quantity and quality of counseling and vocational programming to better equip these women to pursue lives *without* using drugs and crime and *with* their children. We need to make visible the children of incarcerated parents by identifying them and their special needs, if for no other reason than to reduce the likelihood that the children will follow their parents' footsteps to prison. Investing money in programs that improve the life situations of mothers on the margin of legitimate society is an investment in their children and in the welfare of society.

REFERENCES

American Bar Association. (1993). *Children on Hold: What Happens When Their Primary Caregiver Is Arrested?* Washington, D.C.

American Correctional Association, Task Force on the Female Offender. (1990). *The Female Offender: What Does the Future Hold?* Arlington, VA.

Barnhill, S., & Dressel, P. (1991). *Three Generations at Risk.* Atlanta, GA: Aid to Imprisoned Mothers.

Barry, E. (1995). Legal Issues for Prisoners with Children. In: K. Gabel & D. Johnston (Eds), *Children of Incarcerated Parents*, (47–166). New York: Lexington.

Baunach, P. J. (1982). You Can't Be a Mother and Be in Prison . . . Can You? Impacts of the Mother Child Separation. In: B. R. Price, & M. J. Sokoloff (Eds), *The Criminal Justice System and Women: Women Offenders, Victims, Workers*, (155–169). New York: Clark Boardman.

Beckerman, A. (1994). Conditions for Permanency Planning. *Social Work, 39*, 9–14.

Bloom, B. (1995). Public Policy and the Children of Incarcerated Parents. In: K. Gable & D. Johnston (Eds), *Children of Incarcerated Parents,* (271–284). New York: Lexington.

Bloom, B. & Steinhart, D. (1993). Why Punish the Children? A Reappraisal of the Children of Incarcerated Mothers in America. *The IARCA Journal, 6,* 14–17.

Caspi, A., Bem, D. J., & Elder, G. H. Jr. (1989). Continuties and Consequences of Interactional Styles Across the Life Course. *Journal of Personality, 57,* 375–406.

Chesney-Lind, M. (1991). Patriarchy, Prisons, and Jails: A Critical Look at Trends in Women's Incarceration. *The Prison Journal, 71,* 51–67.

Chicago Legal Aid To Incarcerated Mothers (CLAIM). (1997). Women in Prison Fact Sheet. Chicago, IL.

Collins, C. F. (1997). *The Imprisonment of African American Women: Causes, Conditions, and Future Implications.* Jefferson, NC: McFarland.

Devine, K. (1999). The End of the Welfare as We Know It. *CLAIM:* Chicago.

Dressel, P., & Barnhill, S. (1994). Reframing Gerontological Thought and Practice: The Case of Grandmothers with Daughters in Prison. *The Gerontologist, 34,* 685–691.

Dressel, P., Porterfield, J., & Barnhill, S. (1998). Mothers Behind Bars. *Corrections Today, 60,* 90–95.

Fletcher, B. R., Shaver, L. D., & Moon, D. G. (1993). *Women Prisoners: A Forgotten Population.* Westport, CT: Praeger.

Gabel, K., & Girard, K. (1995). Long-term Care Nurseries in Prisons: A Descriptive Study. In: K. Gabel, & D. Johnston (Eds), *Children of Incarcerated Parents,* (237–354). New York: Lexington.

Gabel, K., & Johnston, D. (Eds). (1995). Incarcerated parents. In: *Children of Incarcerated Parents,* (3–20). New York: Lexington.

Gabel, S. (1992). Children of Incarcerated and Criminal Parents: Adjustment, Behavior, and Prognosis. *Bulletin of the American Academy of Psychiatry and the Law, 20,* 33–45.

Gaudin, J. M., Jr., & Sutphen, R. (1993). Foster Care vs. Extended Family Care for Children of Incarcerated Mothers. *Journal of Offender Rehabilitation, 19,* 129–147.

Geen, R., Boots, S. W., & Tumlin, K. C. (1999). *The Cost of Protecting Vulnerable Children: Understanding the Complexities of Federal, State, and Local Child Welfare Spending.* Washington, DC: Urban Institute Press.

Hagan, J., & Palloni, A. (1990). The Social Reproduction of a Criminal Class in Working Class London, Circa 1950–80. *American Journal of Sociology, 96,* 265–299.

Hairston, C. F. (1991). Family Ties During Imprisonment: Important to Whom and for What? *Journal of Sociology and Social Welfare,* 87–102.

Hairston, C. F. (1998). Family Ties During Imprisonment. Do They Influence Future Criminal Activity? *Federal Probation,* 48–52.

Hostetter, E., & Jinnah, D. (1993). Research summary. Families of Adult Prisoners. Washington DC: University Press of America.

Hungerford, G. P. (1996). Caregivers of Children Whose Mothers Are Incarcerated: A Study of Kinship Placement System. *Children Today, 24,* 23–29.

Irwin, J., & Austin, J. (1997). *It's About Time: America's Imprisonment Binge.* Belmont, CA. Wadsworth.

Johnston, D. (1995). Effects of Parental Incarceration. In: K. Gabel & D. Johnston (Eds), *Children of Incarcerated Parents,* (59–88). New York. Lexington.

Johnston, D., & Carlin, M. (1996). Enduring Trauma Among Children of Criminal Offenders. *Progress: Family Systems Research and Therapy, 5,* 9–36.

Koban, L. A. (1983). Parents in Prison: A Comparative Analysis of the Effects of Incarceration on the Families of Men and Women. *Research in Law, Deviance and Social Control, 5*, 171–183.

Martin, S. L., & Cotton, N. U. (1995). Literacy Intervention for Incarcerated Women. *Corrections Today, 57*, 120–123

McGowan, B. G., & Blumenthal, K. L. (1978). *Why Punish the Children? A Study of Children of Women Prisoners.* Hackensack, NJ: National Council on Crime and Delinquency.

Minkler, M., & Roe, K. (1993). *Grandmothers as Caregivers: Raising Children of the Crack Cocaine Epidemic.* Newberry Park, CA: Sage.

Morash, M., Bynum, T. S., & Koons, B. A. (1998). Women Offenders:

Programming Needs and Promising Approaches. *National institute of Justice: Research in Brief* Washington, DC.

Moore, J., & Devitt, M. (1989). The Paradox of Deviance in Addicted Mexican American Mothers. *Gender & Society, 3*, 53–70.

National Black Child Development Institute (1989). *Who Will Care When Parents Can't?* Washington, DC.

National Institute of Justice. (1998). *Women Offenders: Programming Needs and Promising Approaches.* Washington, DC: U.S. Government Printing Office.

National Women's Law Center. (1995). *Women in Prison.* Washington, DC.

Owen, B. (1998). *In the Mix Struggle and Survival in a Women's Prison.* Albany, NY: State University of New York Press.

Phillips, S., & Bloom, B. (1998). In Whose Best Interest? The Impact of Changing Public Policy on Relatives Caring for Children with Incarcerated Parents. *Child Welfare, 77*, 531–541.

Prendergast, M L., Wellisch, J., & Falkin, G. (1995). Assessment of and services for substance-abusing women offenders in community and correctional settings. *Thye Prison Journal, 75*, 240–256.

Rowe, D., & Farrington, D. (1997). The Familial Transmission of Criminal Convictions. *Criminology, 35*, 177–201.

Sampson, R. J., & Laub, J. H. (1993). *Crime in the Making.* Cambridge, MA: Harvard University Press.

Schram, P. J. (1998). Stereotypes About Vocational Programming for Female Inmates. *Prison Journal, 78*, 244–271.

Stanton, A. M. (1980). *When Mothers Go to Jail.* Lexington: D.C. Heath.

U.S. Department of Justice. Bureau of Justice Statistics. (1971). *Sourcebook of Criminal Justice Statistics 1970* Washington, DC: U.S. Government Printing Office.

U.S. Department of Justice. Bureau of Justice Statistics. (1999a) *Sourcebook of Criminal Justice Statistics 1998.* Washington, DC: U.S. Government Printing Office.

U.S. Department of Justice. Bureau of Justice Statistics. (1994). *Women in Prison.* Washington, DC: U. S. Government Printing Office.

U.S. Department of Justice. Bureau of Justice Statistics. (1999b). *Women Offenders.* Washington, DC: U. S. Government Printing Office.

U.S. Department of Justice. Bureau of Justice Statistics. (1998). *The Women's Prison Association: Supporting Women Offenders and Their Families.* Washington, DC. U S. Government Printing Office.

Watterson, K. (1996.) *Women in Prison: Inside the Concrete Womb.* (Rev. ed). Boston: Northeastern University Press.

Zaplin, R. T., & Joyce Dougherty. (1998). Programs that Work: Mothers. In: R. T. Zaplin (Ed), *Female Offenders: Critical Perspectives and Effective Interventions,*, Ch.15. Gaithersburg, MD. Aspen.

A BALANCED AND RESTORATIVE APPROACH TO JUVENILE CRIME: PROGRAMMING FOR FAMILIES OF ADOLESCENT OFFENDERS

Stephen M. Gavazzi, Courtney Yarcheck, Deborah Wasserman and Charles Partridge

Crime is devastating to individuals and families because it upsets two fundamental assumptions: first, the belief that the world is an orderly, meaningful place and, second, the belief that we have a relatively high degree of personal autonomy and control over our lives (Zehr, 1990). Thus, crime is an event that has the potential to leave victims and their families feeling agonized and out of control. Members of these families often find themselves asking questions such as: Why did this happen to a member of my family? What could I have done to prevent this? and What will happen if someone in my family becomes a victim again?

Criminal activity can have a profound impact on the families of offenders as well. Members of these families also are faced with arduous questions, including: How could a member of my family commit such a crime? Did the

Salaries and research support for this study were provided, in part, by The Alliance for Cooperative Justice (OSURF #737264) and The State of Ohio Department of Youth Services (OSURF #738358).

The authors would like to express their deep appreciation for the thoughtful comments provided by Dr. Gordon Bazemore on a previous version of this manuscript.

Families, Crime and Criminal Justice, Volume 2, pages 381–405.

ISBN: 0-7623-0737-4

event happen because of something I did or failed to do? and Is this family member at risk of further involvement in illegal activity?

The need to formulate answers to these and related questions for families of victims and families of offenders compels us to examine how we as a society define and implement justice. Among other things, the model of justice employed by our society is thought to greatly influence the type of programming that can be done in the way of providing assistance to those families who are impacted by criminal activity. As is discussed below, a retributive model of justice has been the predominant framework for dealing with criminal behavior, and individual-focused treatment for the offender has been the prevailing form of programming. Growing utilization of alternative justice models, in conjunction with innovations in the field of family life education, have has given rise to the development of intervention methods that are more family-oriented, especially regarding treatment efforts that engage offenders and their families.

The present chapter focuses on one such family life education programming effort that targets the families of offenders. The Growing Up FAST: Families and Adolescents Surviving and Thriving™ program (Gavazzi, 1995) as developed for use with adolescent offenders and their families (Gavazzi, Rhine & Partridge, forthcoming; Gavazzi, Wasserman, Partridge & Sheridan, 2000) is used as an example of the type of family-based programming that is possible within a justice framework that addresses criminal activity less in terms of 'the laws that have been broken' (the retributive justice model) and more in terms of 'the harm that has been done' (i.e. the balanced and restorative model). A brief overview of these two approaches to juvenile justice is given, and is followed by an explication of the Growing Up FAST program's utilization with the families of juvenile offenders, with particular attention paid to identifying the harm done by the adolescent's criminal activity, what has been done already to repair that harm, and what still must be done to repair that harm.

RETRIBUTIVE JUSTICE

Much of what constitutes our current approach to crime and justice centers on retribution. Most simply put, it is a strategy for justice that centers on offenders receiving their 'just desserts.' Within this approach, punishment (the retributive act of society at large) is administered according to the severity of the offense in combination with the offender's prior involvement in criminal activity. Thus, justice is created through the sentencing procedure, and the subsequent punishment (incarceration, fines, etc.) is meted out in order to force the offender to be held accountable for his/her actions.

Within the retributive model, criminal activity is thought to be controlled through the intimidation of punishment. Further, recidivism is thought to be inhibited by the threat of even greater punitive actions within this model. Interestingly, one outgrowth of this retributive approach to justice has been a series of legislative acts at the state and national level that focus on a 'get tough' stance with regard to juvenile crime. While this 'get tough' posture may play well to certain segments of the public, its effects have been controversial. Among other things, Andrews & Bonta (1994) have noted that sanctioning without any sort of rehabilitative effort leads to increase recidivism rates. Also, Bishop, Frazier, Lanza-Kaduce & Winner (1996) have reported that the transfer of juvenile cases to adult courts, another part of the 'get tough' approach, is linked to increased recidivism as well. Such transfers, often resulting in placement within an adult correctional setting (Schwartz, 1989), also are reported to lead to violent offenders to become even more violent (Mattingly, 1996a, b).

In addition to a lack of supportive evidence for the 'get tough' approach, the retributive model has been challenged on philosophical and ethical grounds as well. For instance, while the offender may be seen as being held culpable in a retributive model, there is no emphasis on the offender being held responsible for their actions (Zehr, 1990), including most importantly how the offense has had a profound negative effect on both the victim and community (Clear & Corbett, 1998). As a result, victims and communities are thought to gain little or nothing in the process of justice being served within the retributive model. While victims and their families increasingly are allowed to speak in court prior to sentencing, for instance, these voices are rarely considered in the actual sanctioning process. By the same token, communities damaged by crime have little or no say in sanctioning those offenders that more than likely will return to their neighborhood at some point in the future. In this way, crime becomes a symptom of community disharmony (Bazemore, 1996) and in turn creates a situation of animosity among communities, victims and offenders that more than likely increased the probability of further criminal activity and increased violence.

BALANCED AND RESTORATIVE JUSTICE

An innovative approach known as the Balanced and Restorative Justice (BARJ) model more recently has been advanced as an alternative framework for dealing with criminal behavior. The basic premise for this model is that crime is not merely about lawbreaking, but rather concerns the harm done by the crime (Zehr, 1990; Van Ness, 1997). The BARJ model stems from the intersection of two newer approaches to juvenile delinquency: the "Balanced Approach" and the "Restorative Justice" movement (Bazemore, 1996). The "Balanced Approach" also is known as the "third alternative" in juvenile justice, and as such represents an attempt to capture a middle ground between proponents of harsh punishment and those who would advocate for more lenient and prevention-oriented responses to juvenile crime (Bazemore, 1997). As is discussed below, a balanced approach attempts to design responses that apportion equal (i.e. "balanced") resources to offender accountability, community safety, and competency development. In turn, the "Restorative Justice" movement emphasizes that the most effective responses to criminal activity focus on healing the harm done by the crime. What is "restored" in this approach are the losses suffered by the victims of criminal activity, as well as the losses suffered by the community at large (Bazemore, 1997). Together, such balanced and restorative efforts adopt an approach that administers to the combined needs of victims, offenders, and communities (Bazemore, 1997; Bazemore & Umbreit, 1995).

The practice of justice within the BARJ model occurs under the umbrella of three goals set forth by juvenile justice systems: accountability, community safety, and competency development. Accountability concerns the ability of offenders to truly make amends for their actions. Accountability is thought to be achieved by the offender taking responsibility for their crimes through an understanding of the impact that it has had on those around them (Bazemore & Day, 1996; Clear & Corbett, 1998). The offender is presumed to be capable of taking steps to make amends and restore losses only after they have internalized their responsibility in the offense and its repercussions (Bazemore, 1997; Clear & Corbett, 1998). Paramount to this process is the input of victims and communities in setting tolerance limits, expressing disapproval of the criminal activity, and providing appropriate consequences for the harm generated by the unlawful behavior (Bazemore, 1997; Clear & Corbett, 1998). Promising practices in the achievement of accountability within the BARJ model include such interventions as victim-offender mediation, restitution, and community service orders.

Community safety and competency development focus on the offender's ability to engage in productive activities within their immediate social context. Offenders are more likely to make a positive contribution to the community when they are equipped with marketable work skills. Also, sustained contact with adults may further increase the chances of an offender becoming successfully reintegrated into society. When competency is defined as the capacity to do something that other residents of the social context find valuable, the standard for achieving success in competency development becomes measured within the community that the individual resides (Bazemore, 1997). Rehabilitation needs are thought to be best met when young offenders make measurable and demonstrated improvements in educational and vocational competency that improve their ability to function as productive and capable adults (Bazemore, 1997). Hence, the BARJ model focuses on efforts such as career-building opportunities and mentor programs that build prosocial skills and social attachments to the community in which the offender resides (Bazemore, 1996). Of course, these social attachments are thought to include family members as well. Hence, family strengthening efforts also have been articulated to be part of competency development efforts within the BARJ model.

Family-Based Work Within the BARJ Model

A concentration on family factors seems to hold particular promise in efforts targeting crime committed by juveniles, since family factors have been shown to have a profound influence both on the development of adolescent delinquent behavior (Gartland & Day, 1991; Larzelere & Patterson, 1990; LeBlanc, 1992; Patterson, Reid & Dishion, 1992; Rankin & Kern, 1994; Simons, Robertson & Downs, 1989). In related fashion, the success rates of treatment efforts targeting the families of juvenile offenders has been impressive (Alexander & Parsons, 1973; Bank, Marlowe, Reid, Patterson & Weinrott, 1991; Barton, Alexander, Waldron, Turner & Warburton, 1985; Gordon, Arbuthnot, Gustafson & McGreen, 1988; Gordon, Graves & Arbuthnot, 1995; Henggeler, Borduin, Melton, Mann, Smith, Hall, Cone & Fucci, 1991; Henggeler, 1997; Quinn, Sutphen, Michaels & Gale, 1994; Klein, Alexander & Parsons, 1977).

At the same time, however, the retributive focus that dominates the modern juvenile justice landscape has fostered a decidedly individual-based orientation to rehabilitative efforts. Most simply put, the individual offender is culpable for his/her actions, and thus the individual is in need of rehabilitative treatment. Bazemore (1996) has argued that the retributive treatment model is both insular and unidimensional because of its restricted focus on the offender and subsequent failure to take into account the diverse interests of victims and communities. Programs within the retributive model often ask little of the offender beyond their participation in individual counseling sessions, remedial services, and/or recreational programs, all activities that encourage the offender to remain self-centered (Bazemore, 1996).

It is thought that the emergence of the BARJ model has created an opportunity for family-based intervention specialists to articulate the connection between their efforts and an approach to juvenile justice that is embedded within the criminology literature. Here, a focus on the needs of the victim becomes expanded to include their family members – both as a resource in the healing of the harm done, and as victims in need of healing efforts as well. Likewise, the focus on the community's needs are further clarified when the indeterminate collection of societal members is reframed as a specific collection of families that have certain identifiable requirements for healing. Finally, an expansion of the focus on the needs of offenders to include the offender's family brings about a similar realization – that the offender's family also may have been a victim of the criminal activity, as well as a potential resource in healing efforts.

Although many family-based programs have been developed as a response to juvenile delinquency [see for instance Kumpfer's (1994) review of over 500 family strengthening initiatives], few have claimed a specific grounding within the BARJ model. One important exception in this regard is the Family Group Conferencing (FGC) model. Originating in Australia and New Zealand, the FGC model attempts to utilize components of an aboriginal value system that emphasizes the gathering of all parties affected by a criminal act in order to gain both an accounting of the harm done and what must be done to repair that harm (Umbreit, 1997).

Juvenile Delinquency and Family Life Education

Since its inception, family life education has sought to address juvenile delinquency. With increasing sophistication of family research, especially in the areas of resiliency and strength, family life educators have recognized the need to upgrade programming by integrating new theory with practice (Arcus, 1995; Hughes, 1994). Based on that need, Arcus (1995) identified four missing components to family life education: (1) family scholarship as a theoretical base to programming; (2) integration of family systems theory into the content of parent education; (3) integration of relevant scholarly advances from other fields; and (4) relief from the underlying value assumptions associated with pre-formed beliefs about the 'normal' family, e.g. two-parent and nuclear.

More specific to prevention programming (including those programs targeting delinquency reduction), Bogenschneider (1996) identified two major approaches that had been used to date: the reducing risk approach (founded largely through an adoption of a medical model), and alternatively a model that involved the building up of protective factors. An integrated third 'ecological risk/protective model' was advocated by Bogenschneider (1996) in order to facilitate an approach to delinquency prevention that combined efforts to reduce risk and build protective factors. The proposed integration of the two major approaches to prevention programming was centered on the identification of these risk and protective factors in the ever-widening social circles of influence as suggested by Bronfenbrenner's (1979) ecological theory of human development.

As described below in more detail, the Growing Up FAST Diversion program emerged in response to the challenges set forth by the family life education discipline. In addition to its theoretical grounding and research base, the program attends to both protective and risk factors at personal, family and community levels; it borrows significantly from scholarly advances in other fields, particularly that BARJ in the criminal justice field; it recognizes that one small change in the family system can affect major change in individual behavior; and it is responsive to the idiosyncratic values of families.

THE GROWING UP FAST PROGRAM

The initiative known as the Growing Up FAST: Families and Adolescents Surviving and Thriving™ Program (Gavazzi, 1991) was developed originally as a primary prevention offering. Implementation of the Growing Up FAST program was conducted as a response to the need for families and communities to take responsibility for the creation of transition markers for their adolescents as they proceeded toward adulthood (Blumenkrantz and Gavazzi, 1993; Gavazzi & Blumenkrantz, 1993).

The original program was based on a combination of literatures concerned with rites of passage, multicultural issues, and solution-focused work (Gavazzi, 1995). In large part, the program's central objective involved the creation of a definition of successful adulthood that articulated a family's ideal response to the typical developmental demands of adolescence and family life (Law & Gavazzi, 1998). The program was designed as a brief one-session family intervention (Gavazzi, 1995), although subsequent program evaluation efforts (Wasserman, Gavazzi & Randall, 1999) prompted the expansion of the family strengthening program into a two-session format. Table 1 contains demographic information about the families of adolescents who have participated in formative evaluation studies of the Growing Up FAST program to date (Law & Gavazzi, 1998; Wasserman et al., 1998).

The first author's increased involvement with the juvenile justice community prompted a second and more comprehensive version of this initiative to be developed as a response to the need for family-based diversion programs (Gavazzi et al., 2000), as well as parole programming options that were oriented toward family involvement (Gavazzi et al., forthcoming). These versions of the Growing Up FAST program are designed to be used with individual families, with the requirement that the adolescent offender, typically age 12 through 16, be accompanied by at least one adult committed to both the young person's well-being and to a parent-like role.

Table 1. Demographic Information and Study Outcomes Regarding the Standard, Diversion, and Parole Versions of the Growing Up FAST Program

	Standard	Diversion	Parole
Target Audience	Any adolescent aged 12–18 and (at least) one significant adult	Adolescents aged 12–18 who have engaged in illegal activity within the last six months, and (at least) one significant adult	Adolescent offenders aged 12–18 released from state juvenile correctional institutions, are on parole, and (at least) one significant adult
Number of Study Participants to Date	53	103	8
Gender	27 males/26 females	52 males/51 females	7 males/1 female
Ethnicity	47 Caucasian 6 Minority	77 CAucasian 26 Minority	1 Caucasian 6 African-American
Mean Annual Income	$55,644	$40,000	$40,000
Mean Age	13.9 years	14.6 years	16.9 years
Family Structure	20 intact 6 stepfamilies 25 single parent 2 other	20 intact 17 stepfamilies 57 single parent 9 other	1 intact 4 single parent 3 other
Study Outcomes to Date	Program satisfaction; sense of accomplishment Thematic analysis of family definitions	Risk and protective factor identification (Level 1 output)	Increased contact between family members and parole officres
Study Citations	Wasserman et al., (1998) Law & Gavazzi, (1998)	Gavazzi et al., (2000)	Gavazzi et al., (forthcoming)

A Family-Based Diversion Program

The family-based diversion program is designed as a linked series of five levels, with each level providing a set of opportunities for the family to achieve some modicum of accomplishment. Success may be experienced at and is measurable upon the completion of each level. Each level builds on the work accomplished in the previous level such that the completion of each successive level is thought to indicate greater overall success. This version of the Growing Up FAST program has been developed and evaluated according to a logic-based evaluation model (Julian, Jones & Deyo, 1995) that organizes program efforts into levels of activities, outputs, and outcomes. Here, program activities (what a program facilitator actually does in a session) are thought to be causally linked to initial, intermediate and longer-term outcomes (what a program participant experiences as a benefit or change that logically relates to the program's purpose) through the generation of program outputs (defined as quantifiable products of program activities).

Growing Up FAST program facilitators engage families in a series of activities that lead to outputs for each program level: a chart of risk and protective factors impacting the family of the offender (level 1); the family's definition of successful adulthood (level 2); application of decision-making steps that prevent recidivistic activity (level 3); repairing the harm done by criminal behavior (level 4); and the mapping and utilization of community resources (level 5). In turn, these program activities and outputs at each level are thought to lead to clearly defined initial outcomes (increased adolescent and adult family member awareness of goals and resources related to the achievement of successful adulthood status); intermediate outcomes (increased family functioning and decreased adolescent risk-taking behavior), and the longer-term outcome of reduced recidivism (Gavazzi et al., 2000). An illustration of the program's activities, outputs, and outcomes is contained in Figure 1.

At the time of writing this chapter, 103 adolescents and their families have participated in this family-based diversion program. Of these adolescents, forty-eight (46.6%) committed delinquent acts that came to the attention of the juvenile justice system, while the remaining fifty-five (53.4%) adolescents committed delinquent acts that did not result in an arrest. Of those adolescents who came to the attention of the juvenile justice system, eighteen (37.5%) were categorized as unruly, twenty-three (47.9%) were charged with a misdemeanant offense, and five (10.4%) were charged with a felony offense. Four (8.3%) of the adolescents who came to the attention of the juvenile justice system used a weapon in connection with their illegal behavior. Thirteen (27%) of these adolescents had at least one prior offense that came to the attention of the authorities. Additionally, sixty-one (59.2%) of these families reported that another member of their immediate or extended family (apart from the adolescent) had experienced a formal arrest. Table 1 contains additional demographic information about the families of these adolescents.

The diversion program currently is thought to be at a formative stage of development. Data gathered on program fit with the needs of its target

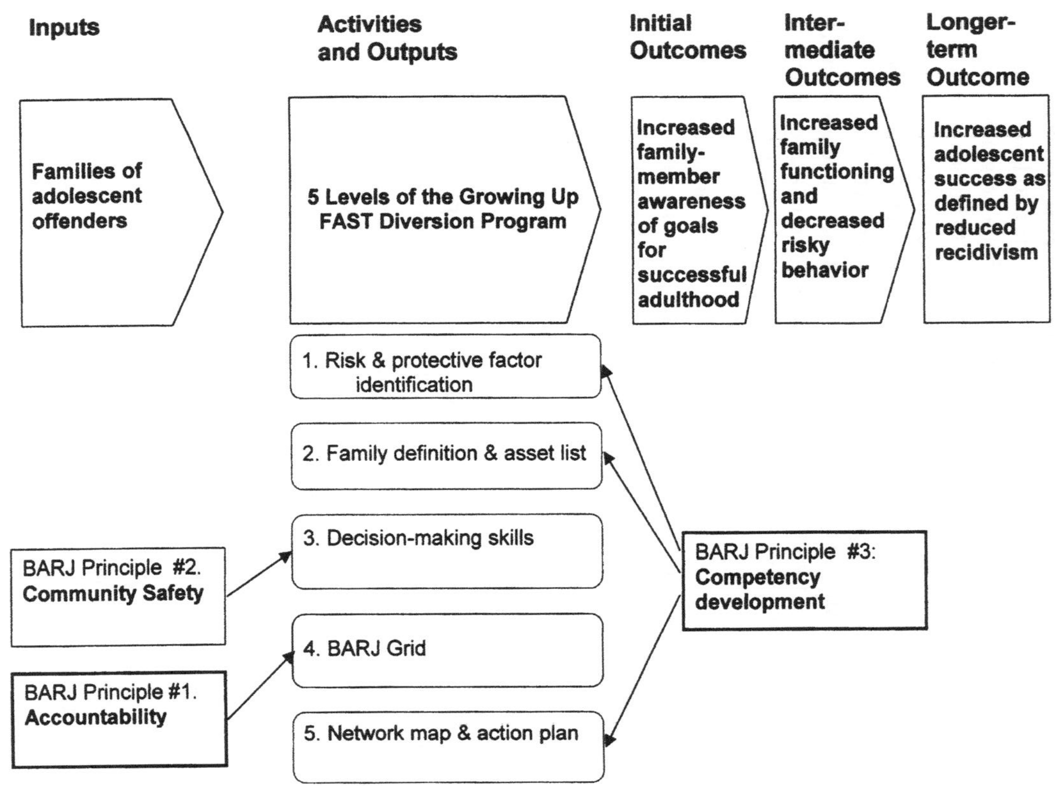

Figure 1. Inputs, Activities, Outputs, and Outcomes of the Growing Up FAST Program as Related to the Three Principles of the Balancd and Restorative Justice Model.

population and the program's saturation of this population currently are being utilized in an ongoing process of program modification (Dumka, Roosa, Michaels & Suh, 1995; Jacobs, 1988). At the same time, some preliminary evidence of the program's effectiveness has been reported (Gavazzi et al., 2000). In particular, initial support for the program's design and its beneficial impact on families has been generated from a focus on the content of family members' responses to the program's activities, including most specifically the construction of program outputs. For instance, the analysis of a pilot group's responses to Level 1 activities (identification of risk and protective factors) yielded information that supported this level's goal of educating families about the impact of their family history on risk-taking behavior (Gavazzi et al., 2000).

This type of formative examination of program effectiveness (Worthen, Sanders & Fitzpatrick, 1997) is consistent with evaluation of other family strengthening initiatives (Hughes, 1994). Data generated from the utilization of a battery of qualitative and quantitative measures will be forthcoming as the six month and one-year post-test evaluations are completed for the first wave of families who have participated in the diversion program. While these empirical activities are designed to document the Growing Up FAST program's impact on problem behavior reduction, efforts also are underway to document the sustainability of this initiative within the community it presently serves (Andrews & Hickman, 1998).

A Family-Based Parole Program

Where the diversion program seeks to reduce recidivism rates through the twin objectives of increasing family functioning and decreasing adolescent risk-taking behavior, the implementation of Growing Up FAST as a parole program is built around the objective of reducing recidivism through activities that directly enhance the relationship between the parole officer and the offender's family. This version of the program was designed to create small but significant successes in a variety of life domains for the youth and his/her family that are reported to the parole officer as part of their regular contact with one another. Through the experience of such modest gains and their subsequently being reported to the parole officer, expectations for further and greater success (including the successful discharge from parole and the prevention of further illegal activity) become fashioned and realized (Gavazzi et al., forthcoming).

Thus, the Growing Up FAST parole program is thought to augment the relationship between the parole officer and the offender's family in such a way as to reduce the likelihood of future reoffending. This approach is thought to be consistent with Rhine, Neff & Natalucci-Persichetti's (1998) call for a new intervention paradigm regarding the supervision of youthful offenders in the community. Within this paradigm, the parole officer's work necessarily must be extended beyond case plan enforcement to include the development of relationships between the offender and those adults, including members of the offender's family, that remain as beneficial components of their lives long after the parole process has ended (Rhine et al., 1998).

The parole programming option was initiated six months prior to this article's writing, and as such is at a much earlier stage of development. To date, 8 adolescents (7 males and 1 female) and their families have been referred to and have initiated their participation in this program following incarceration in a juvenile correctional institution. All 8 adolescents had committed felony delinquent acts that came to the attention of the juvenile justice system, including: serious property offenses (4), drug trafficking (1), carrying a concealed weapon (1), aggravated burglary (1) and a sex offense (1). Four of the adolescents who came to the attention of the juvenile justice system had used a firearm in connection with their most recent illegal behavior, and all of these adolescents had at least one prior offense that came to the attention of the authorities. Additionally, all but one of these families reported that another member of their immediate or extended family (apart from the adolescent) had experienced a formal arrest. Table 1 contains additional demographic details about the families of these adolescents.

Data from these families are being generated from a battery of qualitative and quantitative measures in a program evaluation design that is identical in scope to that of the diversion program evaluation effort. Similar to the diversion program implementation and evaluation project described above, the parole program currently is thought to be at a formative stage of development. Some preliminary evidence of the program's effectiveness has been reported, however, including program effective as evidenced by the increased amount of contact between family members and parole officers generated as the result of family member participation in the Growing Up FAST program (Gavazzi et al., forthcoming).

A BALANCED AND RESTORATIVE APPROACH TO FAMILY PROGRAMMING

The Growing Up FAST program is thought to touch on all three realms of the BARJ model. Certain program activities focus on competency development, including the increased understanding of those risk and protective factors that impact the adolescent and her/his family, the development of the family's definition of successful adulthood, and the development of an action plan for post-program activities that connects the adolescent and family to their social network. Other program activities focus on accountability, specifying what victims and community members need in order to repair harm done by the criminal activity, as well as identifying what actually has been done to date that has met those needs. Finally, still other program activities focus on community safety, including the development of decision-making skills that can be applied to the prevention of future illegal behavior. The connections between each program level and the principles of the BARJ model also are illustrated in Figure 1.

In order to further depict how BARJ principles are embedded in the Growing Up FAST program, attention now turns to a more specific description of the program's fourth level. This level consists of four steps: (1) an introduction to the concept of restorative justice and how it relates to what successful adults do when they make poor decisions; (2) design and completion of a restorative justice grid (illustrated below); (3) extending the family definition to include a component related to making amends for mistakes; and (4) gaining closure in the family concerning discussion of the illegal activity. The steps of Level 4 are summarized in Table 2.

Level 4 Introduction

The introduction to the fourth level of this program serves as an important solution-focused bridge to the activity of making amends, an activity that can be emotionally difficult and challenging even for high-functioning adults. Thus, the introduction to this program level focuses on the following three points:

- All adults make mistakes, and understanding these mistakes and their consequences is a hallmark of the learning process.
- One difference between successful adults and those who are less successful can be defined in terms of what they do when they discover they have made a mistake or a poor decision.
- One way to consider the mistakes made regarding involvement in illegal activity that leads to the greatest success is through consideration of the harm done to various relationships by the criminal act itself.

Hence, the program facilitator introduces certain concepts related to the BARJ model at the beginning of this program level. Most importantly, this includes information that focuses family members on how the process of healing relationships (i.e. repairing the harm done by the criminal activity) involves consideration of harm that may have occurred not only to the victim, but also to one's community, to one's family, and to oneself.

Table 2. Key Steps to Take in Level 4: Creating and using a BARJ Grid

1. Preparation: In Level 3, work is done to gain an understanding of the specific choices and decisions made at the time of the illegal behavior.
2. Introduction of the task: Discussion of how to reach closure about the illegal activity by:
   (1) identifying and repairing the harm,
   (2) committing to a plan for making amends, and
   (3) answering all remaining questions about the illegal activity.
3. Introduction of the 3 BARJ principles (accountability, community safety and competency development): Discussion of how the process of healing relationships depends on repairing the harm done to victims, community, and the offender.
4. Introduction of the BARJ grid with the following question:
   "Do you feel the harm caused by the illegal/incorrigible behavior has been repaired?" and the explanation that the grid will help answer the question.
5. Consideration of who is to be named in each of the four categories of the BARJ grid: victim, family, community, and offender (column 1).
6. Consideration of the harm that was done (column 2).
7. Consideration of what has been done so far to make amends (column 3) and what else might need to be done (column 4).
8. Prioritization (including the creation of a timeline) for the actions listed in column 4.
9. Agreement to gain closure in exchange for answers to all remaining questions.
   1. The program facilitator begins by asking the adult if all their questions about what happened that day are answered. If not, the adults are asked about what other (if any) questions remain.
   2. If there are additional questions, they are asked and answered If family members still cannot gain closure on that incident and move on to the future, they are asked to identify what might be getting in the way.
   3. In those cases where families cannot reach agreement about closure, families are assisted in making a plan for what will happen next, including referrals for other services.
10. Progression to Level 5 work

The Restorative Justice Grid

The program facilitator then proceeds to assist the family in the placement of information in the appropriate spaces of the restorative justice grid (see Figure 2). The grid has been constructed to help families methodically consider each person (or groups of people) who have been affected by the crime, what harm has been done, what has already been done to repair that harm, and what remains to be done in terms of reparative work.

	Who is involved?	What harm was done?	What has been done to make amends thus far?	What more could be done?
Victim				
Community				
Family				
Offender				

Figure 2. The restorative justice grid.

The first step is to define the who and the what of each category of people that may have been harmed. This process begins with the identification of the victim – usually well established by this point – and the specific harm that was done, including its nature and extent (physical, emotional, property, etc.). Also, there may be other victims not considered as such by the official record. For instance, in the case of a physical fight, there may have been a third party harmed by the assault who had left the scene prior to the situation being discovered by authorities; or in the case of a theft, stolen electronic equipment may have included an item that the victim had borrowed from a friend.

Prompts such as "Was there any one else affected by what you did?", along with seeing the grid posted on the wall, assist the adolescent in a consideration of what additional tangential effects their action may have had on people beyond the identified victim. After shoplifting, for instance, the adolescent may consider what happens to a store owner's attitude toward teenagers in general, or after an incident of vandalism, the adolescent might consider the public's limited access to amenities because they were destroyed. In this way, family members are invited to consider what harm has been done to specific community members, as well as the community at large. Also, as a vital part of this community, the family itself addresses what harm has been done to relationships among friends, employers, and acquaintances, as well as within the family itself.

Utilizing the grid, the family lists the relationship and the nature of the breech of trust or other harm that may have occurred. It also is important for the adolescent to acknowledge what harm has been done to self. For instance, adolescents are prompted to answer questions such as "What has happened to your image of yourself?" "What level of trust do you have in yourself?" It is important to remember that the entire process of what may appear to be an act of 'dredging up guilt' is done in the context of a successful individual working to 'make better' what he or she already has come to see as poorly-formed decisions. Hence, the process of identifying harm done is designed to help the adolescent feel empowered rather than chastised. Given the problems inherent to maintaining a solution-focused approach within problem-ridden situations (Gavazzi, 1995), a cautionary note is needed here. As with any other tool, this approach has the potential to be abused by family members and/or program facilitators. Thus, a tremendous amount of sensitivity to family member comfort level and adherence to a solution-focused framework must be monitored and maintained throughout this level.

Following completion of the 'who is involved' and 'what harm has been done' columns, the family works to identify what (if anything) already has been done to repair each harm that has been listed on the grid. Here, any and all actions taken by the adolescent to redress the harm done by the criminal activity are noted on the grid. This might include formal restitution work that had been completed as a requirement of sentencing, for instance, or more basically might include the fact that an apology has been made.

Finally, family members develop ideas concerning what, if anything, might still need to be done in order repair harm. Further actions such as seeking forgiveness, monetary compensation, and other restitution efforts are considered. Of course, in the case of the victim, victim's rights are respected and proxies may be established for receiving the benefits of what the adolescent may want to do in the victim's benefit. One of the important steps in this process is for the adolescent to acknowledge what they have done to repair harm that they have done to themselves, i.e. responding to the question 'What does it take to forgive oneself?'

Extending the Definition of Successful Adulthood

In Level 2, family members were asked to create a definition of successful adulthood according to the various roles, responsibilities, and behaviors that those family members believed were components of what it meant to be a successful adult in their family. Typically, these definitions include such components as 'getting an education', 'being able to support yourself financially', 'being a good family member', and other common themes of this developmental stage (Law & Gavazzi, 1999). Additionally, the program facilitator helps the family identify those behaviors on the part of the adolescent and parents that are already in existence that support the adolescent achieving the various components of the family's definition of successful adulthood (Gavazzi & Law, 1997).

The next step is for the family to review their definition of successful adulthood in light of what has occurred so far in Level 4. More specifically, family members are asked to explore whether or not it is important to include some reference to what has been examined in this program level; i.e. what successful adults do when they find they have made a mistake or a poor decision. Similar to Level 2, the prompt for family members to consider this potential addition is given no value beyond its consideration. Family members are reminded that there are no right or wrong ways to construct and edit their definitions. At the same time, however, family members are reminded that they must achieve consensus regarding the addition of another component to their definition of successful adulthood, an important consideration that is introduced initially in Level 2.

> The use of a consensus-based definition of successful adulthood is reflective of the solution-focused basis (Walter & Peller, 1992) of the program. Since the program utilizes only those definitional components that all family members agree upon, potentially problem-focused issues (e.g. conflict over definitional components that all family members did not agree on) are avoided. This is not to say that those issues are not important; however, they have no place in the structure of this particular program. The goal of reaching consensus thus presents families with the opportunity to work with those issues that they all agree upon. In essence, this exercise serves to highlight an existing strength of the family; namely, that there are certain issues that everyone agrees deserve attention by all family members (Law & Gavazzi, 1999, p. 323).

In the event that all family members agree that a new component regarding 'making amends' should be added to their definition of successful adulthood, the program facilitator then assists the family in identifying behaviors that are supportive of this component. Family members might not reach consensus about such an addition, however, or might believe that such a notion already has been expressed in an existing component of the family's definition of successful adulthood. Following a brief exploration of this possible addition to the family definition of successful adulthood, the program facilitator summarizes the work done within this level as preparation for the concluding step of this program level.

Closure and Moving On

The final step of Level 4 is for the family to consider the possibility of reaching a sense of closure regarding the illegal activity committed by the adolescent. Family members are asked the specific question, "Do you know enough about what happened to put this activity in the past and not feel the need to talk about it ever again?" If the family can reach consensus on this question, in effect coming to an agreement that everything about this illegal activity has been understood, thought through, and that the harm has been identified and repaired to the best possible extent (or there is at least a strong commitment to a plan for doing so), closure is usually possible and achieved.

If the family is unable to reach consensus, then family members are asked to explain what they think may be preventing them from agreeing to put this event in their collective past. Sometimes a final apology or an answer to one last question allows the family to move on. In other cases, unresolved issues about the illegal activity may produce the sense that some other equally unacceptable behavior will occur in the future, despite all of the work done to this point in the program. In our experience, these unresolved issues typically have involved concerns related to psychological problems, including substance abuse, depression, and symptoms of ADHD, or longstanding relationship concerns (i.e. parent-adolescent conflict that had resulted in physical violence) that require clinical attention. In these cases, the inability to answer affirmatively to the question of closure provides the family with opportunity to make a plan for addressing the unresolved issue through the initiation of therapy services.

Case Example of Level 4 Implementation

Heather, a 16 year old adolescent, and Mrs. Q., her mother, were referred to the Growing Up FAST Diversion Program after Heather had been arrested for shoplifting $250 worth of CD's and jewelry from a department store. In Level 1 of the Growing Up FAST program, Mrs. Q. reported a number of significant family-related risk factors that were currently impacting Heather, including the fact that her violent alcoholic father had died from liver disease exactly three years prior to the shoplifting incident. The mother also reported significant peer-related risk factors related to Heather's involvement with a set of anti-social friends. Protective factors reported by Heather and her mother in Level 1 included the helpful involvement of a maternal aunt in the life of Heather, as well as Heather's current participation in a class production of a musical.

In Level 2 of the program, the family had conjointly defined successful adulthood as: 'Being responsible for yourself financially'; 'Doing things on your own and in your own way and making good choices'; 'Knowing when you need help and knowing where to get it from'; and 'Being someone people can depend on'. In Level 3, with this definition as a foundation, Heather and her mother had come to understand the decision to shoplift as a misguided attempt to 'do things on her own and in her own way to help her feel good about herself'. Later in this level, Mrs. Q agreed to Heather's request (prompted by the program facilitator) that they add 'Being someone who feels good about herself' to the family definition of successful adulthood.

Heather was prompted to develop alternative options for 'feeling good about herself' that would come with less cost and create greater benefit (than choosing to involve herself in illegal activities such as shoplifting). Of the many alternatives that she came up with, Heather subsequently chose to focus on one short-term option and one longer-term option for dealing with similar situations in the future. The short term option (what Heather claimed she could have chosen to do in the store that day and thus what she could commit to choosing if she should find herself in a similar situation in the future), was to try on new clothes with the immediate goal of 'feeling good about what she saw in the mirror' instead of focusing on the immediate possession of those clothes. Instead, Heather linked actual clothing purchases to her longer term option, which was her getting a job that would allow her to have her own money to spend on clothes that she desired to own.

While Heather and her mother both expressed a sense of relief regarding their ability to discuss the illegal activity in this open and honest manner, Mrs. Q. had some questions remaining about Heather's continued involvement with her current friends, as well as how simply 'looking in the mirror' would really help Heather 'feel good' about herself and prevent her from shoplifting again. The facilitator noted that Mrs. Q's questions would be useful focal points for the ensuing work that was going to occur, because the goal of Level 4 was to see if it was possible for their family to reach closure on Heather's past involvement in illegal activity.

Following a brief review of the work that had been done in the first three levels of the program, the facilitator introduced the main points of Level 4 (everyone makes mistakes; the most successful adults use their mistakes to learn and grow as individuals; consideration of the harm done to relationships by these mistakes helps to heal the harm done and facilitate the growth process of the successful adult) and the restorative justice grid. The program facilitator noted that Heather and her mother were being invited into a process of looking at relationship harm within the same framework of strength and shared responsibility that they had been using throughout the program. The facilitator stressed that the family was not being asked to take part in a 'blame game', whereby Heather and her mother would be at risk of engaging in unproductive discussions that focused on accusations and condemnations.

Using the grid, Heather first identified the store owners as victims because their goods were taken. The facilitator wrote 'store owners' on the grid, and then asked who else may have been harmed. Heather immediately named her Mom, and discussed the embarrassment and shame she saw in her mother's eyes when she came home from the police station. Heather also identified her younger sister, Angie, who now had a 'negative role model'. Finally, Heather also named her Aunt Sarah, who came to get her when the police called. She sheepishly remembered how disappointed her aunt seemed to be in her, especially because she knew the store owners and one of the police officers.

When the facilitator asked Heather about harm to the community, Heather thought about her friends and how she may have reinforced for them the idea that shoplifting was an option, perhaps even a "cool" thing to do. Finally, Heather expressed surprise when the facilitator asked her to consider the harm done to herself as the offender. She was able to quickly identify that she had harmed herself by having a court record. With some prompting by her mother, Heather also was able to describe the damage that she had done to her self image, now thinking of herself as someone who couldn't be trusted. All of these ideas were added to the appropriate boxes of the grid.

As the next part of this level, the facilitator began to direct the family's attention to what Heather already had done to repair the harm that they had identified, as well as and what remained to be done in the way of restoration. In the case of the store, Heather noted that she had already returned the goods and had given the store owners the amount of money that the clothing had cost. Turning attention to the harm done to her relationship with her mother, both Heather and Mrs. Q. stated that they felt that participation in the program had helped them to complete this part of the healing process. Because the program had been the first time that she had been prompted to think about the harm done to her sister and aunt, Heather stated that she couldn't think of anything that she had done already to repair the harm. In her planning for what she could do in the future, however, she added, "tell my aunt and my sister that I am sorry" and "get a job to show them that I am working on staying out of trouble." She also thought it would be helpful to share with her aunt and her sister what she had found out about herself through the program.

As for her friends, Heather thought that her making a verbal commitment to not shoplift ever again would send a helpful message to them. Heather also thought that her getting a job would help her to become a good role model for them as well. For herself, Heather noted that her participation in the Growing Up FAST program already had helped her to repair her self image, but she added three additional commitments to her plan: (1) to apologize to herself, (2) to forgive herself, and (3) to find herself a therapist to help with some recurrent bouts of hopelessness she had been feeling lately. Mom also made the commitment to helping Heather contact and begin treatment with one of the three mental health providers that the program facilitator already has given them referrals to at the beginning of the program.

When the facilitator asked Heather to review her plan (i.e. the fourth column of the grid) to see if she could make a commitment to it, Heather felt she could commit to all but sharing her feelings and decisions with her Aunt. The facilitator used that resistance as a way to reinforce the decision to enter therapy; i.e. to underscore how helpful it would be for Heather to have the opportunity to come to understand and manage her feelings better before she chose to share them. Heather committed to completing all other steps in her plan. Finally, the facilitator turned to Mrs. Q. and asked if she had any other lingering questions about this shoplifting incident. As she indicated none, the facilitator asked Mrs. Q. if she felt that she could put the incident in the past and not talk about it again. Mrs. Q. responded that to do so would not only be easy but would in fact make it easier for Heather to achieve the plan that she had constructed for herself.

DISCUSSION

This chapter has discussed the Balanced and Restorative Justice model as an alternative method for dealing with criminal behavior that takes its adherents beyond the more traditional model of retributive justice. In turn, this chapter has emphasized that the BARJ model provides a theoretical framework that is supportive of intervention efforts that focus on families as the unit of analysis and treatment. BARJ principles within a specific intervention effort known as the Growing Up FAST: Families and Adolescents Surviving and Thriving™ program were explicated, and a thorough description of the program's fourth level, including a case example, were used to demonstrate how these principles are utilized in actual work with the families of offenders.

The focus on the Growing Up FAST program's fourth level was used to underscore the assertion that families of offenders stand to benefit from efforts to identify the victims who have been harmed by criminal activity and the nature of that harm, as well as efforts to identify what has been done and remains to done in terms of repairing that harm. Most importantly, this program focuses attention on the harm done to various relationships, including of course what has been damaged inside of the family itself.

It must be remembered, however, that such work occurs within the much more extensive framework of a program that contains five comprehensive levels of participation, and as such the program is thought to be implemented most successfully and evaluated most fairly when based on the entire program package. Nonetheless, family-based professionals may find the utilization of the grid a valuable tool in their work with the families of offenders. Other levels of the Growing Up FAST program have been discussed as having similar potential, including most notably what occurs in Level 2 regarding the creation of family definitions of successful adulthood (Gavazzi & Law, 1997; Law & Gavazzi, 1999).

This chapter has underscored the Growing Up FAST program's close adherence to core principles of the Balanced and Restorative Justice model. Gavazzi et al. (forthcoming) have discussed this concentration on discussing relationship harm within a family context as bearing close similarity to the BARJ literature's utilization of a concept known as 'relational rehabilitation.' Bazemore (1999) notes that one core assumption about relational rehabilitation within the BARJ model "is that sanctioning processes are more likely to enhance rehabilitative/reintegration processes when they involve family, victims, and key members of the offender's community" (pp. 168–169). We assert here that utilization of the relational rehabilitation concept within a family-based context would seem to extend itself well beyond the confines of a single program such as Growing Up FAST. In essence, we are encouraging the developers of other family-based intervention efforts to explore the potential benefits of an increased familiarity with the BARJ literature, at least in this specific regard.

More generally, however, the BARJ literature also might help these family-based program developers articulate a clear connection to the criminal justice literature. Unfortunately, one element missing in much of the literature on these family-based programs to date has been the development of a specific linkage to a existent justice model. As such, written material about many family-based treatment models often does not contain information about how such efforts compare and contrast with the treatment modalities that have been generated from a more retributive model. In turn, the BARJ model is seen as providing some necessary conceptual linkages that would allow criminal justice professionals to become more familiar with such family-based work.

At the same time, the BARJ model is not the 'be-all' and 'end-all' of criminal justice, nor does it offer iron-clad assurance of the adoption of a family-based orientation to treatment. A variety of initiatives that have adopted a BARJ approach have been criticized on many grounds, including its use by program developers who target only low-risk offenders, as well as the overall lack of information about the effectiveness of programs that have adopted a balanced and restorative approach (Rhine et al., 1998).

Rhine et al. (1998) have asserted that the full impact of the BARJ model can be realized only through the addition of certain principles stemming from the "what works" literature (Andrews & Bonta, 1994). Programs based on this literature tend to target higher-risk offenders, link program activities to specific criminogenic needs, and stress the importance of risk assessment and program evaluation (Rhine et al., 1998). Of particular note is how the focus on criminogenic needs has included the identification and promotion of factors related to a healthy family environment (Andrews & Bonta, 1994). Thus, at the same time that family-based program developers examine literature based on the BARJ model, concurrently they might want to become more familiar with the 'what works' literature as well.

In sum, the Growing Up FAST: Families and Adolescents Surviving and Thriving program is thought to represent an effort to implement programming for families of adolescent offenders. At the beginning of this chapter, we noted that these families were faced with rather arduous questions (How could a member of my family commit such a crime? Did the event happen because of something I did or failed to do? Is this family member at risk of further involvement in illegal activity?). Grounded as it is in a Balanced and Restorative Justice model, the Growing Up FAST program seeks to generate answers to such questions in ways that emphasize shared family member responsibility for the creation of an environment that supports efforts to heal the various relationships harmed by the criminal act and, in turn, prevents the adolescent from engaging in recidivistic activity.

REFERENCES

Alexander, J. F., & Parsons, B. V. (1973). Short-term behavioral intervention with delinquent families: Impact on family process and recidivism. *Journal of Abnormal Psychology, 81*, 219–225.

Andrews, D. A., & Bonta, J. (1994). *The psychology of criminal conduct.* Cincinnati, OH: Anderson.

Andrews, D. W., & Hickman, G. P. (1998). Family involvement in juvenile diversion and the prevention of problem behaviors. *Family Science Review, 11*, 259–276.

Arcus, M. E. (1995). *Advances in family life education. Past, present, and future. Family Relations, 44*, 336–344.

Bank, L., Marlowe, H., Reid, J. B., Patterson, G. R., & Weinrott, M. R. (1991). A comparative evaluation of parent-training interventions for families of chronic delinquents. *Journal of Abnormal Child Psychology, 19*, 15–33.

Barton, C., Alexander, J. F., Waldron, H., Turner, C. W., & Warburton, J. (1985). Generalizing treatment effects of functional family therapy: Three replications. *The American Journal Of Family Therapy, 13*, 16–26.

Bazemore, G. (1997). What's New About the Balanced Approach? *The Juvenile and Family Court Journal, 48*, 1–23.

Bazemore, G. (1999). After shaming, whither reintegration Restorative justice and relational rehabilitation.' In: *Restorative juvenile justice: Repairing the harm of youth crime*, (155–194). Monsey, NY: Criminal Justice Press.

Bazemore, G., & Umbreit, M., (1995). Rethinking the sanctioning function in juvenile court: Retributive or restorative responses to youth crime. *Crime and Delinquency, 41*, 296–316.

Bishop, D. B., Frazier, C. E., Lanza-Kaduce, L., & Winner, L. (1996). The transfer of juveniles to criminal court. Does it make a difference? *Crime and Delinquency, 42*, 171–191.

Blumenkrantz, D. G., & Gavazzi, S. M. (1993). Guiding transitional events for children and adolescents through a modern day rite of passage. *Journal of Primary Prevention, 13*, 199–212.

Bogenschneider, K. (1996). An ecological/risk/protective theory for building prevention programs, policies, and community capacity to support youth. *Family Relations, 45*, 127–138.

Bronfenbrenner, U. (1979. *The ecology of human development*. Cambridge, MA: Harvard University Press.

Clear, T. R., & Corbett, R. P. (1998). Community corrections of place. *Perspectives*, (winter), 24–32.

Dumka, L. E., Roosa, M. W., Michaels, M. L., & Suh, K. W. (1995). Using research and theory to develop prevention programs for high-risk families. *Family Relations, 44*, 78–86.

Gartland, H. J., & Day, H. D. (1991). Parental conflict and male adolescent problem behavior. *The Journal of Genetic Psychology, 153*, 201–209.

Gavazzi, S. M. (1991). The Growing Up FAST: Families and Adolescents Surviving and Thriving Program. Department of Family Relations and Human Development, The Ohio State University, Columbus, Ohio.

Gavazzi, S. M. (1995). The Growing Up FAST: Families and Adolescents Surviving and Thriving™ Program. *Journal of Adolescence, 18*, 31–47.

Gavazzi, S. M., & Blumenkrantz, D. E. (1993). Facilitating clinical work with adolescents and their families through the Rite of Passage Experience Program. *Journal of Family Psychotherapy, 4*, 47–67.

Gavazzi, S. M., & Law, J. C. (1997). Creating definitions of successful adulthood for families with adolescents: A therapeutic intervention from the growing up FAST program. *American Journal of Family Psychotherapy, 8*, 21–39.

Gavazzi, S. M., Rhine, E. E., & Partridge, C. (Forthcoming). Growing Up FAST and community supervision: Building bridges between field staff and the family of the offender. *Corrections Management Quarterly*.

Gavazzi, S. M., Wasserman, D., Partridge, C., & Sheridan, S. (2000). The Growing Up FAST diversion program: Comparative treatment outcome. *Aggression and Violent Behavior, 5*, 159–175.

Gordon, D. A , & Arbuthnot J. (1987). Individual, group, and family interventions. In H. Quay (Ed.), *Handbook of juvenile delinquency*, (290–323). New York: Wiley and Sons.

Gordon, D. A., Arbuthnot, J., Gustafson, K. E., & McGreen, P. (1988). Home-based behavioral-systems family therapy with disadvantaged juvenile delinquents. *The American Journal Of Family Therapy, 16*, 243–255.

Gordon, D. A., Graves, K., & Arbuthnot, J. (1995). The effect of functional family therapy for delinquents on adult criminal behavior. *Criminal Justice And Behavior, 22*, 60–73.

Henggeler, S. W. (1997). Treating serious antisocial behavior in youth: the MST approach. *Juvenile Justice Bulletin* (May). Washington DC: Office of Juvenile Justice and Delinquency Prevention.

Henggeler, S. W (1996). Treatment of violent juvenile offenders – we have the knowledge: Comment on Gorman-Smith et al. (1996). *Journal of Family Psychology, 10*, 137–141.

Henggeler, S. W., Borduin, C. M., Melton, G. B., Mann, B. J., Smith, L. A., Hall, J. A , Cone, L , & Fucci, B. R. (1991). Effects of multisystemic therapy on drug use and abuse in serious juvenile offenders: A progress report from two outcome studies. *Family Dynamics of Addiction Quarterly, 1*, 40–51.

Henggeler, S. W., Melton, G. B., & Smith, L. A. (1992). Family preservation using multisystemic therapy: An effective alternative to incarcerating serious juvenile offenders. *Journal of Consulting and Clinical Psychology, 60*, 953–961.

Hughes, R. (1994). A framework for developing family life education programs. *Family Relations*, *43*, 74–80.

Jacobs, F. H. (1988). The five-tiered approach to evaluation: Context and implementation. In: H. B. Weiss, & F. H. Jacobs (Eds), *Evaluating family programs*, (37–68). New York: Aldine de Gruyter.

Julian, D. A., Jones, A., & Deyo, D. (1995). Open systems evaluation and the logic model: program planning and evaluation tools. *Evaluation and Program Planning*, *18*, 333–341.

Klein, N. C., Alexander, J. F., & Parsons, B. V. (1977). Impact of family systems intervention on recidivism and sibling delinquency: A model of primary prevention and program evaluation. *Journal of Consulting and Clinical Psychology*, *45*, 469–474.

Kumpfer, K. L. (1994). *Family strengthening in preventing delinquency*. Washington, DC: U.S Department of Justice, Office of Juvenile Justice and Delinquency Prevention.

Law, J. C., & Gavazzi, S. M. (1999). Definitions of adulthood: From the voices of parents and adolescents. *Family Science Review*, *11*, 318–335.

Larzelere, R. E., & Patterson, G. R. (1990). Parental management: Mediator of the effect of socioeconomic status on early delinquency. *Criminology*, *28*, 301–324.

LeBlanc, M. (1992). Family dynamics, adolescent delinquency, and adult criminality. *Psychiatry: Interpersonal and Biological Processes*, *55*, 336–353.

Mattingly, M. (1996a). *Word from Washington. Juvenile Justice Update*, December-January , 6. Washington, D.C.: Office of Juvenile Justice and Delinquency Prevention/National Criminal Justice Reference Service.

Mattingly, M. (1996b). *Word from Washington. Juvenile Justice Update*, April-May, 5–6. Washington, D.C.: Office of Juvenile Justice and Delinquency Prevention/National Criminal Justice Reference Service.

Patterson, G. R., Reid, J. B., & Dishion, T. J. (1992). *Antisocial boys*. Eugene, OR: Castalia Press.

Quinn, W. H., Bell, K., & Ward, J. (1997). Family Solutions for Juvenile Offenders. *Prevention Researcher*, *4*, 10–12.

Rankin, J. H., & Kern, R. (1994). Parental attachments and delinquency. *Criminology*, *32*, 495–515.

Rhine, E. E., Neff, A. R., & Natalucci-Persichetti, G. (1998). Restorative justice, public safety, and the supervision of juvenile offenders. *Corrections Management Quarterly*, *2*, 40–48.

Simons, R. L., Robertson, J. F., & Downs, W. R. (1989). The nature of the association between parental rejection and delinquent behavior. *Journal of Youth and Adolescence*, *18*, 297–310.

Simons, R. L., Whitbeck, L. B., Conger, R. D., & Conger, K. L. (1991). Parenting factors, social skills, and value commitments as precursors to school failure, involvement with deviant peers, and delinquent behavior. *Journal of Youth and Adolescence*, *20*, 645–664.

Umbreit, M. S. (1997). Victim offender mediation in criminal conflict. In: E. Crack (Ed.), *Mediation and Conflict Resolution in Social Work*, 279–296. Chicago: Nelson-Hall.

Walter, J. L., & Peller, J. E. (1992). *Becoming solution-focused in brief therapy*. New York: Brunner/Mazel.

Wasserman, D., Gavazzi, S. M., & Randall, A. P. (1999). Furthering prevention-intervention linkages in family strengthening programs: The creation of a follow-up phase for the Growing Up FAST™ program. *Family Science Review*, *11*, 300–317.

Worthen, B. R., Sanders, B. R, Fitzpatrick, J. L. (1997). *Program evaluation: Alternative approaches and practical guidelines*. New York: Longman.

VanNess, D. (1997) Perspectives on Achieving satisfying Justice: Values and Principals of Restorative Justice *ICCA Journal of Community Corrections*, *8*, 7–12.

Zehr, H. (1990). *Changing Lenses: A New Focus on Crime and Justice*. Scottsdale, PA: Herald Press

Printed and bound by CPI Group (UK) Ltd, Croydon, CR0 4YY

16/09/2026

14963401-0004